Ecology, Capitalism and the New Agricultural Economy

With increasing pressure on resources, the looming spectre of climate change and growing anxiety among eaters, ecology and food are at the heart of the political debates surrounding agriculture and diet. This unique contribution unravels agri-environmental issues at different spatial levels, from local to global, documenting the major shifts in agriculture from a long-term perspective.

The book begins by exploring the changes in the industrialisation and socialisation of agriculture over time, through the lens of institutional economics including The French Regulation School and Conventions Theory. Building on Polanyi's 'Great Transformation', the chapters in this volume analyse long-term and contemporary changes in agriculture and food systems that have occurred throughout the last few centuries. Key chapters focus on the historical changes in provisioning and the social relations of production, consumption, and regulation of food in different socio-political contexts. The future of agriculture is addressed through an analysis of controversial contemporary political claims and their engagement with strategies that aim to improve the sustainability of agriculture and food consumption.

To shed light on ongoing changes and the future of food, this book asks important environmental and social questions and analyses how industrial agriculture has played out in various contexts. It is recommended supplementary reading for postgraduates and researchers in agricultural studies, food studies, food policy, the agri-food political economy and political and economic geography.

Gilles Allaire is an economist and Emeritus Director of Research at the french National Institute for Agricultural Research (INRA). He was trained as an agronomist engineer. His main research topics concern institutions, markets and agricultural policies; more specifically, rural employment, rural development, genetic resources, knowledge systems, organic agriculture movement, geographical indications and alternative food networks. He directed several national and European research projects. He has been a visiting scholar in the Institute of International studies, University of California at Berkeley. He is member of the editorial board of the *Revue de la regulation*.

Benoit Daviron is Senior Researcher at the Centre de Coopération Internationale en Recherche Agronomique pour le Développement (CIRAD) and consulting professor in SupAgro Montpellier, France. He has been a visiting scholar in the Department of Agricultural and Resource Economics of the University of California at Berkeley and responsible for economics and social sciences in CIRAD. For the last 20 years he has worked on agricultural trade and food policies in developing countries with a special focus on the questions of governance and regulation.

Critical Food Studies

Series editor: Michael K. Goodman, *University of Reading, UK*

The study of food has seldom been more pressing or prescient. From the intensifying globalisation of food, a worldwide food crisis and the continuing inequalities of its production and consumption, to food's exploding media presence, and its growing re-connections to places and people through 'alternative food movements', this series promotes critical explorations of contemporary food cultures and politics. Building on previous but disparate scholarship, its overall aims are to develop innovative and theoretical lenses and empirical material in order to contribute to – but also begin to more fully delineate – the confines and confluences of an agenda of critical food research and writing.

Of particular concern are original theoretical and empirical treatments of the materialisations of food politics, meanings and representations, the shifting political economies and ecologies of food production and consumption and the growing transgressions between alternative and corporatist food networks.

Digital Food Activism
Edited by Tanja Schneider, Karin Eli, Catherine Dolan and Stanley Ulijaszek

Children, Food and Nature
Organising Meals in Schools
Mara Miele and Monica Truninger

Taste, Waste and the New Materiality of Food
Bethaney Turner

Ecology, Capitalism and the New Agricultural Economy
The Second Great Transformation
Edited by Gilles Allaire and Benoit Daviron

Alternative Food Politics
From the Margins to the Mainstream
Edited by Michelle Phillipov and Katherine Kirkwood

For more information about this series, please visit: www.routledge.com/Critical-Food-Studies/book-series/CFS

Ecology, Capitalism and the New Agricultural Economy

The Second Great Transformation

Edited by Gilles Allaire and Benoit Daviron

Routledge
Taylor & Francis Group

LONDON AND NEW YORK

First published 2019
by Routledge
2 Park Square, Milton Park, Abingdon, Oxon OX14 4RN

and by Routledge
52 Vanderbilt Avenue, New York, NY 10017

First issued in paperback 2020

Routledge is an imprint of the Taylor & Francis Group, an informa business

British Library Cataloguing-in-Publication Data
A catalogue record for this book is available from the British Library

Library of Congress Cataloging-in-Publication Data
Names: Allaire, Gilles, editor. | Daviron, Benoãit, editor.
Title: Ecology, capitalism and the new agricultural economy : the second great transformation / Gilles Allaire and Benoit Daviron.
Description: First edition. | New York : Routledge, [2019] |
Series: Critical food studies series | Includes bibliographical references and index.
Identifiers: LCCN 2018035704| ISBN 9780815381617 (hbk : alk. paper) | ISBN 9781351210041 (ebk) | ISBN 9781351210010 (mobi/kindle)
Subjects: LCSH: Agriculture–Economic aspects. |
Agriculture and state. | Ecology. | Capitalism.
Classification: LCC HD1415 .E23 2019 | DDC 338.1–dc23
LC record available at https://lccn.loc.gov/2018035704

ISBN 13: 978-0-367-58273-9 (pbk)
ISBN 13: 978-0-8153-8161-7 (hbk)

Typeset in Times New Roman
by Integra Software Services Pvt. Ltd.

To Martino Nieddu

Contents

Illustrations

Figures

Tables

Contributors

Gilles Allaire is an economist and Emeritus Director of Research at the French National Institute for Agricultural Research (INRA). He was trained as agronomist engineer (Dr). His main research topics concern institutions, markets and agricultural policies and, more specifically, rural employment, rural development, genetic resources, knowledge systems, organic agriculture movement, geographical indications, alternative food networks and norms in general. He has directed several national and European research projects. He is member of the editorial board of the *Revue de la regulation*.

Ward Anseeuw studies political economy and the economy of development at the Agricultural Research Centre for International Development (CIRAD). He carried out research programmes in Africa on agricultural and land policies. He is co-founder of the Land Matrix (www.landmatrix.org), to promote transparency in land governance. Presently he is working at the International Fund of Agricultural Development.

Nicolas Béfort is a Researcher in socio-economics, specialising in the emergence of a bioeconomy at the Laboratoire Regards, Université de Reims Champagne-Ardenne. His work is mainly dedicated to the analysis of the institutionalisation of a bioeconomy and how actors are developing their identities in an emerging economic space.

Lawrence Busch is Distinguished Professor of Sociology at Michigan State University. He has worked at the Norwegian University of Science and Technology, Lancaster University (UK), and the *Institut de Recherche pour le Développement (IRD)*. He is co-author or co-editor of 16 books and more than 150 other publications. He was president of both the Rural Sociological Society and the Agriculture, Food, and Human Values Society and is currently a fellow of the American Association for the Advancement of Science, a *Chevalier de l'Ordre du Mérite Agricole* and an elected member of the *Académie d'Agriculture de France*.

Emmanuelle Cheyns is Senior Researcher in Sociology at the Centre de Coopération Internationale en Recherche Agronomique pour le Développement (CIRAD), research unit 'Markets, Organizations, Institutions and Stakeholders

Strategies' in Montpellier. She has been involved in research on food quality and food consumption of urban dwellers in African cities, including the impact of liberalisation in the oil palm value chain in Africa, forms of participation and the construction of the agreement between stakeholders involved in sustainability voluntary standard organisations.

Benoit Daviron is a Senior Researcher at the Centre de Coopération Internationale en Recherche Agronomique pour le Développement (CIRAD) and Consulting Professor at SupAgro Montpellier, France. He has been a visiting scholar in the Department of Agricultural and Resource Economics of the University of California, Berkeley and responsible for economics and social sciences in CIRAD. For the last 20 years, he has been working on agricultural trade and food policies in developing countries with a special focus on the questions of governance and regulation.

Antoine Ducastel is a Social Science Researcher at the Agricultural Research Centre for International Development (CIRAD) in Montpellier, France. His work focuses on the relations between finance and agriculture, especially in sub-Saharan African, from an economic sociology perspective.

Eve Fouilleux is a political scientist, Senior Research Director at the French National Center for Scientific Research attached to the University of Montpellier and associated with the Center for International Cooperation and Research in Agriculture for Development (CIRAD). She works on global policy-making, multi-level governance and policy changes, with a two-fold approach, analysing ideas, discourses, narratives and concretely implemented policy instruments. She has been working on the EU common agricultural policy, food security policies in Africa, global food security debates, and on diverse voluntary sustainability standards.

David Goodman is Professor Emeritus in the Department of Environmental Studies, University of California, Santa Cruz and now lives in London. He has published widely on changes in agriculture and food systems and is co-editor (with Michael Goodman) of the 'Contemporary Food Studies' series published by Bloomsbury Press.

Pascal Grouiez is Associate Professor in economics at Paris Diderot University. He analyses the emergence of new forms of firms in order to comprehend the relation between institutional innovations and capitalist change. His topics include the agro-holding organisations in post-Soviet economies.

Petia Koleva is Associate Professor in economics at Paris Diderot University. Her research topics include institutional economics, post-communist transformation and sustainable development.

Pierre Labarthe focuses on agricultural economics and is Researcher at Inra, UMR-Agir, Université de Toulouse, INRA, INPT, INP- EI PURPAN and Castanet-Tolosan, France.

Julie Labatut is a Researcher in management sciences at INRA in Toulouse, France. Her research focus on the management of animal genetic resources as common goods and the impact of new technologies in generating organisational and institutional changes in the animal breeding industry. Her recent publications examine the impact of genomic technologies in the dairy sheep and cow breeding industry in France.

Sylvaine Lemeilleur holds a PhD in development and agricultural economics from the University of Montpellier and is a Researcher at CIRAD (UMR Moisa). Her research focuses on agri-food industry transformation and the socio-economic impacts of sustainable standards on small farmers in developing countries.

Les Levidow is a Senior Research Fellow at the Open University, UK, where he has been studying agri-food-environmental issues since the late 1980s. His research topics have included the following: sustainable development, agri-food-energy innovation, agricultural research priorities, agbiotech, agroecology, governance, European integration, regulatory expertise, scientific uncertainty, and the precautionary principle. He is co-author of two books: *Governing the Transatlantic Conflict over Agricultural Biotechnology: Contending Coalitions, Trade Liberalisation and Standard Setting* (Routledge, 2006) and *GM Food on Trial: Testing European Democracy* (Routledge, 2010).

Allison Loconto is a Sociologist and Researcher at the French Institute for Agricultural Research (INRA) and co-leads the cluster on 'transitions, emergences and transformations' in the Laboratoire Interdisciplinaire Sciences, Innovations et Sociétés (LISIS). She is a Visiting Expert on Institutional Innovations in the Plant Production and Protection Division of the Food and Agriculture Organization of the UN (FAO). She is President of the Research Committee on Food and Agriculture (RC40) within the International Sociological Association (ISA). Dr. Loconto explores innovation and the governance of transitions to sustainable agriculture with a focus on standards, institutional innovations and questions of responsibility.

Martino Nieddu was Professor of Economics at the University of Reims, and the head of the economics and management research laboratory EA REGARDS. His research focused on green chemistry, the substitution of fossil resources with renewable resources and the bioeconomy. He used the French theory of regulation approach to analyse research programmes in synthetic biology, biotechnology and agronomy in a perspective of double intensification (ecological and economic). Martino Nieddu passed away on 11 June 2018 at the age of 60.

Thomas Poméon is a socio-economist and Research Engineer at the French National Institute for Agricultural Research (INRA) 'Observatoire du Développement Rural' unit. He holds a PhD in rural and agro-industrial economy and is involved in research projects on rural development policies and markets related to food quality schemes, organic farming and the interaction between agriculture and environment. He currently manages an observatory dedicated to official food quality schemes in France, run in partnership with INAO. He has contributed to several projects funded by regional, national and European agencies.

Stefano Ponte is a Professor in the Department of Business and Politics, Copenhagen Business School, Denmark. His research focuses on global value chains, quality conventions, the political economy of sustainability initiatives and transnational environmental governance.

Robert Scollay is an Associate Professor of Economics and Director of the APEC Study Centre at the University of Auckland, New Zealand. His research interests focus on economic integration in the Asia Pacific, including issues raised by the role of agriculture in Asian countries.

Germain Tesnière holds a PhD in management sciences (MINES ParisTech) and carried out his research at the French National Institute for Agricultural Research (INRA) in Toulouse.

Jean-Marc Touzard is an agricultural economist, Director of Research at the French National Institute for Agriculture Research (INRA, UMR Innovation, Montpellier). His research focuses on innovation processes in agriculture and agri-food chains and their contribution to food security and adaptation to climate change. He is coordinating the national programme on adaptation to climate change in the wine industry (LACCAVE) and is contributing to European projects on innovation systems for sustainable agriculture. He is Vice President of the Research Network on Innovation (RRI) that publishes the *Journal of Innovation Economics and Management*.

Franck-Dominique Vivien is Professor of Economics at the University of Reims. He is a member of the REGARDS Laboratory of this university. His work focuses on bioeconomy and sustainable development economics.

D. Hugh Whittaker is Professor in the Economy and Business of Japan at the Nissan Institute of Japanese Studies, University of Oxford. His research interests span entrepreneurship and innovation management, agriculture business models and value chains, and development in East Asia.

John Wilkinson is Associate Professor at the Graduate Centre for Development Agriculture and Society (CPDA) of the Rural Federal University, Rio de Janeiro, where he teaches and researches on the agri-food system and economic sociology with a special interest in Convention Theory. He has recently published on geographical indications, markets and social movement and on the Brazil–China agri-food nexus.

Acknowledgements

Previous versions of all the chapters of the present book (except Chapter 12, the introduction and the conclusion) were published in French in *Transformations dans l'agriculture et l'agro-alimentaire: Entre écologie et capitalisme* (QUAE Collection Synthèse) edited by Allaire and Daviron in 2017. These chapters were revised for publication in English in the 'Critical Food Studies' series with Routledge. Thanks to Michael Goodman for his support of the development of this book into the series.

We wish to thank all the contributors to this volume and to those who contributed to the previous book published in French, for the exchanges we had during the editing process and the seminar *'Renouveler les approches institutionnalistes sur l'agriculture et l'alimentation: La "grande transformation" 20 ans après'*, Montpellier (France), 16 and 17 June 2014. We are particularly grateful to the editorial advisory group who helped us in reviewing and editing the book chapters: Lawrence Busch, David Goodman, Michael Goodman, and John Wilkinson.

Many thanks to INRA and CIRAD, the French research bodies who funded the initial French publication and financed the translation. We thank QUAE for authorising the English edition by Routledge.

We thank *Revue de la Regulation* editorial board for authorising the re-publication of the English translation of one of the QUAE chapters (Chapter 3 in this book) that was previously published as an article in Autumn 2016.

Finally, we wish to acknowledge the sad loss of our colleague and friend Martino Nieddu, who passed away on 11 June 2018 at the age of 60. We lost a friend and a creative economist for our Regulationist community. He is one of the four authors of Chapter 9 of this book. Martino made an original and important contribution to the treatment of the agrarian question in Regulationist approaches. By studying the dynamics of long periods in productivist agriculture and the changes in contemporary French agro-industrial systems, he highlighted the heterogeneity of 'productive configurations' – coordination around a product supporting a strategy in a competition space – and the variety and historical permanence of the strategic logics of various actors. Going deeper, he re-evaluated the concept of heritage to identify 'collective productive heritages' in the various spheres of economic life, extending this concept to industrial and services sectors. He was among the first French Regulationists who mobilising John Commons' institutional economics, including his 'Theory of Reasonable Value', to reassess the distinction between market and non-market goods. He introduced the concept of the 'patrimonial relationship' by noticing that 'non-market goods' refers to commodities that conserve or change the state of various societal domains. These goods produce collective states, which have a plasticity and facilitate market relationships, so they have parallel historical transformations and relate production ecologies to capitalism. He will be sorely missed.

1 Introduction

Industrialisation and socialisation of agriculture, towards new regimes

Gilles Allaire and Benoit Daviron

Landscape ecologies and food production and consumption remain major concerns across global ecosystems. This book assembles research from world-leading scholars to explore these issues from an institutionalist perspective that builds directly on the established French Regulation School and Conventions Theory approach to agri-food governance, through novel empirical and theoretical contributions to contemporary questions of food, agriculture and ecology. Going beyond classical French Regulation Theory, with its focus on national 'regimes of accumulation' and 'modes of regulation', this volume addresses transnational ecological issues that are novel expressions of the global agri-food economy in order to open up new conceptual spaces and theoretical understandings of these issues. Through novel empirical cases and theorisation – namely through engagements with modern economic sociology, political and ecological economy and institutional economics – this book extends institutional approaches and Regulation Theory to capture contemporary changes, impacts and outcomes of global capitalist agri-food and ecological development.

To shed light on the ongoing changes and the future, the book engages with a series of long-term transformations of agriculture and food throughout the 20th century. That century saw wars and violent class struggles, changes in dominant forces and world order, and changes in the nature of capitalism, which today structure the whole world economy. We propose to analyse and discuss these transformations around two processes: the *industrialisation* and *socialisation* of agriculture. While these notions have more general uses, we consider here the industrialisation and socialisation of agriculture as processes characteristic of historical capitalism from the end of 19th century.

Industrialisation is a longstanding systemic transformation process characterised by the massive injection of fossil energy into agricultural production. It is defined both by the use of fossil fuels, synthetic fertilisers, biocides and farm machinery and by the spread of generic technical skills. Historically, agriculture has been more or less mechanised using self-built devices, but industrialisation requires a growing investment in inputs from the chemical, mechanical equipment and energy industries and a corresponding need for financial resources and participation in the market, thereby transforming family agriculture. Industrialisation changes both the agricultural 'metabolic regime' (Krausmann and

Fischer-Kowalski, 2012), and the organisation and the nature of productive work.[1] First, with the use of synthetic and mineral fertiliser, it offers a radical new way to produce and manage soil fertility and to increase yields. Second, it revolutionises manual labour by the use of petrol- or electric-powered engines. Third, and more recently, it is extended by the increasing role of information systems and codified knowledge (progressively from the 1980s), including genetic selection, robotisation and labelling. Public, collective and private investments in these domains have increased from the turn of the 21st century.

Agricultural industrialisation gave humanity the opportunity to escape the curse of Malthus and Boserup: increasing labour productivity was now accompanied by increasing yields. According to Malthus, humans face a given amount of natural resources and technology, and thus have to confront repeated periods of disequilibrium between population numbers and the available food supply that can only lead to demographic collapse (Malthus, 1803). Boserup's (1965) thesis is less famous and she built her analysis in opposition to Malthus. Basing her argument on a large body of diverse cases, she showed that, far from being fixed, the technologies a society masters respond to demographic pressure, for example, by reducing fallow periods. But she also demonstrates that increased yield is only made possible by an increased amount of human labour, and generally involves a fall in labour productivity.

Simultaneous increases in yields and labour productivity (and, in fact, lower inputs of agricultural labour) that have occurred since 1945 represent a major change in the history of humanity. This can only be understood if one considers that agricultural modernisation not only resulted from the progress of science and improved varieties, but from a positive transfer of energy and matter. Thus, Boserup's argument is accurate in terms of systemic energy expenditures: increased yields can only be achieved by an increase of energy inputs. During most of human history, this energy was mainly human labour, possibly supplemented by that of animals. During the second phase of the industrial metabolic regime, fossil energy replaced animal and human energies.

In this book, we use the term *socialisation* to refer to the mediation devices that link individuals and groups or classes to society as a whole. It is a historical dialectic between the irresistible human quest for autonomy and the enlargement of the social structures characterising modernity (Aron, 1969). In a Regulationist perspective, socialisation corresponds to a set of institutional mediations shaping social identities, which allow coherent regimes of growth but which can also fail. Markets function as sites of struggle over values and are in relationship with the creation and evolution of social rights, both collective and personal. From such a Regulationist perspective, we use the notion of the capitalist socialisation of agriculture, referring both to the hegemony of the logic of accumulation and the integration of rural classes in capitalist societies.[2]

The longstanding socialisation process includes the formation of national agricultural policies and international markets. It corresponds to both the integration of agri-food sectors in a logic of capital accumulation and to the restructuring of peasant and rural classes in an urban economy and economic culture, depending on historical context. National agricultural policies create social rights resulting from social security programmes and specific ways to regulate markets, and thus institutionalise the status of family agricultural labour. But socialisation takes different forms depending on the periods and countries, with different roles for the market, nation-state policy, and social class struggles and compromises in shaping institutional mediations.

Standardisation of resources, products and markets, as well as the individualisation with respect to the shifting social organisation of production and forms of consumption, are part of the socialisation process. The socialisation of agriculture is nested in a wider process of social transformation that emerged with the 'modernity' expectations at the end of the 19th century in the USA, Western Europe, and parts of Latin America, with the expansion of the share of monetary earnings in family revenues and of the commodification of food and progressively all of the vital resources.

Peasant classes engaged in socialisation with a claim for autonomy, which translated into both an acceptance of the bourgeois ideology of free enterprise and the important development of agricultural cooperatives movements at several scales and in various domains. This socialisation was primarily supported by farmers' collective action and makes the rural classes part of the whole society but as a distinct component governed by corporatist organisation and a set of specific public regulations encapsulated in national agricultural policies. Transformations in this mode of regulation, which were initiated in the context of the 1970s crisis, have disrupted stable social mediations, both in the global North and the South. In this new context, food demand becomes organised around the differentiation of product chains and food services, and is associated with a more general change in the competition regime in the post-Fordist period (Petit, 1998). It is part and parcel of the movement of agricultural policy liberalisation initiated in OECD countries after the creation of the World Trade Organization (WTO). At the same time, new actors emerge in the socialisation of agriculture when environmental issues enter political agendas. Socialisation today happens in a new media order defined by ambivalent processes creating new individual concerns and new communities. While we can observe significant changes in several aspects of individual and collective life in different segments of the contemporaneous societies, the future is largely uncertain.

Two great transformations in the political economies of food and agriculture

The main historic period addressed by the book covers two 'Great Transformations' of agriculture and agri-food political economy characterised by epochal changes in the industrialisation and socialisation of agriculture. These two Great Transformations, named following Polanyi (1944), can be interpreted as historical social reactions following periods of domination of liberal thought.[3] The first Great Transformation gave birth to the Fordist era. As with the first, the second Great Transformation originates from the reaction to the false promise of a self-governing market, which reappeared with the abandonment of Keynesianism and the justification of deregulation in a neoliberal era. In addition, these two major transformations are also a reaction, but by no means a solution, to the ecological crisis, of, for example, the Dust Bowl, the New Deal (Worster, 1979), and the current energy and climate crisis of what we call the second Great Transformation.

The first Great Transformation of agriculture, which culminated after WWII with the industrial model of the Fordist era (Allaire and Boyer, 1995), boosted production to feed the world, including animal proteins, according to the consumption norm developing with the rise of wages in the global economy, the wealth distribution issue aside. Poverty had certainly not been removed but the advanced societies and the world more generally overcame the Malthusian curse, to a large extent.

As a longstanding transformation of production logics, this process inscribes itself in a change of metabolic regime with the use of fossil biomass (Daviron and Allaire, this volume). The integration of agri-food markets at the national and international levels and national market regulation policies organised the socialisation of agriculture corresponding to the industrialisation model based on the high productivity of mechanical and chemical investment in the agri-food sectors. The accumulation of capital concerned both the physical (the engines but also animals considered as machines) and technical (articulated with the experimental sciences). These investments were 'socialised', meaning they were supported and guaranteed by public policies, according to Keynes.[4] Even though the share of agriculture declined in the global economy, the agri-food complex remains important and accumulation in these areas remains a key component of aggregate demand until at least the 1970s.

Kenney et al. (1989) developed the Regulationist framework to understand transformations in US agriculture during the Fordist era and the agricultural debt crisis of the 1980s. In France, to analyse the agricultural accumulation regime after WWII (Bertrand, 1980) and the mode of regulation installed in the 1960s, Allaire (1988) adopted the same Regulationist framework, though these works were developed independently.[5] Overall, a variety of works were developed in the 1990s to characterise or criticise this first Great Transformation in agri-food economy.

As termed by Bonanno and Busch (2015: 2), in the case of the USA,

> The political economy of agriculture of this Fordist era featured state promoted intervention in a variety of areas, including land redistribution, infrastructure building, publicly sponsored research and development, and price control and commodity programs. The declared goal was to increase production and productivity in order to generate abundant food to feed the growing domestic and world population.

The dynamics of agricultural growth, the policies, and their justification, were similar in Europe but also, with the Green Revolution, in a number of less-developed countries of the Global South, where Fordism involved primarily the application of modernisation policies in peasant-dominated economies (e.g. the Green Revolution in India in the 1960s). Thus, this Fordist-intensive type of agricultural modernisation expanded to countries that traditionally were agricultural exporters.

The political economy of agriculture of this Fordist era also pursued social and economic parity in line with the social benefit and average income enjoyed by the majority of waged society. According to classic French Regulationist authors, the 'wage nexus', or employee–capital relation, is the dominant structural or institutional form determining the general regimes of accumulation. Nevertheless, according to work introduced by Aglietta and Brender (1984), in Fordist economies, all types of enterprises and independent or liberal workers, including peasant categories, are included by various mechanisms in the wage society.[6] Fordist national agricultural policies integrated family agriculture in wage-society mediations. States were socialising investments and instituting mechanisms to stabilise incomes of the farms involved in modernisation (i.e. the social basis for accumulation) and to cover agricultural workers' social risks (i.e. social security or retirement pensions). In this context, the consolidation of medium-sized farms and the migration of the rural population to the cities – including Mexicans to the US or southern Mediterranean peoples to Northern Europe – was part of the social compromise around the industrialisation of agriculture and food production. Until the recent economic crisis, the wage society extended throughout the world, absorbing a substantial portion of rural classes.

Following his election to the US presidency in 1981, Ronald Reagan implemented sweeping new liberal political and economic ideas. Reagan's counter-revolution had four pillars: tax cuts for the rich; a reduction in spending on education, infrastructure, energy, climate change and vocational training; a huge increase in military expenses; and deregulation of the economy, including the privatisation of public services. Thatcher followed the same orientation in the UK, and more generally neoliberalism inspired monetary and economic policies in the capitalist world. The Reagan administration saw the Uruguay Round of the General Agreement on Tariffs and Trade (GATT) as an opportunity to rewrite US agriculture and food policy, and OECD national agricultural policies were progressively liberalised.

The increase in interest rates, costs of agricultural production, and the instability of market prices at several different times from the 1970s and in different countries put farmers in serious financial difficulties, resulting in the closure of farms, and more widely jeopardised the pattern of Fordist socialisation. Conversely, especially in the Global North, the differentiation and reorganisation of food markets opened opportunities for a shift to local markets, and alternative quality schemes developed in recent years.

In addition to the diminished role of the state, the internationalisation of industrial and retail firms contributed to the erosion of national oligopolies by boosting competition, generally with state support, whereby they became world oligopolies, including in finance. A small number of large corporations established quasi-monopolistic control of key commodity markets and distribution channels and exercised power over production processes. But these new market structures also affected Russian and Asian operators. According to Bonanno and Busch (2015: 3; see also Touzard and Labarthe, Chapter 3),

> This shift was aided by the growth of information technology and contain-erisation that opened up numerous new commercial options. State regulation was replaced by third party programs that privatised control, imposed production strategies and prescribed behavior to producers, hired labor and consumers worldwide.

The development of ICT-related changes affected producers in the transformation of professional relationships and skills, and in the restructuring of food chains and food provisioning systems.

The rising level of industrial concentration in sectors located upstream or downstream of agriculture continues today. The case of the seed sector is the most spectacular; in 2015, in the United States the market share of the top four companies reached 85% for corn, 90% for cotton and 75% for soybeans (McDonald, 2016). The phenomenon is also evident throughout the world. In 2009, according to Fuglie et al. (2011: 31), four companies controlled 54% of the global seed market (Monsanto 29%, Dupont 18%, Syngenta 9% and Limagrain 5%). As a result of two mega-mergers, Dow and Dupont and Bayer and Monsanto, the market share of the three larger seed companies reached 62% in 2015 (Clapp, 2017: 9).

After the fall of the Berlin Wall and the collapse of the Soviet Union in the 1990s, a finance-led regime of accumulation as theorised by Regulationist macro-economists was established in the USA[7] and seemed to replace the Fordist regime of accumulation. This novel regime could not be exported to other countries, apart from the UK, given that reforms in Europe and in general in OECD countries aimed to make credit and finance key instruments of economic activity. Financialisation, which was initially understood as a transformation of capitalism stemming from the Anglo-Saxon world, has in recent years completely reconfigured the global macro-economic environment and influenced public policies in different types of economies. In addition to consumer credit, which expands financial inclusion, the financial sector has sought to conquer new territories, including healthcare, education, and climate.

The neoliberal growth model, as installed in the United States, does not lead to improved living conditions of the majority of the population or to a reduction of inequalities. The results of the work of Piketty and his colleagues are unequivocal.[8] In the United States, the cradle of Fordism, the people of the 50% lowest income group have not seen their incomes increase since 1980, whilst the income of the richest 1% has tripled. The national income share for the 1% richest increased from 10% to 20% (Piketty et al., 2016). The development of inequality, both in the North and the South, is apparent in several domains, including access to food. That is revealed with the uneven social development of obesity.

Retrospectively, according to Boyer (2015: 259–260), the decades after WWII seem to mark the apogee of nation-state power, in the historical context of a steady and permissive international system allowing the restructuring of economy on a national basis, notably allowing the building of national agricultural policies (Daviron and Voituriez, 2006), eroding sectoral, regional and local particularities. The 1980s, on the other hand, were characterised by the retreat of the state, with both an internationalising dynamic and a decentralisation movement (Benko and Lipietz, 1992). But, in the middle of the 2010s, according to Boyer, globalisation appears to have been limited and has embraced a variety of forms.

As Jessop (2001: 298) argues, there are

> three newly important contradictions that hinder the search for a stable post-Fordist accumulation regime and mode of regulation. They comprise: first, a dissociation between abstract flows in space and concrete valorisation in place; second, a growing short-termism in economic calculation vs. an increasing dependence of valorisation on extra-economic factors that take a long time to produce; and, third, the contradiction between the information economy and the information society as a specific expression of the fundamental contradiction between private control in the relations of production and socialisation of the forces of production.

In this general context, and in agri-food studies, three concepts have attracted the interest of scholars: globalisation, neoliberalism and financialisation. They have generated numerous contributions, notably in the framework of 'food regime' analysis (see Chapter 5, Chapter 7 and the Conclusion, in Chapter 14).

From the turn of the 21st century, many changes have signalled a second Great Transformation that encompasses change in the accumulation regime (industrialisation forms) and mode of regulation (socialisation forms). Whereas under the Fordist mode of development of agriculture, agri-industry and food corresponded to a process of national integration of agricultural and food markets and was sustained by national agricultural policies, the neoliberal era concerning agri-food organisation is characterised by the global processes of transnational normalisation and financialisation and by a growing concern with global health and ecological issues. It is in this context that we can envisage a new Great Transformation.

Regarding *industrialisation*, many debates since the 1990s have focused on the characterisation of a 'post-Fordist agriculture' (quality or service turns, multifunctional agriculture, etc.), without a real consensus emerging among researchers. The existence of a 'post-productivist agriculture' is contested (Evans et al., 2002) and, for our part, at the global level we do not detect signs of a reversal in the metabolic aspect of the industrialisation of biomass production (see Daviron and Allaire, Chapter 6). The new growth regime based on information technologies, life sciences, and 'non-cost' competitiveness is not a reversal in the level of energy use. But important changes in the mode of regulation have transformed power relations in agri-food ecology. The productivity of capital now depends on intangible assets and the opening up of opportunities for financialisation (see Ducastel and Anseeuw, Chapter 13). Questions also arise in the context of the ecological and information frontiers of this era.

With respect to *socialisation*, liberalisation constitutes a breach both from the point of view of international trade and of the sectoral mode of regulation. Beyond tariff changes, national agricultural policy was reformed with a reduction in the public domain and the weight of sectoral corporatism, albeit with some strong resistance. Whereas the industrialisation of farming and of food was facilitated by upstream and downstream systemic economies of scale, new food markets appear to have moved into the era of what is called the 'new economy', where design models and private norms are leading features. Allaire and Daviron (2006) interpret this change in terms of market devices and competition rules – what they call a 'market regime'. The transformations in agri-food markets during recent decades can be construed as a 'media regime', emerging in response to the new demand for food services that was generated by the individualisation of resources and behaviours. In this new configuration, downstream forces gradually take over from upstream ones in directing innovation dynamics with the development of a service logic.

This shift has brought new food sectors into being, not corresponding to the classic commodity complexes or agri-food chains, but to transversal quality standard markets. This includes food safety standards promoted by international retail firm consortia and so-called alternative food labelling. These new economic sectors found their outlets in behaviours sustained by transnational conceptions of control that offer opportunities for several scales of government (cf. Staricco (2016) on fair trade and Levidow et al., Chapter 9, on organic food). Allaire and Boyer (1995: 26), stressing that 'competition now covers a set of forms of organisation and institutional devices and not just anonymous products', questioned the viability of a set of national sectoral regulation mechanisms in the context of a transnationalisation of standards. This issue is now reopened, involving different levels of analysis, taking into account the restructuring of capital, the role of finance, the transformation of forms of competition, the restructuring of international trade of commodities, and the role of transnational social movements.

Considering food safety and private 'green' or social labels, the multilateral level (e.g. WTO) is not proactive and efficient, not only because of a paralysis of the multilateral organisations manifest at the time of the 2008 crisis (HLPE, 2011), but also because these issues are beyond the scope of existing multilateral rules. In this context, the roles of multinational firms and transnational social movements in the regulation of the vast arena of quality standards should be noted. The increase in multi-party international fora of normalisation of standards (Daviron and Vagneron, 2012; Cheyns et al., 2016), despite hardships faced by nations in the process of globalisation, does not, however, lead to the absence of the nation-state in regulating the rules of exchange.

Conceptions of social progress compatible with the globalisation of economies and new technologies have been oriented, in post-Fordist economies, towards the service sector, notably health services (Boyer, 2015) and various supports for personal development and entertainment. Agriculture with differentiation of the immaterial qualities of food is part of this shift (Allaire, 2004, 2010). Primitively the increase of service demands in the USA or Western Europe was hindered by high prices and quality uncertainty. But these sectors rapidly developed in conjunction with information and communication technologies and the setting up of a transnational standardisation regime (Busch, Chapter 5).

It is important not to confound the so-called quality turn with the alternative organisation of local food systems incorporating moral values. In the context of the implementation of liberal reforms of agricultural policies in OCDE countries, the 1990s model involved a transformation of product markets and of food services focused on the differentiation of qualities. The spread of this model occurred through sectoral crises and the reorganisation of global supply chains by intangible assets (Allaire, 2002). While vast networks of common-pool resources, such as technical professional knowledge, genetics (e.g. the Holstein cow, Labatut and Tesnière,Chapter 8) and cooperative trade organisations, for example, were built up during the first Great Transformation, the dominant tendency of the second transformation is to privatise these resources as part of a differentiation process in which alternative initiatives also participate.

The so-called 'alternative organisations of local food systems' or 'alternative food networks' (AFNs) are nested in global markets (Van Der Ploeg et al., 2012), in a multi-polarised world (see Wilkinson and Goodman, Chapter 7). Alternative food initiatives emerged in many places and they were interpreted by many observers as opposing the globalising food system and creating alternative systems of food production that are environmentally sustainable, economically viable, and socially just. Whether they share such a political agenda has been challenged (e.g. Allen et al., 2003). AFNs, depending on the context and the proclivities of the observer, can be presented as springboards for the construction of more equitable and sustainable food systems or a contrary diversification of markets, introducing duality in the industrial quality regime and a social re-stratification of consumption (Verhaegen, 2012).

Wolf and Bonanno (2014: 284) conclude their book *The Neoliberal Regime in The Agri-Food Sector: Crisis, Resilience and Restructuring* by stressing the 'paradox of interpretive flexibility', which assigns resilience to neoliberalism notwithstanding signs of crisis. For these authors, one of the reasons for this is that the cultural and political origins of these initiatives are linked to the individualisation of society, which supports neoliberal ideologies (Bonanno and Wolf, 2017).

While the second Great Transformation is generally understood as a global expansion of the economic rationality of market logics, several authors, focused on socialisation, see new potentialities. Addressing the 'politics of globalisation in the global north', Zincone et al. (2000: abstract) argue that

> recent trends towards a more global economy can be seen in a different light: as entailing a potential positive transformation in the geographical organisation of political rights and citizenship on a par with the first Great Transformation in economic life and citizenship signalled by the coming of market society.

In addition, Howard-Hassmann (2005: 3) argues that 'finally and paradoxically, there is the global social movement against globalisation, which forces some reflection upon its deleterious consequences'.

The feasibility of a mode of regulation geared towards social progress was supposed 'to depend above all on the type of political mediation' (Aglietta, 1998), unless social progress is identified by more goods, more services, and more access to information flows. Economic growth has been, and is still, a goal defended by the Left because of its contribution to the improvement of the living conditions of the working classes and the reduction of inequalities (without having to impose a revolutionary redistribution of income and wealth). But, today what is a reasonable capitalism compatible with a social progress concerning humanity as a whole? The future of the second Great Transformation might very well be addressed by controversial and ambivalent political claims and strategies addressing the sustainability of agriculture and food. In short, the continued and narrowed focus on scientific inputs could potentially continue the domination of nature or its conservation. Taken together, the different chapters of this book show the ambivalence of social change and claims regarding our ties with food and land.

Origin and content of the book

The French approaches to agri-food studies known as '*Théorie de la Régulation*',[9] or Regulation Theory, and '*Economie des conventions*', or Convention Theory, are infused throughout the collective book *La grande transformation de l'agriculture* (Allaire and Boyer, 1995), which was not translated but was presented to an Anglophone audience by Wilkinson (1997). At that time, research relations were highly tenuous between Anglophone communities in Europe and America and the French community gathered in the 1995 book. Different research traditions had developed in Anglophone food and rural studies (for an overview, see Buttel, 2001), including Food Regime Analysis (Friedmann et al., 2016; Wilkinson and Goodman, Chapter 7), which at that time was not widely known in France. Subsequently, closer exchanges developed,[10] as shown by the existence of this volume.

The present book contains some of the chapters (translated and revised) published in a newly edited French book (Allaire and Daviron, 2017) which is an extension, although after a 20-year hiatus, of the essays published in the 1995 book, which drew on the Regulationist and Convention approaches to analyse the crises, new formations and perspectives of the 'agriculture of Fordism'.[11] Since the 1990s, intellectual landscapes have changed while new major concerns about the future of land and food are on the social and political agendas. Regulation Theory and Convention Theory have generated several research lines and are mixed in with other approaches in agriculture and food studies (see Touzard and Labarthe (Chapter 3) and Cheyns and Ponte (Chapter 4)).

Four theoretical references structure the present book and enlarge the theoretical perspective of the 1995 book (i) institutional political economy, according to Commons (1934); (ii) economic sociology of standards and markets; (iii); global ecological economics (Krausmann and Fischer-Kowalski, 2012); and (iv) Braudel's world economy construction (Arrighi, 1994). Through these lenses, this volume develops novel contributions in three domains: (i) new perspectives on the global history of agri-food ecology in relation to social structures; (ii) an institutionalist approach to global agri-food ecology; and (iii) an overview of the shifting organisation of agri-food capitalism and the diversity of agriculture across global structures through studies of specific issues addressed in various national contexts. Without claiming to be exhaustive, the selection of areas and themes is designed to provide instances of relevant features that characterise these domains.

The first section of the book, entitled 'Novel approaches and theories of global agri-economies in the second Great Transformation', explores different theoretical frameworks addressing global ecology, and food and agriculture in ways that build on 20 years of experience in applying institutional approaches to agri-food economies. The common perspective of the four first chapters is to situate the first and second Great Transformations in a global theoretical analysis, referring to institutional economics, Regulation Theory, Convention Theory, and economic sociology.

Two chapters survey approaches to agri-food studies inspired by Regulation Theory (Touzard and Labarthe, Chapter 3) and by Convention Theory (Cheyns and Ponte, Chapter 4), after the book edited by Allaire and Boyer (1995). These two approaches were initially developed by heterodox economists seeking a political economy perspective and advocating reconciliation with the other social sciences, including geography, sociology, history and political science. These were debated in the Anglophone research in sociology in post-Fordist debates, human geography focused on the new division of labour, environment, and global ecology, and political science and international political economy engaging in analyses of varieties of capitalism. From the beginning, Convention Theory was concerned with 'quality conventions' in particular in the realm of food (e.g. Letablier and Delfosse, 1995) while Regulation Theory addressed the specificity of capitalist and Fordist agriculture.

The second section of the book, entitled 'Ongoing transformations of the agri-economy', examines new and emerging shifts in biomass, genetic resources, and bioeconomy and agroecology across global agriculture and food regimes. It addresses not only recent changes related to privatisation (Labatut and Tesnière,

Chapter 8) and the bioeconomy agenda (Levidow et al., Chapter 9) but also offers a long historical perspective at the world level considering biomass and food (Daviron and Allaire, Chapter 6 and Wilkinson and Goodman, Chapter 7).

The third section, entitled 'Cases studies: competition in markets and policies', focuses on specific empirical case studies in global agroecologies and structures across different topics and locations, such as organic agriculture, agricultural policy in Japan, the post-socialist transition, and financial investment in agriculture in order to examine facets or opposing trends and their political and institutional implications. Chapters in this section draw specifically on national cases: France (Poméon et al., Chapter 10), Japan (Whittaker and Scollay, Chapter 11), Russia and Bulgaria (Grouiez and Koleva, Chapter 12), and South Africa (Ducastel and Anseeuw, Chapter 13). All the cases show phenomena that have greater global implications, such as the political economy of organic food or the financialisation of agriculture.

Novel approaches and theories of global agri-economies in the Second Great Transformation

Property and the socialisation of agriculture in the Second Great Transformation

As developed here, the idea of socialisation includes not only the emergence and the development of social and economic policies (with their sectoral specificity), but also the way in which markets are organised and governed, and overall a transformation of the form of property, with the emergence of intangible property as a distinctive feature of modern capitalism. According to Commons (1934), the economy is a set of bargaining, managerial, and rationing transactions that organise the transfer of property rights on goods (corporeal property), debts (incorporeal property), and expected future incomes (intangible property). Intangible property, as considered by Gilles Allaire in Chapter 2, includes intellectual, collective and social rights. He mobilises the concept of 'intangible property' as introduced by Veblen and Commons in two different perspectives to analyse the historical process of socialisation of agriculture and food (and more generally capitalist socialisation) as an ambivalent one.

This chapter shows the power and actuality of the concept of intangible property and, more generally, of J. R. Commons' works for a Regulationist research programme. New institutional economics attracted some agri-food French scholars to analyse 'hybrid forms of organisation', which are numerous in agri-food markets, such as the production and marketing contracts, that have rapidly increased in recent decades both in the United States and in Europe (Ménard and Klein, 2004)[12] or to address the multi-level 'governance of the quality of products' (Gonzalez-Diaz and Raynaud, 2007). But old institutional economics and the works of J. R. Commons have influenced some Regulationist authors (Théret, 2001; Dutraive and Théret, 2017) on the relationship between money and sovereignty), as did, more recently, Ostrom's institutional analytical framework (Chanteau et al., 2013). This includes agri-food studies. In Chapter 12, Pascal Grouiez and Petia Koleva draw on Commons' institutional framework to focus on conflicts about ownership, quality production issues and social objectives in the dairy sector in Russia and Bulgaria (see also Grouiez, 2018).

French Regulation Theory: trajectory and agri-food studies

French research contributions related to Regulation Theory and applied to agriculture and food, focusing on the growth and crisis of the French Fordist agri-economy, were analysed and explored in Allaire and Boyer (1995). In Chapter 3, Jean-Marc Touzard and Pierre Labarthe survey the research from the last two decades and show the new issues that have emerged and the continuation and renewal of the theoretical perspectives in this domain. They show how this body of research extends the analysis in different dimensions of food production and provision, and of rural ecologies, and thus progressively combine with other research programmes. They propose a new research agenda, arguing that there are opportunities to revive Regulation Theory research on agriculture and food.

Convention Theory

According to Eymard-Duvernay et al. (2005: 25), who are the developers of the 'Economy of Conventions' approach, Convention Theory involves an articulation between three fundamental issues of political economy: 'the characterisation of the agent and his/her reasons for acting; the modalities of the coordination of actions; and the role of values and common goods.' In this view, any market coordination

> is uncertain insofar as it brings into play heterogeneous actors, takes place over time, and focuses on a product (or service) that is never entirely predefined. Overcoming that uncertainty requires the conventional construction of products, services and expectations that are the media of the commercial interaction and productive activity of firms.

In Chapter 4, Emmanuelle Cheyns and Stefano Ponte examine applications of Convention Theory in the Anglophone agri-food studies literature and review 51 relevant contributions. They highlight how Convention Theory has helped explain different modes of organisation and coordination of agri-food operations in different places, and how it has provided new avenues to understand food quality.

A tripartite standards regime

In Chapter 5, Lawrence Busch distinguishes four components of this 'Brave New World' in which we now live: a new de facto internationalised standards regime;

the extension of assembly line technologies perfected by Ford a century ago to much of the agri-food chain; a New Taylorism; and the rise of Big Data, which permits all of this to become real. Busch considers what the consequences might be for democratic governance.

A key ingredient in supply chain management is that every actor in the supply chain is expected to conform to a wide and ever-growing variety of (more and more international) legal and de facto standards. This concerns both large and local chains. The strength of these standards can be related to conceptions based on the individuation of individuals out of family, community, and profession. In today's media environment of markets, becoming more transnational in what sometimes is an odd cultural mix, these standards reflect values hierarchies of organisations, regions, products, and knowledge.

Conformity to de facto standards is enforced through audit systems that extend from the behaviour of CEOs to janitors, from farm supply companies to farmers to processors to retailers. They are organised according to a tripartite standard regime, developing in the last few decades as specific institutions of the neoliberal quality regime, named the 'triple transformation', linking (i) standards, (ii) certifications, and (iii) accreditations (Loconto et al., 2012). Not only has this regime developed in a transversal way within different industries, but also by transnational organisations. What the TSR does is to establish a private global system of governance that extends far beyond that of individual firms. Analysis of the historical development of the TSR shows that it involves more than the uniformity of standards specifications, but also leads to a substantial standardisation of procedures, which has structural impacts.

Busch argues that modern markets, and especially the global markets constructed over the last several decades, demand the creation of multiple massive bureaucracies at different governance levels; they are part of neoliberal forms of intangible property as explored by Allaire (Chapter 2). This system of governance is largely lacking in accountability to any particular government, and its procedures are largely invisible to the ordinary consumer. But this system is sensitive to an ambivalent media regime (Allaire and Daviron, 2006), as shown by several 'food scares' and the loss of public confidence in Europe since the 'mad cow' (BSE) crisis and in China because of recent food scares around milk and baby formula.

Ongoing transformations of the agri-economy

Sources and uses of biomass through the ages and industrial growth

In Chapter 6, Benoit Daviron and Gilles Allaire examine the modes of production and consumption of biomass in historical perspective and look at ongoing transformations in the world of agriculture and, more widely, in the sources and uses of biomass. The chapter opens with a presentation of the concept of metabolic regime put forward by scholars of the Vienna Institute of Social Ecology (Krausmann and Fischer-Kowalski, 2012). This perspective in ecological economics is related to the 'general economy' analysis proposed by the French philosopher Georges Bataille (1967), who was interested in how the energy received by earth circulates and the ways in which societies throughout history consume an energy surplus that he calls the 'cursed part'. Then, referring to Arrighi, the chapter aims to demonstrate that historically specific conditions of production and consumption of biomass can be linked to the different hegemonic configurations that have appeared since the 16th century. In this respect, the position put forward in this chapter contrasts with the first French Regulationist works on agriculture, which concentrated on national Fordist economies. While Chapter 5 refers to world system analysis, it departs from Regulation Theory and other political economy frameworks, including Food Regime Analysis, by introducing a global ecology approach with the concept of metabolic regime and building on the ignored contribution of Bataille to economic thinking.

Food regimes: a critical assessment

Friedmann and McMichael (1989: 25) advanced the concept of *food regimes*, with reference to world system analysis and sought to 'link international relations of food production and consumption to forms of accumulation broadly distinguishing periods of capitalist transformation since 1870'. While these authors draw on Aglietta's distinction between extensive and intensive regime of accumulation in order to contrast the two first food regimes that correspond to British and then US hegemony, respectively, they ignored Regulation Theory (Friedmann et al., 2016).

Food Regime Analysis has provided a macro-historical framework for Anglophone agri-food studies, and it became the conceptual perspective of choice for many scholars analysing the transformations underway in the world agri-food system. The attempts to identify and formalise a new food regime since the 1990s stimulated various debates around the respective roles of finance, transnational companies, greening policies and social movements.

In Chapter 7, John Wilkinson and David Goodman argue that the food regimes account of historical developments in agriculture and food systems is flawed and discuss how its analytical framework compromises an understanding of the changing international agrarian political economy in the early 21st century. They suggest that the limitations of the Food Regime Analysis stem from an over-emphasis on systemic rupture and processes of hegemonic succession. This has led to the corresponding neglect of multipolarities in the evolving world capitalist system, as well as the historical continuities in accumulation strategies based on agriculture and food sectors pursued by other ascendant economies. Accordingly, they argue that it is necessary to recognise a plurality of regional food orders, which increasingly share a common scientific and technological frontier and institutional arrangements rather than retain the concepts of hegemony and an international food regime.

This international 'frontier', which shapes but does not by itself determine the variety of food orders, is designed and maintained by the institutions of what Larry Busch calls the 'tripartite standards regime' and, more generally, by intellectual property rights institutions. But an internationalisation of food and nature stimulated by social movements is also influencing the transformations experienced by diverse local food orders. Both these ambivalent tendencies are associated with the development of the intangible property regime discussed in Chapter 1.

Calibration of animals for capitalist production

Holstein cattle were the mainstay of dairy production in the period of Fordist industrialisation. The recent breeding strategy adopted in the USA, as in Europe, and the transformation of the market in genetic material are the result of technological and political changes that vividly illustrate the global transformation of norms of production. Animal genetic resources are in the public domain and, according to most national regulations, animal breeds are common-pool resources that belong to a community of dairy farmers. The Holstein breed is a particularly relevant case study for analysing the paradox of management of a common-pool resource faced with growing commodification of the products generated by that resource. This issue is becoming increasingly important in the context of recent changes in the animal genetics market: growing globalisation, deregulation, withdrawal of government funding from genetic selection programmes, and the emergence of new technologies that dramatically accelerate genetic progress. In recent years, scientific development in the domains of bioinformatics and genomics have resulted in an epochal restructuring of animal breeding activities and have enhanced the exploitation of living nature. In Chapter 8, Julie Labatut and Germain Tesnière analyse the relationship between common-pool resources and markets within genetic selection plans. The authors discuss scientific and political developments, including the evolution of life sciences and deregulation, which have led to a segmented industrialisation of the cow breeding sector.

Ambivalent knowledge-based bio-economies

The dominant political discourse advocating innovation and investment for the growth of agricultural production has been reversed. Until recently, the discourse used to justify constantly increasing capitalist agricultural intensification – even if 'green' – was the need to 'feed the world'. But it is now constituted by the new 'Eldorado' of biomass resources, including biological wastes and an ideal type of new capitalist growth regime. According to the OECD, a bioeconomy denotes 'the aggregate set of economic operations in a society that uses the latent value incumbent in biological products and processes to capture new growth and welfare benefits for citizens and nations' (OECD, 2006: 1). Priorities for a bioeconomy vary across countries (Levidow, 2015). In Europe, the Knowledge-Based Bio-Economy (KBBE) on which the European Union bases its competitive strategy has been framed in various ways and has gained prominence as an agricultural research and development agenda.

In Chapter 9, Les Levidow, Martino Nieddu, Franck-Dominique Vivien and Nicolas Béfort analyse this European agenda and show conflicting visions. In the dominant liberal agenda, natural resources offer renewable biomass amenable to conversion into industrial products via a diversified biorefinery complex. Environmental sustainability becomes dependent upon markets to stimulate technological innovation. In an alternative agenda, diverse sources of knowledge inform agroecological methods as the basis for a very different knowledge-based bioeconomy.

The KBBE attracts rival visions, each favouring a different diagnosis of unsustainable agriculture and finding its remedies in agri-food innovation. Each vision links a technoscientific paradigm with a quality paradigm (distinguished according to Allaire and Wolf, 2004): the dominant Life Sciences vision combines converging technologies with decomposability, especially for industrial uses of non-edible biomass, while a marginal vision combines agroecology with integral product integrity for quality food. From these divergent visions, rival stakeholder networks contend for influence over EU research policies and priorities. Although a Life Sciences vision remains dominant, agroecological approaches have gained a presence, thus overcoming their general exclusion from agricultural research agendas. In their own way, each rival paradigm emphasises the need for collective systems to gather information for linking producers with users, as a rationale for funding distinctive research priorities. Both are examined in the chapter in relation to the media system and current political debates.

So far, the promoters of the bioeconomy typically take the existence of dormant, under-used or misused biomass resources as their point of departure. This perspective recalls one that flourished in elite discourses at the end of the 19th century concerning the development of empty and wild land. However, the promoters of the agroecological alternative are well-aware of this and the focus of agroecological research has moved to territorial food systems (Francis et al., 2003).

Cases studies: competition in markets and policies

Tensions in organic products marketing (France case)

In Chapter 10, Thomas Poméon, Allison Loconto, Eve Fouilleux, and Sylvaine Lemeilleur analyse the debates and tensions that characterise the field of organic agriculture. These tensions concern the principles, but perhaps more often, the practices and systems put in place to implement organic agriculture, which can lead to an important gap between the discourses and the facts. They analyse the specifications and the mechanisms of certification and accreditation in organic food and explore the three poles of the TSR mentioned above. Two models are compared to reflect on these tensions: the 'official' organic label, supported by public policies, and private standards, such as Nature and Progrès. Although these private standards are less significant when measured by volume and turnover, they are promoted by proactive actors in different social and political forums.

The analysis presented in this chapter – on the organic TSR, the rise of private standards and participative guarantee systems – traces the weakening and reactivation of the critical dimension of the organic movement. Beyond simply coordinating supply chains and market segmentation, private standards might be vehicles for both defending the pluralist message and delivering a structural critique of the current agri-food system. Along with participatory guarantee systems (PGS), which are built into some of these standards, they represent a critical structural element of the current movement that could enable actors to reclaim the identity of organic agriculture. PGS puts forward a vision that is coherent with the founding principles of the organic move-ment that revelled in the diversity of contexts, rather than the reductive homogeneity that is mostly linked to the governing model of the public organic standard. Yet while the attractiveness of PGS as a regulatory tool controlled by the movement itself is enticing as a solution, it is important not to forget that this tool alone cannot bring about systemic change. Regulatory regimes and agricultural policy in general are just as important. Long-term success and sustainability cannot be achieved without a profound reform of European and national agricultural policies, the creation of a system of incentives that are far more favourable to agriculture, and food practices that are more respectful of the need to balance both ecological and socio-economic needs of future generations.

The future of the Japanese exception

Hugh Whittaker and Robert Scollay, in Chapter 11, have two main objectives. Their first is to explore Japanese agriculture, notably its endurance but gradual erosion from the 1970s, and recent upheavals that appear to mark a break in that continuity. The second is to tentatively frame Japan's agriculture in a broader agri-food framework that extends to consumers and agri-food importers and their overseas interests.

This chapter presents Japan's post-war agriculture regime, its role in Japan's economic 'miracle' and its current upheavals. It considers how the status quo was maintained in a changing environment, especially during the Uruguay Round, which effectively prolonged the post-war order by two decades. The main focus, however, is on the state of agriculture and agricultural policy since 2010, the unfolding debates, and currents of change. New participants are now entering agriculture, with diverse agendas. Industrial and market order forces jostle for position, while alternative networks that partly shun these orders have also experienced a groundswell. Japanese agriculture in 2030 will look hugely different to that in 2010, with or without a 'big bang' shock from the Trans Pacific Partnership.

Re-industrialisation of dairy industry in post-Soviet economies of Bulgaria and Russia

After the collapse of the Soviet Union in 1989, and during the 1990s and early 2000s, the countries of Central and Eastern Europe experienced a major wave of institutional change: systemic transformation and the process of accession to the EU, barring Russia. Policies adopted by Eastern European and Russian leaders at the end of socialism were a variant of the Washington Consensus, which in the same timeframe oriented the interventions of the IMF and World Bank in Africa and Latin America, summed up by the triptych of stabilisation, liberalisation, and privatisation. Stabilisation means the implementation of a restrictive policy mix in fiscal and monetary terms. Liberalisation implies openness to international competition and capital movements. Privatisation is intended to destroy the old system of social relations quickly and definitively. The strategy of 'shock therapy' was implemented by Poland in 1990, Bulgaria and Czechoslovakia in 1991, and Estonia, Latvia and Russia in 1992. Proponents of this therapy seemed to be unaware that institutions of Anglo-Saxon capitalism were not built in a day. The lessons learned and theorised by political economy scholars of the consequences of these measures have highlighted the multiplicity of socio-economic regimes that succeeded the Soviet regime. The institutional and organisational legacy of the old socialist system, underestimated in the dominant paradigm, played a decisive role in the determination of national trajectories (Chavance, 2008; Magnin, 2016). This legacy, which varies according to country, must be considered in its interactions with the strategies of national elites and the influence of external forces such as the EU, international organisations and multinationals.

In Chapter 12, Pascal Grouiez and Petia Koleva analyse the transformation of the agricultural industry in the post-Soviet economies of Bulgaria and Russia. They draw on Commons' institutional framework to focus on path-dependent and path-shaping processes leading to conflicts about ownership, quality and market power issues. The socialist era's legacy left the agricultural sectors of Russia and Bulgaria highly concentrated around collective farms and state-owned farms. The comprehensive agricultural sector reforms (e.g. land restitution, dismantling of existing organisational structures, opening up to competition, etc.) challenged the inherited compromises between actors and opened up a field for strategic action leading to new compromises about the production, distribution and consumption of agricultural goods. This chapter endeavours to analyse the issue of the integration of the Russian and Bulgarian agricultural sectors into the globalised agri-food industry by focusing on the dairy industry.

The main actors 'diffusing institutional rules' are identified to understand their role in conflict resolution at different scales (national or local). Over the last two decades in Russia, the variety in local agricultural policies has led to significant disparities in the transformation of the dairy industry, accentuated by the access to the sector granted to international firms. The strong market power gained by international businesses left little room to local operators. However, the latter were able to implement different niche strategies. In Bulgaria, domestic actors had to deal with several norms imposed by the European Union. While these norms were considered by some actors as favouring standardisation of milk products to the benefit of the mass-market, other actors used different EU norms to defend their strategy based on the development of niche markets. But, in Bulgaria, the upgrading of the dairy industry has been impeded by the effects of land restitution reform; in spite of the European policies implemented since Bulgaria's accession to the EU, this path-dependency could certainly continue in the near future.

Large-scale land investment and agricultural financialisation in South Africa

Financialisation is not a new phenomenon in agriculture. Agricultural commodity futures markets have a long history and the financialisation of peasant agriculture began with the use of credit to invest and produce for the market. However, there are new phenomena. Following the crisis of 2008, the financial industry is committed in its quest for alternative assets which are not correlated with the stock market and that provide protection against inflation, including investment funds specialising in agricultural land. In Chapter 13, based on two case studies of financial funds invested in land in South Africa, Antoine Ducastel and Ward Anseeuw investigate the phenomenon of financial land grabbing 'from below' by framing their analysis in terms of agri-financial *filières*.

As a first step, the authors undertake a critical review of the literature to establish the relevance of the *filière* approach. The analysis identifies the entire chain of intermediaries engaged in the framing and implementation of the financialisation of the land: investors, managers of assets, consultants, national and local administrations, and farmers. In their case studies, they follow the movement of capital from the investors to the farms. The financial institutions involved implement a double operation of securitisation and of delegation of the management of the acquired agricultural funds. It is an investment formula to establish the remuneration of the capital invested in one or more farms based on the financial profitability and the 'social profitability' of the company, capitalised by various forms of certification. A plurality of factors, including entrepreneurial and fiscal rules, information systems on environmental and social assessment procedures and practices, are considered in order to understand the underlying coding specific to this financial product. Finally, the authors examine the transformations in South African agriculture induced by the development of these new investments.

Conclusion

Taken together, the chapters in this book throw light on the Great Transformations of global agri-food ecology and economy during the 20th century. The roots of these transformations are to be found in the transition to the industrial metabolic regime, the modern capitalist nature of property, intangible property and the conflicting organisation of the world system. The contributions assembled here give an overview of the shifting organisation of agri-food capitalism and the diversity of agriculture and food systems across time and space. They explore various issues that dominate contemporary intellectual and political landscapes: financialisation, transnationalisation and privatisation of standards, alternative food networks, liberalisation and the reshaping of agricultural policies. They also address the future.

Various economists and political scientists have observed that the changes initiated in the 1990s represent a second Great Transformation, that, in the Polanyian sense of an epochal transformation of capitalism, takes into account the creation and circulation of knowledge, new forms of personal identity and of socialisation, and the extension of social rights. Yet in the domain of agri-food ecology, all the chapters here describe and analyse what appear to be very ambivalent historical processes. While they stress noticeable changes in the modes and relations of production, a second Great Transformation is characterised by the agri-food system's relations with nature, in food distribution and international trade, by standardisation processes, and property rights regimes and market governance structures. Yet, while a series of changes are identified in the chapters of this volume, they do not constitute a unique and stabilised political and economic model worthy of stasis and essentialisation. On the contrary, the second Great Transformation is characterised by competition between different strategies, which leaves issues open for future research and future politics.

Notes

1 While we consider here Anglo-Saxon and Occidental Europe capitalisms, the historical industrialisation also concernsl former socialist countries. Planning and the authoritarian State in the Soviet bloc permitted a very high rate of accumulation (absolute surplus in Marx's terms). In the 1930s, there was organised exchanges between American and Russian agronomists around the motorisation and large-scale farming, and importation of American tractors in the USSR (Fitzgerald, 1996).

2 B. Jessop introduces the term 'societalization' with the objective of exploring not only the regulation of the accumulation process but also 'that of capitalist societies as a whole through specific modes of mass integration and the formation of an "historic bloc" which unifies the economic "base" and its political and ideological superstructures' (Jessop, 1990: 7). This objective is associated with the North American approach of the 'social structure of accumulation'. In the same perspective Jessop mentioned the growing body of work concerned with the political geography of accumulation, and urban and rural restructuring.

3 There are diverging understandings on what Polanyi (1944) called the 'Great Transformation'. While for some authors it is the triumph of the market economy and ideology (and thus the second transformation being the globalisation which from the 1990 expand capitalism in all parts of the world), we follow the understanding of L. Dumont: '*La Grande Transformation, c'est ce qui est arrivé au monde moderne à travers la grande crise économique et politique des années1930–1945, c'est-à-dire, Polanyi s'emploie à le montrer, la mort du libéralisme économique*' (Dumont, 1982,

preface in 'La Grande Transformation (K. Polanyi)'). Another debate related to Polanyi's work is about its vision of the future. While it can be presented as an anticipation of the post WWII welfare states, he was not thinking that this will be the definitive form of capitalism and nor a re-embedding of the market economy; and he was not according credit to social capitalism. Instead, the economic rationality continued to expand (Lacher, 1999).

4 In the end of *The General Theory of Employment, Interest and Money*, Keynes (1936), one of the stated recommendations is the socialisation of investment, as oriented, supported and secured by public policies as well as by collective industry's programmes; but at that time this concept was not always correctly understood. This meaning is close to the institutionalist concept of the socialisation of ownership introduced by American institutionalists (see Chapter 1).

5 Regarding agriculture, the first works referring to Regulation Theory relied on the identification and specification of the institutional forms that allowed for the stabilisation of the agricultural growth regime characterised as Fordist. The development of Regulation Theory in this field faced a double challenge: (i) in regard to the notion of 'petty commodity production', which was dominant thought in the 1970s in rural sociology and political economy where agriculture forms of production were still considered separate from capitalism, and (ii) in regard to Regulation Theory itself, as a macroeconomic approach. Thus Regulation Theory was completed by the analysis of sectoral and regional dimensions of national economies (Chapter 3).

6 'For these authors, the cohesion of a society is not to be found in an abstract and uniform general law. It is to be found in the uniqueness of its local structures, in the complexity of relationships created between various behaviours, ways of producing and living; it is fundamentally to be found in the differentiation of its members. Capitalism may thus be the dominating movement of several types of society that succeed each other over time or coexist in space ... We call a wage labour society one in which the main differences are to be found within the wage labour class.' (our translation)

7 According to Boyer (2013: 98), in this regime,The central variable is none other than stock prices, since it is the stock market that is observed simultaneously by the companies in their choice of governance and their investment decisions, by the employees in the management of their savings and debt decisions, and by the central bank in the pursuance of its objective of averting financial instability. Peasants on their part, not only the large entrepreneurs observe future markets prices. In the bars of any rural area of Argentina, including the poorest, soybean prices are displayed on TV throughout the day.

8 See http://wid.world/.

9 There is a difficulty regarding the translation of the French concept of 'régulation' which does not have the meaning of regulation in English, which refers to public rules, while the concept of mode regulation refers to complex sets of institutions. In the collective book (Boyer, Saillard 2001) drawing a state of the art approach, the French spelling of the word was maintained in the book title.

10 These exchanges are reflected by the international participation in the 2014 Montpellier seminar and in publications that followed: Allaire, Daviron (2017); Revue de la régulation n°20, autumn 2016 (Allaire et al., 2016).

11 References to the 1995 book often credit it for introducing the concept of the 'economy of quality'. Karpik (1989) was in fact the first to use this expression to designate the extension of markets governed by mechanisms for judging quality. Allaire and Boyer (1995, Introduction) see in this a transformation in the forms of competition. The contributions to this collective book covered various issues that cannot be summarised in this expression and the authors later explored different theoretical paths (Chapter 2). Another collective book on food and agriculture from the same year (Nicolas and Valceschini, 1995) corresponds better to this expression.

12 Contracting is not a new practice and developed in various sectors and contexts at least since the end of the 1940s. Watts (1994: 23) argues that it 'represents one fundamental way in which the twin processes of internationalisation of agriculture and agri-industrialisation are taking place on a global scale'.

References

Aglietta M., 1998. Capitalism at the Turn of the Century: Regulation Theory and the Challenge of Social Change. *New Left Review*, 232(232), 41.

Aglietta M., Brender A., 1984. *Les métamorphoses de la société salariale*. Paris, Calmann-Lévy.

Allaire G., 1988. Le modèle de développement agricole des années 1960. *Economie rurale*, n° 184-185-186, 171–181.

Allaire G., 2002. L'économie de la qualité, en ses territoires, ses secteurs et ses mythes. *Géographie, Economie et Société*, 4(2), 155–180.

Allaire G., 2004. Quality in Economics: A Cognitive Perspective. In Harvey M., McMeekin A. and Warde A. (Eds), *Qualities of Food*. Manchester, Manchester University Press, pp. 61–93.

Allaire G., 2010. Applying Economic Sociology to Understand the Meaning of 'Quality' in Food Markets. *Agricultural Economics*. 41, 167–180.

Allaire G., Boyer R. (Eds), 1995. *La grande transformation de l'agriculture. Lectures conventionnalistes et régulationnistes*. Paris, INRA/ECONOMICA. p. 444.

Allaire G., Daviron B., 2006. Régimes d'institutionnalisation et d'intégration des marchés: Le cas des produits agricoles et alimentaires. *Les nouvelles figures des marchés agri-alimentaires*. Montpellier, 23–24 mars 2006. Symposciences. pp. 101–114. www.symposcience.org/exl-doc/colloque/ART-00001973.pdf.

Allaire G., Daviron B., (coordinateurs) 2017. *Transformations dans l'agriculture et L'agri-alimentaire. Entre écologie et capitalisme*. Versailles, France, QUAE Collection Synthèse, p. 429.

Allaire G., Nieddu M., Labarthe P., 2016. Régulations agricoles et formes de mobilisation sociale (présentation du dossier). *Revue de la régulation*, [En ligne], mis en ligne le 17 janvier 2017, consulté le 17 janvier 2017. http://regulation.revues.org/12190.

Allaire G., Wolf S., 2004. Cognitive representations and institutional hybridity in agrofood systems of innovation. *Science, Technology and Human Values*, 29(4), 431–458.

Allen P., FitzSimmons M., Goodman M., Warner K., 2003. Shifting plates in the agrifood landscape: The tectonics of alternative agrifood initiatives in California. *Journal of Rural Studies*, 19(1), 61–75.

Aron R., 1969. *Les désillusions du progrès. Essai sur la dialectique de la modernité*. Paris, Calmann-Lévy. p. 375.

Arrighi G., 1994. *The Long Twentieth Century: Money, Power and the Origins of Our Times*. London, Verso, p. xiv, 400.

Bataille G., 1967. *La part maudite: précédé de La notion de dépense*. Paris, Éditions de Minuit.

Benko G., Lipietz A., 1992. *les régions qui gagnent: districts et réseaux: les nouveaux paradigmes de la géographie économique*. Paris, Presses universitaires de France.

Bertrand H., 1980. Le régime central d'accumulation d'après-guerre en France et sa crise. *Économie rurale*. 138, 16–21.

Bonanno A., Busch L., 2015. The International Political Economy of Agriculture and Food: An Introduction. In *Handbook of the International Political Economy of Agriculture and Food*. Edward Elgar Publishing.

Bonanno A., Wolf S.A. (Eds.), 2017. *Resistance to the Neoliberal Agri-Food Regime: A Critical Analysis (Introduction)*. Routledge.

Boserup E., 1965. *The Conditions of Agricultural Growth. The Economics of Agriculture under Population Pressure*. London, Allan and Urwin.

Boyer R., 2013. The Global Financial Crisis in Historical Perspective: An Economic Analysis Combining Minsky, Hayek, Fisher, Keynes and the Regulation Approach. *Accounting, Economics and Law*, 3(3), 93–139.

Boyer R., 2015. L'essor du secteur de la santé annonce-t-il un modèle de développement anthropogénétique? *Revue de la régulation*, [En ligne], 17, consulté le 18 octobre 2015. http://regulation.revues.org/11159.

Boyer R., Saillard Y. (Eds), 2001. *Régulation Theory, the State of the Art*. London & New-York, Routledge.

Buttel F.H., 2001. Some Reflections on Late Twentieth Century Agrarian Political Economy. *Sociologia Ruralis*. 41, 165–181.

Chanteau J.P., Coriat B., Labrousse A., Orsi F., 2013. Autour d'Ostrom: Communs, droits de propriété et institutionnalisme méthodologique. Introduction. *Revue de la régulation*, 14 (2), revue en ligne https://regulation.revues.org/10516 (consulte le 28 mai 2016).

Chavance B., 2008. Formal and Informal Institutional Change: The Experience of Post-socialist Transformation. *The European Journal of Comparative Economics*, 5(1), 57–71.

Cheyns E., Daviron B., Djama M., Fouilleux E., Guéneau S., 2016. La normalisation du développement durable par les filières agricoles insérées dans les marchés internationaux. In Estelle B., Alain R. and Denis L. (Eds.), *Développement Durable Et Filières Tropicales*. Versailles, Ed.Quae, pp. 275–294.

Clapp J., 2017. *Bigger Is Not Always Better: The Drivers and Implications of the Recent Agribusiness Megamergers*. Waterloo, ON, Canada, University of Waterloo.

Commons J.R., 1934. *Institutional Economics: Its Place in Political Economy*. New York, Macmillan.

Daviron B., Vagneron I., 2012. Standards, risques et confiance dans le commerce à longue distance à destination de l'Europe: Une lecture historique à partir de Giddens. In Alphandery P., Djama M., Fortier A. and Fouilleux E. (Eds.), *Normaliser au nom du développement durable*. Versailles, Editions Quae, pp. 25–41.

Daviron B., Voituriez T., 2006. Régimes internationaux et commerce agricole. In Berthaud P. and Kebadjian G. (Eds.), *La question politique en économie internationale*, Paris, La Découverte.

Dumont L., 1982. Préface. In Polanyi K. (Ed.), *La Grande Transformation*. Paris, Gallimard, pp. 1–8.

Dutraive V., Théret B., 2017. Two Models of the Relationship Between Money and Sovereignty: An Interpretation Based on John R. Commons's Institutionalism. *Journal of Economic Issues*, 51(1), 27–44.

Evans N., Morris C., Winter M., 2002. Conceptualizing Agriculture: A Critique of Post-Productivism as the New Orthodoxy. *Progress in Human Geography*, 26(3), 313–332.

Eymard-Duvernay F., Favereau O., Orléan A., Salais R., Thévenot L., 2005. Pluralist Integration in the Economic and Social Sciences: The Economy of Conventions. *Post-Autistic Economics Review*, 34(30), 22–40.

Fitzgerald D., 1996. Blinded by Technology: American Agriculture in the Soviet Union, 1928–1932. *Agricultural History*, 70(3), 459–486.

Francis C., Rickerl D., Lieblein G., Salvador R., Gliessman S., Breland T.A., Wiedenhoeft M., Creamer N., Allen P., Harwood R., Altieri M., Salamonsson L., Flora C., Helenius J., Poincelot R., 2003. Agroecology: The Ecology of Food Systems. *Journal of Sustainable Agriculture*, 22(3), 99–118.

Friedmann H., Daviron B., Allaire G., 2016. Entretien Avec Harriet Friedmann; "Political Economists have been Blinded by the Apparent Marginalization of Land and Food". *Revue De La Régulation*, 2. http://regulation.revues.org/12145.

Friedmann H., McMichael P., 1989. Agriculture and the State System: The Rise and Decline of National Agricultures, 1870 to the Present. *Sociologia Ruralis*, 39(2), 93–117.

Fuglie K.O., Heisey P.W., King J.L., Pray K.E., Day-Rubenstein K., Schimmelpfennig D., Ling Wang S., Karmarkar-Deshmukh R., 2011. Research Investments and Market Structure in the Food Processing, Agricultural Input, and Biofuel Industries Worldwide. ERR-130. U.S. Dept. of Agriculture, Econ. Res. Serv. December.

Gonzalez-Diaz M., Raynaud E., 2007. La gouvernance de la qualité des produits. *Économie rurale*, 299, 42–57.

Grouiez P., 2018. Understanding Agri-holdings in Russia: A Commonsian Analysis. *Journal of Economic Issues*, 52, 4, 1012–1037.

HLPE, 2011. *Price Volatility and Food Security*. A Report by the High Level Panel of Experts on Food Security and Nutrition of the Committee on World Food Security. RomeCFS-HLPE.

Howard-Hassmann R.E., 2005. *The Second Great Transformation: Human Rights Leap-frogging in the Era of Globalization*. Laurier, Canada, Political Science Faculty Publications. Paper 24. http://scholars.wlu.ca/poli_faculty/24.

Jessop B., 1990. Regulation Theories in Retrospect and Prospect. *Economy and Society*, 19(2), 153–216.

Jessop B., 2001. What follows Fordism? On the Periodization of Capitalism and Its Regulation. In Albritton R., Itoh M., Westra R. and Zuege A. (Eds.), *Phases of Capitalist Development*. London, Palgrave Macmillan, pp. 283–300.

Karpik L., 1989. L'économie de la qualité. *Revue française de sociologie*, 30(2), 187–210.

Kenney M., Lobao L.M., Curry J., Goe W.R., 1989. Midwestern Agriculture in US Fordism: From the New Deal to Economic Restructuring. *Sociologia Ruralis*, 29(2), 131–148.

Keynes J.M., 1936. *The General Theory of Employment, Interest and Money*. London, Macmillan.

Krausmann F., Fischer-Kowalski M., 2012. Global Socio-Metabolic Transitions. In Singh S.J., Haberl H., Chertow M., Mirt M. and Schmid M. (dir.), *Long Term Socio-Ecological Research: Studies in Society-Nature Interactions accros Spatial and Temporal Scales*. Heidelberg, Germany, Springer, pp. 339–365.

Lacher H., 1999. Embedded Liberalism, Disembedded Markets: Reconceptualising the Pax Americana. *New Political Economy*, 4(3), 343–360.

Levidow L., 2015. Eco-efficient biorefineries: Techno-fix for resource constraints?, *Économie rurale* [En ligne], 349–350: 31–55, septembre-novembre 2015, http://economierurale. revues.org/4729 and https://economierurale.revues.org/4718, mis en ligne le 15 décembre 2017.

Letablier M.T., Delfosse C., 1995. Genèse d'une convention de qualité. Cas des appellations d'origine fromagères. In Allaire G. and Boyer R. (Eds.), *La Grande Transformation De L'agriculture*. Paris, INRA-Economica.

Loconto A., Stone J.V., Busch L., 2012. Tripartite Standards Regime. In *The Wiley-Blackwell Encyclopedia of Globalization*. Vol. 4, New York, John Wiley & Sons, pp. 2044–2051.

Magnin É., 2016. *La grande transformation des pays d'Europe centrale et orientale : tous les chemins (r)évolutionnaires mènent-ils au capitalisme dépendant?* Paris, Les Dossiers du CERI, Centre d'études sur le développement et la coopération internationale, Sciences Po.

Malthus T., 1803. *An Essay on the Principle of Population; Or, a View of Its past and Present Effects on Human Happiness; with an Enquiry into Our Prospects Respecting the Future Removal or Mitigation of the Evils Which It Occasions.* London, Printed for J. Johnson.

McDonald J., 2016. *Mergers and Competition in Seed and Agricultural Chemical Markets.* Washington, DC, Amber-Waves (USDA) Economic Research Service.

Ménard C., Klein P.G., 2004. Organizational Issues in the Agrifood Sector: Toward a Comparative Approach. *American Journal of Agricultural Economics*, 86(3), 750–755.

Nicolas F., Valceschini E. (Eds.), 1995. *Agri-alimentaire: une économie de la qualité.* Versailles, Éditions Quae.

OECD, 2006. *The Bioeconomy to 2030: Designing a Policy Agenda Scoping Paper.* Paris, OECD.

Petit P., 1998. Formes structurelles et régimes de croissance de l'après-Fordisme. *L'année de la régulation*, 2, 169–196.

Piketty T., Saez E., Zucman G. 2016. *Distributional National Accounts: Methods and Estimates for the United States.* Working Paper 22945. Cambridge, NBER, p. 58.

Polanyi K., 1944. *The Great Transformation: The Political and Economic Origins of Our Times.* New York, Farrar & Rinehart.

Staricco J.I., 2016. Towards a Fair Agri-Food Regime? A Regulationist Reading of the Fairtrade System. *Revue de la régulation*, [En ligne], consulté le 17 décembre 2017. http://journals.openedition.org/regulation/12148.

Théret B., 2001. Saisir les faits économiques: La méthode Commons. *Cahiers D'économie Politique*, 2001/2(40–41), 79–137.

Van Der Ploeg J.D., Jingzhong Y., Schneider S., 2012. Rural Development through the Construction of New, Nested, Markets: Comparative Perspectives from China, Brazil and the European Union. *Journal of Peasant Studies*, 39(1), 133–173.

Verhaegen E., 2012. Les réseaux agroalimentaires alternatifs: Transformation globales ou nouvelle segmentation du marché?. In Nizet J., Stassart P., Van Dam D. and Streith M. (Eds.), *Agroécologie, entre pratiques et sciences sociales.* Dijon, France, EDUCAGRI ed..

Watts M.J., 1994. Life Under Contract: Contract Farming. Agrarian Restructuring and Flexible Accumulation. In Little P.D. and Watts M.J. (Eds.), *Living under Contract: Contract Farming and Agrarian Transformation in Sub-Saharan Africa.* Madison, University of Wisconsin Press, pp. 21–77.

Wilkinson J., 1997. A New Paradigm for Economic Analysis. *Economy and Society.* 26, 305–339.

Wolf S.A., Bonanno A. (Eds.), 2014a. *The Neoliberal Regime in the Agri-Food Sector: Crisis, Resilience, and Restructuring.* London & New-York, Routledge.

Wolf S.A., Bonanno A., 2014b. The Plasticity and Contested Terrain of Neoliberalism. In *The Neoliberal Regime in the Agri-Food Sector: Crisis, Resilience, and Restructuring,* London & New-York, Routledge, 284–296.

Worster D., 1979. *Dust Bowl. The Southern Plains in the 1930s.* New York, Oxford University Press, p. 277.

Zincone G., Agnew J., 2000. The Second Great Transformation: The Politics of Globalisation in the Global North. *Space and Polity*, 4(1), 5–21.

Part I

Novel approaches and theories of global agri-economies in the second Great Transformation

2 The ambivalence of the capitalist socialisation of agriculture

Gilles Allaire

Introduction

What does socialisation mean?[1] This chapter addresses the framing of agriculture by markets, social movements, and public policies, in the course of capitalist development. The unusual expression 'socialisation of agriculture' has a double origin, in sociology and in law. And it refers to two interrelated historical processes: the emergence of institutions conditioning the insertion of agriculture in markets and the creation of specific social rights as a result of agricultural policies and public market regulation. In a general anthropological understanding, socialisation can refer to all types of social structures from language to commodities and division of labour; and thus concerns any society in history. Here, I have in mind the mediations that regulate societies dominated by capitalism.

The choice of the expression 'socialisation' refers to the notion of 'social property' as developed by some French scholars. This form of property developed during the 20th century in the form of social security and labour rights (Castel, 2002). Yet the direct translation of this concept from French is not easily understandable in English; the political and social science literature speaks about the 'welfare state' to design such policy regime (and not 'social state', as is current in French social sciences, e.g. Ramaux, 2012). According to Castel, the concept of social property includes public services and economic policy benefits. It differs from that of the welfare state in stressing the social rights corresponding to this form of property, which should not be confused with the material properties of public institutions, but the rights to access various types of social benefits guaranteed by the State and delivered to individuals according to their status. The state recognises the existence of a 'social debt' in relation to these rights, of which it is the debtor of last resort (Théret, 2013).

In the same period, coupled with the construction of social property, i.e. the creation of social rights and the institutionalisation of the social debt, a socialisation of private property (ownership) occurred, which was theorised by early American institutional economics. In his seminal book, R. T. Ely (1914) argued that '*private property does not evolve just based on individual interests, but just as much based on interests for the society as a whole*'. Private property is gradually 'socialised' through law. This socialisation is not collectivisation, but rather a limitation of the rights of the owner through rules of responsibility that carry legal weight. Beyond the limitation of private property rights by patrimonial and moral duties controlled by communities, in modern capitalism the socialisation of property comes to be controlled by law and jurisprudence and transforms the property *'from being a right over things to a right over rights'* (id.). According to Ely (1914), the history of property was characterised above all by this shift. The mechanisms of the socialisation of property depend on the jurisdiction and the historical context. Limitations on property rights are justified by fair competition rules, security, national objectives and various social and environmental issues.

According to J.R. Commons (1934), ownership has gradually been extended from corporeal to incorporeal assets (debts) and then to *intangible assets*, where capital is not valued in physical terms but according to expected future revenue depending on future transactions. Recognition of intangible property played a central role in the transition to the juridical form of sovereignty and the development of credit money (Dutraive and Théret, 2017). As rights over expected future revenue, intangible property corresponds to goodwill, patents and other intellectual property rights. Rights over common-pool resources and rights generated by the development of social property are also intangible property as they concern future benefits.

The aim of this chapter is to show the power and actuality of the concept of intangible property for a Regulationist research programme. I argue that the conception of property developed by early American institutionalism enables one to account for the socialisation of agriculture, if one distinguishes the historical forms of intangible property in its various dimensions from the 19th century to the present day, in the light of the different global political and economic contexts, with a specific focus on the Fordist and the neoliberal eras.

The concept of socialisation as argued here not only refers to the emergence and development of social and negotiated economic policies and thus both social rights and the socialisation of ownership; it also refers to the ways markets are organised and governed according to specific bundles of institutions. Social integration through markets is essentially a result of rationalisation and standardisation processes, involving not only a standardisation of techniques and products, as in a utilitarian view, but also the standardisation of economic and political representations and practices engaged in the functioning of market governance structures. This socialisation materialises in a global standardisation regime (Busch, Chapter 5). Socialisation by market extension and the logic of accumulation does not mean, however, that the market is incompatible with the maintenance of human values – just the opposite (Zelizer, 1988). Quality standards allow both for economic market differentiation, and the continuance of attachments to the 'moral economy' (Busch, 2000). As in the case of public policy, the forces that influence markets, at least the less obscure ones, justify their control through social goals: food security, biodiversity, protection of social and natural heritage, public health, social equity, and so on.

This capitalist socialisation of agriculture is nested in a wider process of social transformation that emerged with the expectations associated with 'modernity' at the end of the 19th century in the USA, occidental Europe, and parts of Latin America, and involving the expansion of monetary earnings in family revenues, together with the commodification of food and more generally resources for human reproduction. It corresponds to the organisational and institutional transformations that took place over a long period concerning labour, nature, and production structures. These transformations are generally described as the passage from so-called 'traditional' to so-called 'modern' forms of agriculture. While the former was based on local social relations, the latter are based on market forces and public policy. Socialisation takes different forms depending on the period and the country, with different roles for the market, for the state, and for social movements.

The concept of socialisation does not prejudge the direction of history.[2] The risk of using concepts such as social property is their almost automatic conflation with visions of social progress or democracy. Yet social rights were claimed and supported by progressive camps while the reactionaries fought against them. The emergence of intellectual property in the 19th century, and that of social property (the welfare state) in the 20th century, as options for the efficient common management of certain resources as proposed by Elinor Ostrom, propounds were met with both political and intellectual conflicts. Hirschman (1991) shows that the arguments of the 'reactionaries' – whether on civil rights in the 18th century, universal suffrage in the 19th century or the welfare state in the 20th century, and I would add, the possibility of common property underpinned institutionally – use the same rhetoric, calling attention to their perverse effects, the impossibility of their implementation ('futility'), and the risks involved ('jeopardy'). He further adds that the 'rhetoric of progress' uses similar arguments that are merely inverted; for example, the argument of imminent danger for society. For political debate to take place, it is necessary to recognise the '*profoundly ambivalent nature of historical movements*' (ibid.). According to Bonanno and Wolf (2017: 284), 'concepts such as sustainability, multifunctionality, or organic have become mainstream references in significant part because [their] interpretative flexibility'.

The structure of the chapter is as follows. In the next section, I introduce the concept of intangible property stressing its social ambivalence and discuss its relevance for a historical perspective. The rights over future values that characterise intangible property have different modalities, depending on whether private incorporeal rights (intellectual property), access rights to collective resources (common property) or social rights (social property) are concerned. In the later sections, I address these differing facets of intangible property. The last section is devoted to agricultural policy and forms of socialisation, with a focus on European agricultural policy. The conclusion addresses research issues.

The two sides or ambivalence of intangible property

More than a century ago, American institutionalist writers described a transformation in property and wealth, and the emergence of a new form of property.

Veblen and Commons diverged on the significance of this transformation for the future of capitalist societies. According to Commons, the ownership of rights over rights (Ely, 1914) became the ownership of rights over future values, which he recognised in the emerging concept of 'intangible property'.

For Commons, a transaction does not just take place in a single moment of time, but over a stretch of time. It includes the time between the transfer of property rights, and real possession of the object of exchange through a process of consumption or valorisation involving managerial transactions. Commons points out that this is not a simple material matter (e.g. the non-conformity of products or services delivered), but a matter of economic valuation, because possession leads to actions in the future that depend on future developments and arbitrages. It is in this sense that property, *'in its modern form of intangible property'* becomes *'rights on future values.'* The conflicts over rights became battles over the power of valorisation, involving decisions on 'reasonable values'. But how reasonable are such authoritative decisions, issued from courts, from the sciences or from customary precedents? From an historical perspective, the ambivalence of intangible property has to be seriously taken into account.

Property and property rights according to early American institutionalism

According to Commons:

> two theories of modern intangible property have been developed since the 1890s. One is Veblen's 'theory of exploitation', and the other the theory of reasonable value of the courts. Both are consistent with the new conception of property as the present value of future profitable transactions.
>
> (1934: 640)[3]

Intangible property emerged as a legal theory based on court decisions when the US Supreme Court, in a decision in 1890, greatly reassessed the value of the stock of a railroad company, and through this set the precedent for resolving such questions. The reasonable value of the courts involves accounting not just for the value of corporeal capital property, but also of incorporeal and intangible property that corresponds to the value one can reasonably expect today from profitable future transactions. Veblen (1908), in basing his assertions on the opinions of business leaders, also discovered that this intangible property was based on flows of future profits. He saw the exploitation of labour, which through its skill and commitment would produce future value, as being at the origin of these flows. Commons distinguished himself from Veblen's extortion theory and constructed his own theory of intangible property on the concept of 'reasonable value'.

Reasonable values of course does not just cover the value of shares taking into account expected future profits, but also the valuation of patents, of goodwill, as well as of salaries, and more generally of any transactional institutional arrangement. A reasonable value is not an 'average' value, but rather refers to a reasoned judgement that leads to the choice of which is of greatest benefit to the community as a whole. Yet, it depends on the representations of what is 'great', in the French Theory of Conventions sense. Commons recognised this way of reasoning both in the decisions of courts of justice, and embedded in custom. For Ely and Commons, a salaried worker holds a claim on a value proportionate to the expected value of the product when it will be ready for sale; this claim is also a form of intangible property. Here reasonable values are determined through the negotiation and the resolution of disputes on wage levels. In his references to reasonable values, Commons places great importance on 'custom' in a particular industry. Reasonable values are determined in different social spaces through 'rationing transactions'.[4]

In line with his conception of reasonable value, Commons sees the intangible property regime becoming effective when social actors (e.g. employers and wage workers, and also farmers) are organised in mass movements such as unions. Collective negotiation results in social rights that are claims on future wealth. Thus, intangible property covers both social property and the socialisation of ownership.

Commons distinguishes between property and property rights: 'property is not just a right but also a conflict of claims to whatever is scarce, while rights of property are the concerted action, which regulates the conflict' (1934: 303). In the same passage, Commons introduces a distinction between 'analysis' and 'justification': analysis deals with the relations between scarcity, property and property rights, while justification of property rights refers to the reasons put forward to preserve or change these rights. The justification of actions in favour of change are based on emerging representations of what is good or great, claiming to be 'good per se'.[5] Critical movements intervene in the building or restructuring of reasonable values. In the case of the AIDS movement (Dodier, 2005), for example, the change resulted in the establishment of new rights for patients as related to information and transparency. Studying critical movements with the right analytical distance makes it possible to identify these socialisation dynamics behind the rhetoric.

Intangible property as a form of wealth should not be confused with property rights that bring it into the realm of private ownership. Depending on the different dimensions of the former, the nature of the latter differs. In the case of common property and social property, we are dealing with access rights to common resources. A distinction must also be made between the justification of social property and the efficiency of public policies that ensure their management. In the case of intellectual property and goodwill, property rights are temporal resource use monopolies. They are private rights and are valued as incorporeal assets, but they are justified, instituted and contestable in the name of the common good or of social interest.

Here, rather than trying to decisively settle the debate between Veblen and Commons, I propose retaining both their visions of intangible property while accounting for the profound changes in context a century on because the two issues of exploitation and of the formation of social agreements or compromises referring to progress remain at the heart of understanding capitalism. These two visions characterise the two faces of intangible property and its ambivalence.

Historical forms of intangible property: are 'reasonable values' really reasonable?

What is happening to the intangible property regime in the 21st century? The welfare state emerged in the 20th century, and then on certain fronts regressed in the liberal era. However, real or formal social rights – such as the right to food security – have continued to make progress on a global scale. With the development of regulatory law, standards have proliferated, especially since the 1990s with the liberalisation of international trade. This development outstrips what was imaginable in the 1920s when the process of standardising industries began after lessons learned from the First World War (Busch, Chapter 5). This process was supposed to generate savings on 'transaction costs'; instead it generated a growing market for standards that is constantly being restructured. Capitalism expanded under the intangible property regime, but this regime has taken different forms and cannot be associated to one single economic regime.

The use of Commons' framework for an historical analysis of intangible property raises two difficulties: that of avoiding anachronisms and that of generalising beyond the legal-political framework of the United States. The second difficulty can be resolved by considering the distribution of sovereign authority in varying constitutional frameworks. Indeed there are numerous sources of production of 'reasonable values' on both a local and global scale. They result from the arbitration of social conflicts at various scales, and the institutionalisation or socialisation of market structures and social property that give them resilience.

To avoid anachronisms, we need to examine the historical context in which the theory of intangible property emerged. One issue here is the very significance of the intangible property as a developing form of wealth. Is it an immaterial wealth, a fictitious capital? Is it a wealth in terms of information and knowledge? In which ways or according to which criteria is the rationing and rationalisation process that rests behind the notion of 'reasonable value' reasonable?

Veblen and Commons lived in an age in which energy was becoming abundant, in which large-scale industry was rationalising production activities, and in which technical knowledge could spread quite easily thanks to the significant expansion in education and to the generic nature of technical experimental knowledge. The process of industrialisation seemed unquestionable, leaving aside the denunciation of the capital exploitation of labour power. Both authors were admirers of 'technology' and the prowess of 'industry'. In Commons' view, industry reached its *'climax with the rapid and large-scale movement of goods, and with the instant transmission of knowledge and negotiations, on a world scale'* (1934: 774). There was certainly a considerable management of data collecting and analysing techniques, benefitting various powerful actors. Today, the digitalisation of information flows is developing even further new forms of accumulation and competition (for example through genomics, see Labatut and Tesnière, Chapter 8).

In the same period, in 1930, Keynes published a short essay, '*Economic Possibilities for our Grandchildren*', reprinted in *Essays in Persuasion* (Keynes, 1963). He attributed the impressive rise in the average standard of living in Europe and the United States in spite of the growth in the population as the result of the industrial and scientific revolutions, which made important advances from the beginning of the 19th century within a dynamic of capital accumulation. Despite the pessimist thinking of the time, he predicted that in 100 years (2030) this standard would be significantly increased, working time reduced and economic scarcity solved. This essay received a great number of comments about the accuracy of Keynes' predictions, which in a global perspective are still relevant even if the world remains profoundly unequal. At this time, Keynes, as Commons, was not concerned with the physical limits of technological development resting on the use of fossil energy (see Daviron and Allaire, chapter 6).

Commons' vision of the links between science and reasonable values is questionable today. Commons bestowed upon science an essential role in the general development of reasonable values, as a sort of superior evaluator. This applied both for the 'physical sciences', which eliminate a big part of technological uncertainty to rationalise production, and for the reflective 'social sciences', which serve to render values reasonable. This is an ideal vision of science, which seems to ignore social inventiveness and imaginative power (Castoriadis, 1975). However, at the same time, other American institutionalists, such as Ayres (1927), were more suspicious about scientific developments. Public investment in scientific research during the 20th century was certainly legitimised by a quest for social progress, but it was also motivated by a quest for military power. While up until 1970 the issue of a reasonable conception of abundance did not occupy the media scene, criticisms of the 'consumer society' and warnings about the 'ecological crisis' have very much diffused since then.

Scarcity, reasonable values and 'radical monopolies'

Commons associated the intangible property regime with the historical, economic stage of 'stabilisation' following that of 'abundance'. In this context, intangible property enables the limitation of the abundance of certain things (for example, that of AOC champagne or products of a specific brand) so as to influence prices. But, it creates new markets for information contributing to financial capital accumulation: first the development of financial services by the banking system, and then the emergence and development of service industries exploiting numeric and communication technologies and based on intangible assets (audit, design, knowledge production, and so forth). Intangible property thus permits the development of new types of services concerning education or social security as new forms of alienation, as revealed by the recurrent criticisms of consumer society and of the mass entertainment industries.

For Commons, 'scarcity operates, as Hume says, both as self-interest and as self-sacrifice, and an economics based on Hume's scarcity permits a union of economics, ethics and jurisprudence' (1934: 143); in other words a reasonable capitalism is possible in a context of abundance. In this sense, scarcity is not an assessment of an objective lack regarding an objective need, but rather a moral justification of the distribution of limited opportunities. Veblen, differs on this point by suggesting that scarcity can be maintained by sustaining an infinite need to consume.

Keynes (1963 [1930]), following the Commons' distinction of three historical periods, scarcity, abundance and stabilisation, considered that, while *'the economic problem, the struggle for subsistence, always has been hitherto the primary, most pressing problem of the human race'*, henceforth it is not. That is an epochal change, characterising the abundance era. It corresponds with the period where the consumption of the public is becoming an engine of the development of capitalism. It is the industrialisation of agriculture and exploitation of fossil energy that helped defeat the Malthusian curse.

According to Keynes, while it *'is true that the needs of human beings may seem to be insatiable'*, they

> fall into two classes—those needs which are absolute in the sense that we feel them whatever the situation of our fellow human beings may be, and those which are relative in the sense that we feel them only if their satisfaction lifts us above, makes us feel superior to, our fellows. Needs of the second class, those which satisfy the desire for superiority, may indeed be insatiable; for the higher the general level, the higher still are they. But this is not so true of the absolute needs-a point may soon be reached, much sooner perhaps than we are all of us aware of, when these needs are satisfied in the sense that we prefer to devote our further energies to non-economic purposes.
>
> (Keynes, 1963: 366)

The elimination of economic scarcity could be expected if we were able to control relative needs, following all or any moral recommendations, and depart from the 'disgusting morbidity' Keynes attributed of the possession of money. Keynes here is mainly thinking about material needs, but there are also services such as education, knowledge production, or heath care, which are ambivalent regarding this classification. These are not ostentatious, as they are relative needs even if their consumption contributes to social status and their need increases more than decreases or stabilises with the pursuit of the good life. The share of these services in the global economy effectively has increased in later phases of abundance such as after WWII, relying on intangible property in its various forms. Moreover, the time gained by working people is used for activities, sport, entertainment, tourism, fitness, foodism, music, etc., rapidly entering into the monetary spheres developed thanks to and as part of the intangible property regime.

Keynes was not really an optimist. The change brought by abundance is frightening because

"if the economic problem is solved, mankind will be deprived of its traditional purpose" and "for the first time since his creation man will be faced with his real, his permanent problem-how to use his freedom from pressing economic cares".

(Keynes, 1963: 368)

The obstacle to control the expansion of needs is not, thus, fundamentally the desire for superiority, but

this purposiveness with which in varying degrees Nature has endowed almost all of us ... The 'purposive' man is always trying to secure a spurious and delusive immortality for his acts by pushing his interest in them forward into time.

(Keynes, 1963: 368)

In the *Affluent Society*, reflecting on the era of abundance beginning in the USA in the early 1940s, J. K. Galbraith (1998) articulated a 'dependence effect' by which '*wants are increasingly created by the process by which they are satisfied*'. Packard drew attention to the mental and psychological manipulations of 'the Hidden Persuaders' and argued that wastefulness has become a part of the American way of life and that to cope with 'glut' is the momentous problem of the nation (Packard, 1960). Since then, various observers have pointed out the absurdities of consumer society, including junk food. Bataille, in an article first written in 1932, denounced utilitarianism as a principle that distorts economic thinking: '*There is indeed no good way, given the set more or less deviating current designs, which allows to define what is useful to humans*' (Bataille, 1967: 25).

Overabundance has a material dimension: product obsolescence, industrial and urban waste, fossil energy consumption, greenhouse gas emissions, etc. It has also immaterial dimensions given the importance of the intangible capital mobilised in the services of armies of creditors, evaluators, monitors, auditors, scientists, entertainers and the media (Allaire and Daviron, 2006). It is in fact a decisive aspect of the development of the intangible property regime and has become an important part of the global economy. What can, then, reasonable value mean in a society of excess? While intangible property develops in the era of abundance and financialisation, are the established reasonable values suffi-ciently reasonable to stabilise an orientation to the good life for mankind?

The economy of abundance relies on institutions and markets that create scarcity. I find it analytically useful to oppose the moral definition of scarcity that Commons attributes to Hume, with the dependence effects and the notion of 'radical monopoly' put forward by Illich (1975) in his radical criticism of the transport, health and education systems that developed after WWII.[6] We are not dealing here with classical monopolistic power through the restriction of supply, but rather with an obligation to consume due to a technical or organisational monopoly. As analysed by Illich, these radical monopolies are beyond the control of reasonable values, although they may be legitimised by discourses referring to social goals. They are denounced in the name of a conception of the good life. In these particular cases, an analytical distinction has to be made between human rights (education, health, and mobility), the related public policies, and the capitalist systems for delivering these services.

Radical monopolies can be observed in industrialised agriculture, creating needs for inputs as a result of dominant conceptions of agricultural efficiency. Whereas we can identify a synergy of innovations enabling the growth of Fordist agriculture, which are justified by the goal of food security, we can also see, when other societal issues are considered, a series of radical monopolies that link upstream industries, research and state services to increase consumption of fertilisers and pesticides. The fundamental right to food security must not be confused with the forms that the socialisation of agriculture has taken.

The notion of radical monopolies denounces more than just a financial logic, but extends also to technical, technoscientific and bureaucratic logics. In general, this line of critique calls on forms of collective property as a solution for autonomy and as an escape from the infernal logic of development. How much of this is utopia is not clear. Institutions that create scarcity are ambivalent; they can be founded on the institutionalisation of a particular right as intellectual property, which is justified and contestable, or can be the result of an alienating capitalistic logic.

Intellectual property

Intellectual property rights, which cover patents, trademarks, copyright and geographical indications,[7] are justified in neoclassical economics as enabling private investment in the production of knowledge and innovation. They confer upon holders a legal and regulated monopoly on the use of an invention, a creation or a distinctive signal. Intellectual property is the ownership of rights over future values resulting from this monopoly. It ensures both the protection of a recognised right to individual (or collective) property of intellectual works and the dissemination of knowledge through the publicity of registered property rights.

The historic origins of intellectual property can be seen, for instance, in the monopolies that were given in medieval times to guilds, craft groups and traders who exerted strict control over techniques and the dissemination of craft knowledge. The authority given to these groups made them arbitrators of reasonable values. It was in the name of the restriction of the freedom to do business that these monopolies were contested and in most cases eliminated by the Bourgeois Revolutions of the 18th century. A philosophical debate then arose on whether intellectual property was a component of human rights. As Marie-Angèle Hermitte points out, '*Intellectual rights were created late, between the end of the 17th and 19th century, when they became international*' (Hermitte, 2016: 17). What is at play is the social nature of innovation and the role of the state in the oversight of innovation: 'laissez-faire' versus regulation.

> Modern intellectual rights were born when public administration abandoned the idea of utility and the pursuit of the common good contained in an invention to focus exclusively on its novelty and industrial character. It was accepted that the common good would automatically follow from the free play of the market ... The abandoning of the ideal of seeking the common good in the attribution of patents set the conditions for the creation of an economy of innovation in the 18th century. Today, after having created a neutral mechanism, we are trying to reintroduce to it the common good.
>
> (Hermitte, 2016: 19–20)

Patents or geographical indications are both property rights on the form of immaterial assets and institutions or property regimes that govern innovation. Food quality standards related to food security or the environment, including organic and fair trade, are also part of the intellectual property regime. These assets have two faces: they are legitimised by doctrines that claim social values and codes that specify required practices submitted to certification guaranteeing economic valuation (Allaire, 2010). The latter are obligations with respect to an (incorporeal) property right; and respect for these obligations can be legally controlled (liability), whereas the doctrines that justify the standards are subject to public criticism (accountability). The discourses justifying or contesting intellectual property rights fundamentally concern their field of application: which type of information, of production of the human mind, of art, of living entity can be monetised?

In principle, by opening a market for knowledge, the system of patents is supposed to encourage innovation. But when holding patents becomes part of competitive strategy, this justification no longer holds up, and the system creates the risk of the 'tragedy of the anti-commons' which arises when scarce resources are under-utilised because too many holders of exclusive rights block each other. This phenomenon was brought to light in the case of biomedical research in the United States (Heller and Eisenberg, 1998) and denounced with regard to the greater good expected from innovation in this field. This denunciation led to reactions such as the mutualisation of pools of patents, which limits their exclusion effects and reduces their competitive value while releasing their social value. The emergence of patents in the field of plant selection was also criticised for its anti-commons role with regard to biodiversity conservation (Thomas, 2015). The key issue here is the ambivalence of the patent system.

Intellectual property alone is not capable of ensuring the dissemination of knowledge, unless one assumes a completely codified world. Madison et al. (2010: 691–692) identify three types of 'commons' concept that have emerged within the regime of intellectual property: i) mechanisms for cooperation, to create specific environments for sharing intellectual resources for specific projects, as in the case of open licences or voluntary quality standards (including geographical indications, organic and fair trade); ii) those constructed in resistance to the privatisation of resources of which in different historical contexts several forms exist in a variety of domains related to knowledge and to natural resources; iii) those that enable the supplementation of knowledge disseminated under the form of property rights or standards, the case of epistemic or academic communities.

Common property

Immaterial resources, other than intellectual property, that are mobilised by economic activity fall under the regimes of common property, which can be defined as regimes of shared access to resources reserved to a possibly large community of users. They are reflected in the user's exclusive rights to realise part of the economic value that result from the mobilisation of these resources. The systems at the origin of these resources were called 'commons' or more precisely 'common-pool resources' by Elinor Ostrom. This concept was first applied to 'natural' resources, such as a lake or a forest, whose indivisible nature meant that private utilisation could threaten the existence of these resources if reasonable rules were not in place for their management. In the wake of Ostrom (see 2010), the concept was extended to designate 'immaterial', 'cultural' or 'intellectual' commons linked to information and knowledge, as well as to characterise complex systems of the immaterial resources that structure the economy (Allaire, 2013). It is these commons I have here in mind, in linking common property regimes and modern intangible property. Regarding agriculture, I do not intend to address the survival or the renewal of community practices in local resource management, but rather the constitution of vast intangible resources systems, such as the Holstein cow shaped by genetic selection (Labatut and Tesnière, this volume). Any resource system, even if immaterial, is dependent on technical, ecological and social infrastructures in one way or another and has multi-polarised nesting connections (Ostrom, 2005).

Ostrom has regularly criticised three semantic confusions that obscure the nature of the commons: i) between common property and open-access regimes; ii) between common-pool resources (socio-ecological or cultural system) and common property regimes (set of property rights and governance structures); and iii) between a resource system and the flow of resource units. She points out that while the system that produces resources falls under common property, the units of the resources delivered are generally appropriated for private (productive or non-productive) use. If correctly managed, the resource system may deliver a lasting flux of resource units with uncertainty related to their quality. This flow is valorised by future mobilisation. The private use of services of these resource units is part of a lasting production or consumption process. The benefit of this appropriation is realised in the future, depending on the outcome of this process. Common property regimes manage rights of resource withdrawal that are privately valorised in future profitable transactions. While the maintenance of the resource system is made possible by collective action, individual access rights are valuated as the present value of these future transactions; in short, they are intangible assets.

One must make the distinction between common property as a regime (institution) and property rights over a commons. Schlager and Ostrom (1992) (see also Hess and Ostrom, 2003, for information knowledge) cite different types of rights, i.e. access, withdrawal, management, contribution to the production of resource units, and exclusion. These determine the position of the holders. Each position can be occupied by actors of varied status (an individual or enterprise, an intermediary body, a public authority), and by different or the same social groups. The benefits accruing to rights holders in a common property regime depend not only on the distribution of the different rights between different types of actors, but also, and above all, on the future strategies of other holders of one or other type of rights. Rights holders constitute all the direct beneficiaries and other stakeholders who exert, in a more or less consensual manner, control over common property. At stake in this control is the ability to influence the rules of management and the justification of control. Disputes always arise over this, and these may lead to existential crises and bringing to the fore political and ethical issues related to cooperation and competition.

Madison et al. (2010) use the expression 'commons in the cultural environment' to designate 'environments for developing and distributing cultural and scientific knowledge through institutions that support pooling and sharing that knowledge'. They cite patent pools, open source systems, news agencies and modern universities. This concept can be extended to 'collective productive heritages' specific to industry sectors or local productive systems (Nieddu et al., 2010). Hess and Ostrom point out that information in general has 'complex tangible and intangible attributes: fuzzy boundaries, a diverse community of users on local, regional, national, and international levels, and multiple layers of rule-making institutions' (Hess and Ostrom, 2003: 132). On their side, markets as specific information systems are fragmented and the knowledge system specific to a given market is maintained through a series of private and collective investments, whose inter-dependencies create complex systems of common resources.

Changes in technology and lifestyles are responsible for a double phenomenon of extinction and globalisation of the commons. Domains of communities of practice have changed and complex integrated technical systems have developed. The future existence of these large systems of resources depends on rules at differing levels, while the valorisation of resource units is linked to opportunities that depend on markets and technologies that are increasingly beyond the control of local user groups. The industrialisation of agriculture everywhere in the world has led to the marginalisation of commons previously indispensable to agriculture activity; and rests, at the same time, on the construction of vast socio-technical systems – which now encompass 'alternative' systems such as the organic one – and the development of a new cultural environment for agricultural activities. The resulting globalization of agriculture is reshaping common property and has led to the re-emergence of the political question of what the goals of collective action are.

For some critics, the vast socio-technical systems that configure biological resources and knowledge, while falling under a common property regime, are close to constituting radical monopolies. The governance of these common resources systems is based on the power to influence technical and scientific choices that can block the development of alternatives (Vanloqueren and Baret, 2009).

Social property

The transformation of civil society through the development of markets and salaried work was accompanied by the development of insurance markets and the institutionalisation of social rights, constituting parts of intangible property. The social state aims to be 'fundamentally an insecurity reducer' (Castel, 2008). From the 19th century, a new legal and political rationality was developed which, in Europe and America, led to the regulation of insurance companies and to a range of social measures, including making insurance obligatory, forming a complex system of social protection that characterises the social states. Castel (2008) sees in this new rationality a response to the question of the economic security of individuals, a precondition for the equality that the constitutionalisation of private property by the Bourgeois Revolutions of the 18th century was supposed to have created:

> At the end of the 19th century a new solution became conceivable and started to be put in place, a system which respected private property while encouraging a certain type of equality between men by constructing a completely new type of property, property for security, which we may call social property.
>
> (ibid.: 171–172)

Social property 'results from the registration of obligations in a legal system, which in return produce rights and resources' (ibid.). These rights are access rights to collective goods and to services with a 'social goal' to ensure the security of members of a society as well as to 'reinforce their interdependence in such a way that they continue to "be a society"' (ibid.). In this regard, the social state is justified as a way of maintaining the economic order of the day.

According to Théret (2013), the social state recognises the existence of a social debt of which it is the debtor of last resort. Citizens, the beneficiaries of national systems of social protection, are the creditors of a mutual debt that has become public. The social debt covers the whole set of social rights of citizenry as a whole, including in my view (following Castel) those created by economic policies. While the neoliberal state cannot eliminate completely existing social debt, it tries to place it under the control of finance. Through a comparison of pension schemes based on distribution and those based on capitalisation, Orléan (2000) articulates an opposition between a regime in which the social debt is guaranteed by the state, and a new regime that introduces 'the logic of private debt, of contracts and of individual responsibility'. The individual 'sees their [social] rights affirmed in the form of negotiable securities. The exercise of these rights no longer depends on their citizenship, or on their membership to such or such an enterprise, but on their trading liquidity'. Where the welfare state was under public and corporatist control, however, such a shift was limited. Regarding agriculture policies, while neoliberal reforms aim to leave more room for so-called 'market oriented policies', food markets for their part, are oriented in new ways because the construction of technical and nutritional standards are no longer confined to the competence of industry, or even to national regulation, but became a global process in which social movements actively participate (Wilkinson, 2006).

New liberal public management destabilises social property and the state qua state. This change occurs progressively through a series of reforms confronted by the resilience of conventions on which social property is founded. Finance tends to devour social property when the value of the debts of states, of wheat, of land and of pension points depends on financial markets, whereas under the stabilising regulation of the Fordism era these values depended on reasonable values determined in negotiations guided by public authorities. Movements that oppose the neoliberal revolution are seeking new reasonable values for distribution. As we will see in the case of agriculture, this means rebuilding the social debt.

Socialisation and agricultural policies, new stakes for the European agricultural policy

Agricultural policies are one component of the social property regime. They create rights and social statuses for producers. The range of these rights, just like social rights of salaried workers, varies greatly depending on national and historical contexts. This political regime is legitimised by the importance of agriculture for national independence and by the social goals of modern states to ensure economic security, in particular food security. After the 1929 crisis, it was based on instruments for controlling and protecting domestic markets and for stabilising farm family incomes. Social statuses and rules of exchange (i.e. land tenure, inheritance rules, and minimum quality standards) were also defined at the time. In the European Union, the Treaty of Rome (1957) took up the principles of stabilisation contained in national agricultural policies and implemented them in the context of the 'common market' (Barthélemy and Nieddu, 2002).

Théret's (2013) framework is applicable to the part of social debt linked to the stabilisation of agricultural markets contributing to food security, exemplified, here, by the case of the European agricultural policy. Two forms of social rights can be distinguished in this case. While national stabilisation policies constitute direct intervention in markets in pursuit of social goals, they do not create any specific holders of legal rights: they represent rights of a public character. For example, all cereal farmers benefit from a guaranteed price of wheat, and all consumers from stabilisation of the wheat market or a subsidised price of bread. It is not easy to modify these public rights due to social agreements jealously protected by unions. Economic stabilisation policies can also create access rights to social debt for specific beneficiaries, as was the case with the establishment of milk quotas in 1984 in the European Union, and later with the Common Agricultural Policy (CAP) reform from 1992 onwards. Public instruments of market control (e.g. public stocks, import taxes, etc.) were deactivated or reduced in scope, and they were replaced by direct payment to farmers.[8] This made it necessary to designate who the beneficiaries – holders of a right to receive public payment – were, and this required long and still ongoing debate. This individualisation of access does not detract from the social character of this policy, but changes its significance. The original CAP had two objectives: farmers' income parity and reasonable prices for popular food products. The second objective is still affirmed, but it is now disconnected from agricultural social debt.

The social protection of salaried workers and labour rights were not integrated into the construction of the European Union. Social debt remained a state affair, while the market changed in scale, and this exacerbated internal competition, and, in the words of Théret (2013), rendered the Euro a 'single' currency rather than a 'common' currency. On the contrary, the creation of a 'common' agricultural policy was based on a transfer of agricultural social debts from states to the European Union. In terms of principles, this was manifest in 'community preference' for supply and in the affirmation that family farming was the social foundation for food autonomy. The liberal dismantling of this policy did not eliminate this agricultural social debt, but rather created a disparity between globalised and financialised agricultural markets, and the political level of expression of this debt. But the renewed political compromise to limit relative reductions in agricultural spending in the European budget through the multi-year financial frameworks confirms the resilience of this debt.

The CAP has, however, changed its base. Deregulation was not limited to the removal of the common organisation of agricultural markets. The agricultural exception, in terms of competition policy, was undermined, in particular by the prohibition of trade association price-fixing agreements. Individual and collective strategies adapted in two directions: first towards a greater attention to market signals (which was expected), difficult to interpret in instable markets, and second, towards an avoidance of competition on mass markets, now characterised by economic insecurity, through local valorisation initiatives and the establishment of long-term contractual relationships. Since the 2007–08 financial and food crisis, a high instability in production prices has spread to national markets and has changed certain viewpoints. The new CAP adopted in 2014 took into consideration an imbalance of market powers between farmers and the big retail companies that have taken control over food chains; the organisation of producers has once again been encouraged and the local control of markets is even possible in the case of geographical indications. Nevertheless, neoliberal ideology, which aims to substitute social rights with market solutions, remains strongly present.

The political heritage of the CAP – its contribution to community sovereignty – cannot be deleted with the stroke of a pen, but the question is, how can it be articulated with common objectives today? Such a perspective needs to address responsibility towards the global world and be rooted in local collective projects, and at both local and global levels it must respond to the need for an energy and environmental transition. Any path capable of giving the CAP renewed direction must go beyond an exclusive focus on agriculture, and once again address, the destiny of society as a whole in its orientation towards an ecological transition.

Conclusion

While the economic regimes and types of state in which the industrialisation of agriculture developed through history varied, they have depended on a two-fold structure: one provided by standards that came to prevail as prerequisites for market access, and the other provided by the state through agricultural policies. This chapter has proposed the concept of the socialisation of agriculture to analyse this two-fold structure, which has its roots, in particular, in the 19th century in the United States, and which grew considerably in the 20th century throughout all industrialised countries, following the First World War and the 1930s crisis. This socialisation of agriculture was linked to the emergence of new forms of economic and political regulation.

John R. Commons' foundational works on institutional economics highlight the trend towards the socialisation of private property and the development of a new form of wealth called intangible property. In this chapter, differing dimensions of this are distinguished (i.e. intellectual property, common property and social property) and for each, an analysis of the transformation of justifications and contestation or restructuration of the related rights is presented. Financial capitalism is consolidating itself and expanding under the intangible property regime, but this regime has taken different forms depending on the period, the country and influence of various political battles, and therefore it cannot be associated to one single economic regime.

To be able to analyse 20th century economic regimes, a joint analysis of the historical development of different forms of intangible property would be necessary. This chapter's ambitions do not go that far but have tried to share this theoretical perspective by underscoring the ambivalence of the historical development of the socialisation of agriculture.

This perspective requires a development of the institutionalist foundations of the theory of regulation, in particular regarding competition regimes and how the regulative institutions (i.e. the structures of the mode of regulation) change. The arguments I have developed here suggest that these changes result from changing 'reasonable' values.

If we give a scientific or moral foundation to reasonable values that cannot be contested or revised we would enter the realm of utopia or madness. History, on the contrary, is ambivalent and the pursuit of the good does not occur without conflicts, which, as we know, are not limited to ideas. Highlighting this ambivalence is not a refusal to take sides, but a way of drawing attention to the necessity for justification. Today, the argument of imminent danger is used to justify the funding of agricultural research and continued support for agricultural intensification, including an 'ecologically intensive' variant. On the other side, there is increased reflection on the capacity of agroecology to meet the challenge of feeding the world and also on the concept of the right to food (de Schutter, 2010). To be able to respond correctly to the issue of sustainable and equitable abundance, we need to abandon the argument of imminent danger. The assessment of the forms of socialisation of agriculture and of its industrialisation dynamics can help us do this.

Notes

1 I thank Larry Busch, John Wilkinson, Bruno Théret, Jean-Jacques Gislain, and Benoit Daviron for useful comments on earlier versions of this text.
2 This chapter is essentially based on and applicable to the history of Western societies. It may, however, be extended, because the idea of the socialisation of agriculture of course is relevant to formerly socialist countries who had differing historical trajectories and is also linked to the concept of development put forward by international organizations (FAO, World Bank, etc.).
3 The term transaction is to be understood in the sense intended by Commons. Future transactions do not mean resale with commercial margins but refers here in particular to the 'managerial' transactions which enable capital to be put to value in a production process. It also concerns future transactions on speculative capital markets.
4 According to Commons (1934), the economy is a set of bargaining, managerial, and rationing transactions that organize the transfer of property rights on goods (corporeal property), debts (incorporeal property), and expected future incomes (intangible property).
5 'Good in itself' (in French '*bien en soi*'), or 'good per se', is a concept introduced by Dodier (2005) when studying Aids activist organizations. The idea of good per se is drawn from sociological surveys illustrating the crucial role in public argumentation of referring to good, which demonstrates two characteristics: people consider that this good is worth something in itself and they believe that the group has a duty to reserve a certain place for it. It is to this manner of exercising a critical sense that the notion of the good in itself points. For example, the preservation of human life, or health, is now part of this category" (Dodier, 2005: 22).
6 'When industry takes on the right to satisfy, alone, a fundamental need, which had previously been met through individual response, it produces such a monopoly' (Illich, 1975).
7 These rights are today recognized by the World Trade Organization under the TRIPS (Trade-Related Aspects of Intellectual Property Rights) agreements.
8 Equivalent reforms appear in other OECD countries at the time of the neoliberal era, depending on the form of the welfare state and of the national agricultural policy. On the Japan case, see Chapter 10. In this case, workers' social rights were managed in line with the particular form of Japanese capitalism (Aoki and Dore, 1994). Agricultural policy contributed to stabilising the revenues of the rural labour power working in big industries. In a completely different setting, in the USSR and other socialist countries, workers' social rights were also managed through the organization of the production units. Some of these rights are still active in some regions of Russia after the transition to liberal capitalism, as is shown by Grouiez (2017) and in Chapter 11.

References

Allaire G., 2010. Applying Economic Sociology to Understand the Meaning of "Quality" in Food Markets. *Agricultural Economics*, 41(1), 167–180.

Allaire G., 2013. Les communs comme infrastructure institutionnelle de l'économie marchande. *Revue de la régulation*, 14(2), revue en ligne: https://regulation.revues.org/10546 (consulté le 19 octobre 2016).

Allaire G., Daviron B., 2006. *Régimes d'institutionnalisation et d'intégration des marchés: le cas des produits agricoles et alimentaires*. Les nouvelles figures des marchés agro-alimentaires. Montpellier, 23–24 mars 2006. Symposciences. pp. 101–114. www.symposcience.org/exl-doc/colloque/ART-00001973.pdf.

Aoki M., Dore R.P. (Eds.), 1994. *The Japanese Firm: The Sources of Competitive Strength*. Oxford, UK, Oxford University Press.

Ayres C.E., 1927. *Science, the False Messiah*. Indianapolis, Ind., Bobbs-Merrill.

Barthélemy D., Nieddu M., 2002. PAC et multifonctionnalité de l'agriculture. *Oléagineux, Corps Gras, Lipides*, 9(6), 383–389.

Bataille G., 1967. *La part maudite: précédé de La notion de dépense*. Paris, Éditions de Minuit.

Bonanno A., Wolf S.A. (Eds.), 2017. *Resistance to the Neoliberal Agri-Food Regime: A Critical Analysis (Introduction)*. London and New York, Routledge.

Busch L., 2000. The Moral Economy of Grades and Standards. *Journal of Rural Studies*, 16, 273–283.

Castel R., 2002. Emergence and Transformations of Social Property. *Constellations*, 9(3), 318–334.

Castel R., 2008. La propriété sociale: émergence, transformations et remise en cause. *Esprit*, 8–9, 171–190.

Castoriadis C., 1975. *L'Institution Imaginaire de la société*. Paris, Le Seuil.

Commons J.R., 1934. *Institutional Economics*. New York, Macmillan.

de Schutter O., 2010. *Rapport du Rapporteur spécial sur le droit à l'alimentation*. New York, Nations Unies. consultable en ligne: www2.ohchr.org/english/issues/food/docs/A. HRC.16.49_fr.pdf (consulté le 19 octobre 2016).

Dodier N., 2005. L'espace et le mouvement du sens critique. *Annales. Histoire, sciences sociales*, 60(1), 7–31.

Dutraive V., Théret B., 2017. Two Models of the Relationship between Money and Sovereignty: An Interpretation Based on John R. Commons's Institutionalism. *Journal of Economic Issues*, 51(1), 27–44.

Ely R.T., 1914. *Property and Contract in Their Relations to the Distribution of Wealth*, tome 1. New York, Macmillan.

Galbraith J.K., 1998 [1958]. *The Affluent Society*. Fortieth Anniversary Edition. New York, Houghton Mifflin Company.

Grouiez P., 2017. Les organisations agricoles dans la Russie de Vladimir Poutine: une lecture commonsienne. In Allaire G. and Daviron B. (Eds.), *Transformations agricoles et agroalimentaires: entre Ecologie et Capitalisme*. Versailles, QUAE, pp. 245–256.

Heller M.A., Eisenberg R.S., 1998. Can Patents Deter Innovation? The Anticommons in Biomedical Research. *Science*, 280(5364), 698–701.

Hermitte M.-A., 2016. *L'emprise des droits intellectuels sur le monde vivant*. Versailles, Éditions Quæ.

Hess C., Ostrom E., 2003. Ideas, Artifacts, and Facilities: Information as a Common-Pool Resource. *Law and Contemporary Problems*, 66(1–2), 111–146.

Hirschman A.O., 1991. *Deux siècles de rhétorique réactionnaire*. Paris, Fayard.

Illich I., 1975. *Énergie et équité*. trad. de l'anglais par Luce Giard. Paris, Seuil.

Keynes J.M., 1963. Economic Possibilities for Our Grandchildren (First Published 1930). In Norton & Co (Eds.), *Essays in Persuasion*. New York, W. W. Norton & Co., pp. 358–373.

Madison M.J., Frischmann B.M., Strandburg K.J., 2010. Constructing Commons in the Cultural Environment. *Cornell Law Review*, 95(4), 657, 659.

Nieddu M., Garnier E., Bliard C., 2010. L'émergence d'une chimie doublement verte. *Revue d'économie industrielle*, 132, 53–84.

Orléan A., 2000. L'individu, le marché et l'opinion: Réflexions sur le capitalisme financier. *Esprit*, novembre, 51–75.

Ostrom E., 2005. *Understanding Institutional Diversity*. Princeton, Princeton University Press.

Ostrom E., 2010. Beyond Markets and States: Polycentric Governance of Complex Economic Systems. *American Economic Review*, 100(3), 1–33.

Packard V., 1960. *The Waste Makers*. London, Longmans, p. 197.

Ramaux C., 2012. *L'État social: pour sortir du chaos néolibéral*. Paris, Fayard/Mille et une nuits.

Schlager E., Ostrom E., 1992. Property-Rights Regimes and Natural Resources: A Conceptual Analysis. *Land Economics*, 68(3), 249–262.

Théret B., 2013. Dettes et crise de confiance dans l'euro: Analyse et voies possibles de sortie par le haut. *Revue française de socio-économie*, 12, 91–124.

Thomas F., 2015. Droits de propriété industrielle et 'communs' agricoles. Comment repenser l'articulation entre domaine public, biens collectifs et biens privés? In Vanuxem S. and Guibet Lafaye C. (dir.), *Repenser la propriété, essai de politique écologique*. Marseille, Presses universitaires d'Aix-Marseille, pp. 171–190.

Vanloqueren G., Baret P.V., 2009. How Agricultural Research Systems Shape a Technological Regime that Develops Genetic Engineering but Locks Out Agroecological Innovations. *Research Policy*, 38(6), 971–983.

Veblen T., 1908. On the Nature of Capital: Investment, Intangible Assets, and the Pecuniary Magnate. *The Quarterly Journal of Economics*, 23(1), 104–136.

Wilkinson J., 2006. The Mingling of Markets, Movements & Menus. In International Workshop: Globalization, Social and Cultural Dynamics, Rio de Janeiro.

Zelizer V.A., 1988. Beyond the Polemics on the Market: Establishing a Theoretical and Empirical Agenda. *Sociological Forum*, 3(4), 614–634.

3 Regulation Theory and transformation of agriculture

A literature review

Jean-Marc Touzard and Pierre Labarthe

Introduction

From an early stage in their history, studies based on the French school of Regulation Theory (RT) used the sectoral scale to better understand the diversity of capitalism and its transformations (Boyer and Saillard, 1995). Such sectoral approaches were developed in the agriculture and the agri-food sectors (Boyer, 1990), which contributed, in particular, to forging the concept of economic regime at the sectoral scale (Bartoli and Boulet, 1990). The book edited by Gilles Allaire and Robert Boyer (1995) attests to the vivacity and perspective of these studies in the mid-1990s. Research on agriculture and the agri-food industry subsequently continued to refer to Regulation Theory, posing new questions, for example, with regard to environmental issues, regional development or new agricultural policies (Laurent and du Tertre, 2008). It highlighted the emergence of new economic regimes in the agricultural and agri-food sectors in a more liberal capitalist world. Yet these developments have not been fully synthesised.

The question that we raise in this chapter concerns the role of Regulation Theory in such a synthesis and in a comparative analysis of economic regimes and their transitions in agriculture and the agri-food sectors. The revival of political debates on the role of agriculture and food in society, along with recent food crises, justifies the renewal of Regulation Theory as a framework for investigating this sector.

We first examine Regulation Theory's historical contribution to the analysis of agricultural sectoral regimes and the way in which this research has combined with other approaches in institutional economics and with other social sciences. We then argue that the recent trends in the international context afford an opportunity to revive Regulation Theory research on agriculture and food. Drawing on our own research and on a literature review, we look at two perspectives for this purpose: we propose to better integrate Regulation Theory's contributions into the analysis of transition studies of agriculture and then into that of the diversity of agri-food models and food regimes.

Regulation Theory's historical contribution to the analysis of agricultural regimes

From the early 1980s (Perraud, 1985), the application of Regulation Theory to the analysis of agricultural development in the post-WWII years strongly contributed to the establishment of a scientific fact (Fleck, 1980): the institutionalisation of a specific sectoral regime of agriculture in industrialised countries during the Fordist period. At the time, the agricultural sector in these countries was experiencing unprecedented growth in terms of production and productivity. Agriculture was based on a technological paradigm oriented towards the goal of enhanced productivity through labour specialisation and intensification (Allaire, 1988a). Agriculture was supported by national agricultural policies, and gradually integrated into markets and with industrial firms, whether private or cooperative, supplying inputs and buying from farmers. Economic analyses embedded in Regulation Theory sought to identify and specify the institutional forms that facilitated the stabilisation of this agricultural regime: forms of competition, wage and labour relations, role of the State, and international relations (Nefussi, 1987; Allaire, 1988b; Debailleul, 1990; Laurent, 1992). To this end, this research mobilised the methodological principles of Regulation Theory: historical exploration, analysis of the evolution of macroeconomic and structural variables; identification of the key institutions and underlying political compromises; and international comparison (France, USA). It was also interdisciplinary, through collaboration with sociologists, political scientists and agronomists.

From describing agriculture in the Fordist period (and its crisis) to a major contribution to the analysis of sectoral regimes

Initially the identification of the sectoral regime of agriculture in the Fordist period drew on the analysis of the institutional forms underpinning it. It relied on the 'canonical' forms of Regulation Theory used to analyse national economic transformations (Boyer, 1989) while calling into question their sectoral specificity.

In examining the wage and labour social relations, a key notion of Regulation Theory, researchers sought to describe their specific evolution in the agricultural sector. These were characterised not only by certain general features of Fordism (specialisation, intensification and simplification of work, creation of norms and standards, dependence on agro-industrial firms) but also by the permanence of family farms and the emergence of an organised farming profession with associated status (Allaire, 1988a, 1991; Lacroix and Mollard, 1990). Empirical studies were undertaken with sociologists to illustrate and examine the specific features of the labour social relations in agriculture (Lacroix et al., 1995). These revealed how the institutionalisation of a model of professional farms and farmers took place – to the detriment of other forms of farming and social groups (Coulomb et al., 1990; Laurent et al., 1998).

Other research has focused on the role of the State, and on the institutionalisation of compromises with social groups of farmers. In France, these compromises have been embodied in public policies based on joint management between the State and farmers' unions. This form of 'neo-corporatism' facilitated the modernisation of agriculture by orienting access to productive resources to certain professional farmers. For instance, the State and unions of professional farmers jointly managed organisations controlling access to credit (role of the mutual agricultural credit,[1] of bookkeeping associations[2]), to land (role of the SAFER[3]), and to knowledge (role of the chambers of agriculture). These institutional arrangements also contributed to the regulation of the economic contradictions that crystallised around the distribution of the means of production (land, capital) and the definition of the publics targeted by the national agricultural support policies (Coulomb and Delorme, 1987; Allaire, 1988b).

Finally, these early Regulation Theory studies of agriculture also focused on market institutions, the new forms of competition, and relations between supply and demand on a national and European scale, in relation to the emergence and subsequent evolution of the Common Agricultural Policy (CAP) (Bartoli, 1985). Beyond the decrease in the prices of agricultural products, which reflected the marketisation and industrialisation of agriculture in the national economy (Servolin, 1989), research explored i) the institutions and social compromises that created opportunities for the emerging quality-based differentiation of agrifood markets (Bartoli and Boulet, 1989), and ii) the evolution of relations between farmers and upstream and downstream firms (Chevassus-Lozza et al., 1999), involving various types of contract.

The development of such research rapidly challenged the possibility of transposing to a sectoral level the concepts of Regulation Theory that were initially constructed on a macro-economic scale. Three risks were identified. The first risk lies in a homothetic transposition of macro-economic periodisation to the sectoral level, which would eliminate any sectoral specificity. The second risk involves the adoption of an overly functional point of view in which agriculture would be only considered in terms of its (partly specific) contributions to the national development model, such as providing workers. Conversely the third risk stems from regarding sectoral dynamics as being autonomous from global economic dynamics (Bartoli and Boulet, 1990; Boyer, 1990). These difficulties were shared with empirical research carried out in other sectors, notably the car industry (Boyer and Freyssenet, 2000). Research on the agricultural sector contributed significantly to theoretical reflections on sectoral regimes, using the example of wine in particular (Bartoli and Boulet, 1990). It spawned generic concepts, a method, and an approach to sector-based economic analysis (Boyer, 1990). Bartoli and Boulet's work thus formalised the notion of sectoral economic regime, that is, '*the economic mechanisms that enable the reproduction of the sector*', and that of sectoral 'institutional arrangements', that is,*' all the institutions that produce the norms, processes and interventions framing and orienting actions and economic flows at the sector level'* (Bartoli and Boulet, 1990: 19). The articulation between sectoral dynamics and the national development model was then envisaged in a less mechanical way, by reasoning in terms of the constraints and opportunities afforded by national models, and the repercussions that sectors had on macro-economic dynamics and its institutions. The first analysis of coexistence between diverse economic regimes within agriculture (e.g. basic wine vs quality wine) also contributed substantially to shape a more comprehensive view of this articulation.

The first studies undertaken within the Regulation Theory framework also examined the crises and transformations of the sectoral regime of agriculture institutionalised during the Fordist period. Several aspects of the crisis were already evidenced in the 1980s: market crises; overproduction and the first reforms of the CAP (milk quotas, subsidies for removing vineyards, etc.); political protest and emergence of alternative agricultural systems (e.g. organic agriculture); and so on. The analysis of these developments also showed that the emergence of new societal issues was being taken into consideration in the political arena in Western European countries (Laurent, 1992; Allaire and Boyer, 1995; Touzard, 1995; Nieddu and Gaignette, 2000), on the North American continent (Debailleul, 1990) and in countries of the South (Losch, 2000; Bosc and Losch, 2002; Anseeuw, 2004). The book *La grande transformation de l'agriculture* by Allaire and Boyer (1995) was a major attempt to synthesise Regulation Theory and institutionalist studies describing the growth and crisis of agriculture in the post-Fordist period. The research gathered in this book signalled both the rise of these new issues and the attendant economic and political processes: the construction of new qualities of food products; the evolution of labour relations in farm households; the growing complexity of agricultural policies; recognition of the role of agriculture in regional development and environmental issues; and so on. By stressing the coexistence of a plurality of forms of agriculture positioned differently with regard to these issues, and the collaboration needed to strengthen institutionalist approaches to capture these trends, the book highlighted a promising new research agenda for Regulation Theory research on agriculture.

An extension of research to different dimensions of the great transformation of agriculture

Subsequent to the publication of this book (and many seminars and discussions on the topic), researchers inspired by Regulation Theory undertook diverse in-depth explorations of the various dimensions of agricultural and agri-food transformations. These studies extended, deepened and completed earlier work.

Research on the organisation and social relations of labour continued to highlight the particularities and evolution of the farming profession (Laurent, 1995; Lacombe, 1998; Mundler and Laurent, 2003), taking into account the tension between specialised tasks and the 'non-agricultural' work in farm households. The analysis was broadened by the concept of 'systems of activity' (Laurent et al., 1998), seen as a component of the diversity of agricultural models. It contributed to the understanding of institutional arrangements framing the sector, and of their crisis (Delord et al., 2000).

The analysis of the forms of competition and market organisation focused primarily on the question of the quality of agri-food products, and on its central role in the definition of various sectoral regimes (or quality regimes) and the differentiation of markets (Boulet and Touzard, 1995; Allaire and Sylvander, 1997; Allaire, 2002).

The economic analysis of public support for agriculture showed the importance of the articulation of the various scales of public action: European, national and regional (Berriet-Solliec, 1999; Delord et al., 2000; Delorme, 2002). It highlighted the growing role of regional institutional arrangements in the dynamics of the sector (Genieys and Smith, 2001).

The insertion of agriculture in local or regional development dynamics thus constituted a new field of study that facilitated the understanding of intra- and inter-sectoral mechanisms of coordination, as well as the conditions of emergence of alternative forms to the dominant agri-industrial development model (Laurent, 1995; Touzard, 1995; Nieddu and Gaignette, 2000; Pecqueur, 2001; Gilly and Wallet, 2005).

The relations between agricultural activities and environmental issues were also a new direction for research. These studies showed the importance of new social compromises about nature and environmental issues, and suggested that agriculture was taking on an important role in the emergence of a political ecology (Le Roch, 1993; Lacroix et al., 1995; Lipietz, 1995; Becker and Raza, 2000; Barthelemy and Niedu, 2003).

The question of innovation and the construction of knowledge in the agricultural and agri-food sector was also studied specifically in reference to Regulation Theory. Researchers sought both further understanding of the processes of technological change in the sector and to critically examine the evolution of agricultural development institutions (Allaire, 1996; Byé, 1997; Touzard, 2000).

There has also been growing interest in agricultural transformations in countries other than France, evidenced mainly in various PhD theses, particularly in Latin America, Europe and North America (Goodwin et al., 1995; Quemia, 2001; Labarthe, 2006; Lopez, 2006; Trouvé, 2007; Grouiez, 2010; Lataste, 2014), as well as several studies of African economies (Hugon, 1993; Griffon, 1994; Chastel, 1995; Losch, 2000; Anseeuw, 2004; Marzin, 2006).

In this abundant literature, it seems that two dimensions have played a particularly important role for Regulation Theory. First, a growing body of research on the transformations of agri-food systems and agricultural regimes has analysed new issues around product quality (organoleptic, sanitary, environmental, origin-related quality, *etc.*) and new forms of organisation of production. It has contributed to integrating a systemic vision and exploration of the diversity of agri-food models and agricultural innovation systems (Touzard et al., 2014). Second, the integration of the territorial dimension into Regulation Theory has been another significant development, focusing on local conditions of agricultural crisis and integration of agriculture in regional governance (Berriet et al., 2006).

It has led to very rich academic debate associated with the new contradictions between the sectoral and territorial implications of agriculture that appear when the multiplicity of objectives associated with agriculture are recognised (Laurent, 1995). These studies were discussed and developed within the French Regulation Theory scientific community, particularly at INRA (French National Institute of Agricultural Research) and CIRAD (French Agricultural Research for Development), as well as through the working group '*Régulation sectorielle et territorialle*' (Laurent and du Tertre, 2008). Interaction grew between Regulation Theory research in agriculture and in other industrial or service sectors.

In the 1990s, Regulation Theory research applied to agriculture concentrated on the factors accounting for the profound crises and shifts taking place in the agricultural sector, particularly in France. These crises related to the upsurge of problems of overproduction and competition, and to those concerning the environmental, sanitary and social consequences of the intensification of agricultural production. Yet no consensus emerged from this research regarding a new sectoral regime of agriculture and a solution to the crisis. It is surprising that research stemming from Regulation Theory does not play a bigger role in today's academic debate on the identification of these new economic regimes of the sector, notably for purposes of international comparison.

Comparisons and combinations of Regulation Theory with other research programmes analysing transformations of agricultural regimes

A particularity of Regulation Theory is that its development takes place partially through exchanges with other institutional economics research programmes and other social sciences. The development of Regulation Theory research applied to agriculture is a typical illustration.

Studies that drew on Regulation Theory to analyse the organisation and social relations of farm labour were similar to approaches in rural or critical sociology, and to systemic agronomy (farming systems, livestock breeding systems). This collaboration was extended (Mundler and Laurent, 2003; Madelrieux et al., 2010), and helped i) to stabilise the generic notion of social relations of activity (Laurent and Mouriaux, 2008), and ii) to explore the new development of large-scale, enterprise-based farms[4] in France (Olivier-Salvagnac and Legagneux, 2012) and Latin America (Requier-Desjardin et al., 2014).

Research on the regimes of 'specific quality' products was associated with the dynamics of other approaches, particularly Convention Theory or economic sociology (Boulet and Touzard, 1995; Allaire and Sylvander, 1997; Karpik, 2007; Allaire and Daviron, 2008; Chiffoleau et al., 2008; Allaire, 2010, 2013). Internationally, this research coincided with studies in political science (Bonanno and Busch, 2015), particularly with that on quality audited supply chains and international negotiations of food standards (Friedmann, 2005).

Research on agricultural policies, rural development, and European policy (Labarthe, 2005; Berriet-Solliec et al., 2008; Trouvé, 2009) has woven collaborative relations together with political science approaches (Smith et al., 2007; Shucksmith, 2010). Collaboration has contributed to a better understanding of the construction of these policies, according to new compromises, which include new social groups (consumers, NGO, different bureaucracies, etc.).

The evolution of agriculture in territorial dynamics has been a subject of new research related to the economics of proximity (Pecqueur, 2001; Pecqueur and Zimmerman, 2004; Gilly and Wallet, 2005), rural sociology and geography, and to research on localised agri-food systems (Muchnik et al., 2007; Albaladejo, 2012). These analyses of agricultural transformation at a local scale have provided empirical evidence of the new territorial governance of agriculture, especially driven by regions, cities or some of the actors in the tourism sector (Torre and Filippi, 2005).

More recent Regulation Theory studies relating agriculture to environmental issues, ethics or innovation have likewise involved collaboration with other academic fields: sustainable development and ecological economics (Zuindeau, 2007), evolutionary theory and transition approaches (Nieddu et al., 2010; Labarthe, 2010; and even history and law (Tordjman, 2008). These collaborations converge in highlighting the renewal of agricultural sectoral specificities, as far as innovation systems, provision of public goods or professional knowledge are concerned (Touzard et al., 2014).

The encounter between Regulation Theory-based approaches and other institutionalist approaches or even other disciplines has unquestionably been of great heuristic value in the understanding of transformations of agriculture. Conversely, the agricultural sector has been a breeding ground for empirical research that has contributed to developing or clarifying new concepts for Regulation Theory (sectoral regimes, quality regime, social relations of activity, territorial governance, etc.). However, these studies have not led to the elaboration of a new synthesis that might serve to characterise or map new sectoral regimes of the agricultural sector, from global, historical and comparative perspectives.

There are various possible explanations for this situation, some of which concern the specific, internal dimensions of Regulation Theory, particularly the need to combine this approach with others in order to connect micro-economic to meso- or macro-economic analyses. Regulation Theory provides a general framework anchored in macro-economics that can be applied to meso-economic approaches. This framework encourages the use of advances from other theoretical approaches that are better equipped from a micro-or meso-economic point of view, but provide no guarantee of a systematic return to Regulation Theory to finalise a more synthetic approach. The operational dimension has also been mentioned. The analytical and particularly retrospective contributions of Regulation Theory have been widely recognised, but the predictive, normative or even maieutic nature of Regulation Theory is a subject of debate. The fact that Regulation Theory is based on sound empirical work (statistics, case studies) is not challenged, but its operational use is (Favereau, 1995; Pecqueur, 2007). One reason may also lie in the difficulty of organising international comparative studies on (national) agriculture and agri-food systems. Attempts have nevertheless been made in this respect, especially in PhD research, for example Anseeuw (2004), Labarthe (2006), Trouvé (2007), Grouiez (2010) and Lataste (2014). There have been various international partnerships and Regulation Theory has been used by international scholars beyond France (Marsden, 1992; Gibbs, 1996; Wilkinson, 1997). But a collective effort to reach a synthesis is still lacking. This situation also relates to the institutional constraints and power relations that have become less favourable to Regulation Theory in the academy of economics. Geographical dispersion and the small number of researchers working on agriculture and Regulation Theory explain this, and the difficulties of recruiting new researchers and securing resources for institutional economics are very real.

The purpose of this article, however, is not to explore in detail the reasons for these trends. Instead, we would like to draw on our personal research experience to propose two avenues to strengthen Regulation Theory's contribution to the analysis of the diversity of agriculture and agri-food systems. To this end, we believe that it is necessary to reinforce the integration of Regulation Theory approaches in international and pluri-disciplinary communities, and so to foster a synthesis within Regulation Theory. In the following two sections, we present two lines of research that call for a contribution by Regulation Theory to comparative projects and syntheses of agricultural economic regimes: one concerning ecological transitions, and the other, food regimes and agri-food models. We believe that the contribution of Regulation Theory to these syntheses would enhance the analyses carried out in these multidisciplinary communities. In turn, Regulation Theory could benefit from debate within these communities, essentially to continue the characterisation of institutional forms that bring to the fore the core mechanisms of sectoral regimes (the economic mechanisms of access to knowledge, quality standards, etc.). It could also help to link the micro-, meso- and macro-economic scales in this analysis.

Strengthening Regulation Theory's contribution to international research on ecological transitions

As noted in the preceding section, since the late 1980 RT research has highlighted the environmental dimension of the crises faced by agriculture in industrialised countries. The analysis of agricultural and agri-food systems' transitions towards a better integration of societal issues, including environmental ones, is therefore an important research thrust for RT and, more broadly, for institutionalist approaches. This is evidenced in the current development of the ecological economics community (Plumecocq, 2014).

Since the early 2000s there has been an upsurge of research on ecological transitions. This movement did not stem predominantly from economics; instead it has been driven by political science and science and technologies studies (STS). Some authors have proposed a multi-level framework, including 'niches', where ecological innovations appear, and 'socio-technical systems', where a sectoral regime that integrates these innovations (or not), is institutionalised (Geels, 2002). The starting point of these approaches lies partially in a critique of the failure to integrate the demand side into earlier formulations of sectoral systems of innovation (Geels, 2004), embedded in the field of innovation economics, and notably evolutionary economics (Malerba, 2002). Transitions studies thus incorporate new modalities of social construction of food product quality and of innovation (Geels, 2010), and highlight new forms of technological and institutional lock-in (Vanloqueren and Baret, 2009).

Yet these transition studies have also been criticised (Bui, 2015), notably for their conceptualisation of socio-technical regimes, where innovation support is institutionalised (Berkhout et al., 2004; Genus and Coles, 2008; Holtz et al., 2008; Markard and Truffer, 2008). Criticisms highlight the failure to analyse the role of certain actors (firms, intermediate organisations, etc.) in transitions, and potential conflicts between these actors (Smith et al., 2005; Shove and Walker, 2007; Genus and Coles, 2008). Transition studies therefore leave a blind spot around the mechanisms regulating access to the resources needed for the transitions. Regulation Theory research can play an important part in that respect, by making it possible to analyse the institutions shaping this access. For instance, the conditions of access to relevant and reliable knowledge for different actors of the agricultural sector (farmers, advisers, researchers, policy makers) have direct impacts on the speed and direction of technological change (Landel, 2015; Laurent et al., 2010, 2012).

In other words, Regulation Theory research could better contribute to debates around ecological transitions by analysing how the institutional arrangements between different social groups regulate access to certain resources (and notably access to knowledge), and can in turn facilitate or, on the contrary, impede transitions (Labarthe, 2009): '*the economic performance of a technical system depends heavily on societal factors [. . .]: they are institutionalised compromises that define socio-technical trajectories marked by phenomena of reversibility and irreversibility*' (Boyer, 1989). This could contribute to renewing the dialogue between Regulation Theory and other heterodox approaches, notably evolutionary economics (Dosi and Coriat, 1995), about the analysis of path-dependency and lock-in mechanisms (Boyer et al., 1991).

In turn, Regulation Theory participation in debates on transitions could reinforce the collective effort needed to adapt the conceptualisation of Regulation Theory's institutional forms to contemporary transformations of economic regimes of sectors such as agriculture. The canonical institutional forms of Regulation Theory, or the concepts derived directly from it, still have a high heuristic value for enriching our understanding of transformations underway in agriculture. They constitute one of the methodological bases of Regulation Theory that we have to develop further collectively in order to analyse new components of contemporary capitalism and sectoral regimes (Petit, 1998).

It therefore seems necessary to continue the analysis of social relations related to labour in the agricultural sector, in view of the profound changes taking place: radical decline in the number of farms; an increase in the share of salaried work in agricultural employment, with a growing proportion of migrants (Laurent, 2013, 2015); new forms of collective organisation of work or of integration of services in agriculture (Nguyen and Purseigle, 2012); and evolution of competences and qualifications related to new technologies or the demands of food industries (e.g. managing traceability), and so on.

Likewise, the role of the State has changed profoundly, as the focus of public intervention has moved away from direct investments in the sector and towards regulation with standards (on the environment, health, *etc.*) or to strengthening institutional capacities and coordination at local and regional level (e.g. promotion of partnerships and platforms for innovation or commercialisation). These policies correspond to new forms of institutionalised compromises between farmers, land owners, environmental organisations, local/regional authorities, and the State. They differ from one country to the next and can come into conflict with one another in European policy-making (Trouvé and Berriet-Solliec, 2010; Lataste et al., 2012).

The analysis of new forms of competition is also at the heart of transformations in farming. This is due to the rapid development of new quality criteria, particularly with regard to sanitary concerns (Saulais, 2015), and to the transformation of international regimes, with the upsurge of new actors on the demand side, such as China (Chaumet and Pouch, 2012), the supply side in Brazil, for example, (Fèvre and Pouch, 2013), storage in China (Courleux and Depeyrot, 2014), and in agricultural land grabbing (Anseeuw et al., 2013), especially by large multinationals.

Greater participation in debates on transitions could enhance the theoretical debates within Regulation Theory about the articulation between the micro-, meso- and macro-economic levels (Lamarche et al., 2015). This articulation is at the heart of the multi-level perspective approach proposed by transitions studies (Geels, 2002). These studies also provide an interesting framework for building collaborations, including forums, seminars, working groups and research projects, and possibly new scientific alliances at the international level to undertake historical studies and international comparisons – approaches that lie at the heart of Regulation Theory. Research applied to agriculture has already explored the potential of this approach, for instance, in the case of the development of green chemistry (Nieddu and Vivien, 2015, 2016). These debates constitute an opportunity for Regulation Theory-based research to pursue the analysis i) of the crisis of the intensive agri-industrial model, ii) of the deployment of a new alliance between science and capitalism (new doctrines around public-private partnerships for innovation, etc.), and iii) of the development of alternative production systems or even lifestyles.

Another possible, complementary way to accentuate and enrich the contribution of Regulation Theory to the analysis of agricultural regimes is to reinforce the dialogue with approaches and debates in terms of agri-food models and food regimes.

Strengthening the contribution of Regulation Theory to the analysis of the transformation of agri-food models and food regimes

RT – or similar – studies of the transformation of agriculture show an overlap between spaces of regulation and those of transition, with the maintenance of the national level and the importance of international dependencies revealed by the food crisis in 2007–08 (Touzard, 2009). Such studies also point out the confrontation of various economic regimes and 'models', at least alternative vs conventional (Goodman et al., 2012), within the agri-food sphere, both in the North and in the South (Allaire and Daviron, 2008). The global scale, the better analytical integration of national regulations, international relations and the diversity of models could be enriched by two concepts: 'agri-food models' and 'food regimes'.

Agri-food models refer to the observation of regularities and consistencies in patterns of production, trade and consumption of food, according to structural and cognitive principles. They were proposed in early research on 'food systems' (Malassis, 1979), by Regulation Theory research on economic regimes (Touzard, 2009) and by contributions inspired by conventions theory integrating consumption models (Fonte, 2002) and conventions on food quality (Ponte, 2016). A characterisation of agri-food models has been proposed based on the Regulation Theory method and on research on the economy of quality, derived from conventionalist approaches (Fournier and Touzard, 2014). Whether these models stem from the persistence of a diversity inherent to agriculture or from the fragmentation of the former agro-industrial model, they stand out for their specific combinations between i) economic dynamics and logic, ii) institutional arrangements already recognised by Regulation Theory (organisation of work, form of competition, relation to the State), and iii) their specific relations and conventions on technology, territory and, above all, food quality, which should be integrated more fully into Regulation Theory.

The *agro-industrial model*, based on the logic of profit maximisation for agri-industrial firms and retailers, aims primarily at homogeneity and low costs for mass consumption, promoting market and industrial conventions on food. It still prevails through two variants (Rastoin and Ghersi, 2010): i) a model that integrates family agriculture with agri-food firms (both upstream and downstream), and that can benefit from sustained government support and the influence of corporatist organisations formally controlled by farmers; and ii) a wage and financial agribusiness model that was reaffirmed in North and South America (Wilkinson, 2002) around genetically modified (GM) maize and soy, wheat, sugar cane and the pursuit of food industrialisation, and in Russia after decollectivisation (Grouiez, 2013). It is also being revived in Europe (e.g. in intensive breeding) and in Africa with land grabbing and agri-industrial complexes (Anseeuw et al., 2013).

In interaction with this agro-industrial model, several others appeared, which seem to be renewed versions of earlier models. The *domestic model*, built around food consumption at the level of the family unit of production and transformation still provides the subsistence of a large proportion of the population in southern countries. It is also being revived in both the North and South, including urban areas (family or community gardens, urban agriculture, etc.). The *proximity model* is characterised by few intermediaries (short supply chains, direct selling, etc.) and direct contact between producers and consumers, building trust through domestic and market conventions. It is being renewed in countries of the North in local food supply chains and over longer distances via information and communication technologies (ICT). The *artisanal commodity model*, which for a long time enabled the trading of food products over medium distances through trade networks with many intermediaries, is still very much present in countries of the South.

Research on the economy of quality has highlighted above all *differentiated quality models*, based on economic regimes that combine higher costs and prices, on institutions that guarantee the product's attributes (signs of quality), and on specific conditions of consumption (social differentiation, festive or cultural conditions). These models have actually existed in various forms for a very long time, but have received extensive media coverage due to their critique of the agri-industrial model and their claim to better meet certain societal challenges concerning agriculture and food (Colonna et al., 2013). The *origin based model* is thus promoting the differentiation of quality according to origin, and the value attributed to the heritage of a place, like products under geographic indications. The economic regime of this model relies on a set of institutions and conventions that links a community to the geographical space and its public goods (Belletti et al., 2017). The *naturalist quality model* highlights practices (agricultural, post-harvest) that respect the environment and that the organic or agroecology supply chains claim to use. The *ethical quality model* groups together products differentiated in terms of respect for an ethic, whether it be social, religious or community. It generally displays support for a category of people (small farmers or disabled farmers, for example).

The coexistence of these models in most countries calls into question sectoral approaches to regulation, essentially by examining their articulation on a national and international scale. It is de facto the generalisation of this coexistence that seems to be a current feature of globalisation, and that invites us to put Regulation Theory back at the heart of our analysis of the various regimes regulating agriculture. On this global scale, two antagonistic movements play a major role: a 'market liberalisation' and disengagement of the State movement that leads to a commoditisation of agriculture and food (desectorialisation); and a movement of political reinvestment in value chains and in different arenas of negotiation (from local to international), which can lead to the reassertion of the sectoral particularity of agriculture on a global scale. Thinking and analysing these trends together in the framework of Regulation Theory requires us to take into account principles that, on an international scale, ensure compatibility between activities of production, exchange and consumption of food, and thus the coexistence of agri-food models. The notion of food regime, derived essentially from the work of Friedmann and McMichael (1989), seems to be useful here.

The food regime approach is above all a historical approach aimed at linking the international relations on production, exchange and consumption of food, with the evolution of the forms of capitalism, underlying the role of hegemonic (global) actors. Beyond recent debates between McMichael, Bernstein and Friedman (McMichael, 2009; Friedman, 2016), the food regime approach identified three food regimes during the two last centuries. We argue that this approach is also useful in relation to Regulation Theory, i) to understand the coexistence (and contradictions) of various sectoral regimes and agri-food models at international scale, and i) to explore scenarios for agriculture and food transition at this scale.

After the *colonialist food regime* (1870–1914), the food regime of the Fordist period has been called either *'mercantile-industrial'* or 'State-regulated' (Friedmann, 2005). It can be characterised ex post by international markets steered by States (GATT, use of food as a weapon, national or regional sectoral regulations, the CAP, etc.) under US hegemony, coupled with the promotion of the intensive agricultural development model (the Green Revolution) and the development of mass food consumption (industrialisation, urbanisation, standardisation, concentration, etc.) progressively benefitting agri-food multinational firms. It led to more regional specialisation of agricultural production, while maintaining overall the modernisation of family farming within a diversity of agri-food models, dominated by the State-regulated agri-industrial model. Sectoral particularities, oriented by national agricultural modernisation policies, were affirmed overall.

The evolution of this regime has led to a *corporate and environmental food regime* (Friedmann, 2016) driven by multinational firms developing in a neoliberal and media form of capitalism (Allaire and Daviron, 2008). This new regime developed from the end of the 1970s and was characterised by the withdrawal of States, a financialisation of agri-food markets, and a political economy of standards driven by multinational firms' strategies, partially integrating some environmental issues (Fouilleux and Goulet, 2012; Loconto and Fouilleux, 2014). This regime is also characterised by the growing importance of ICTs and the promotion of new biotechnologies, under a more fragmented world in terms of political control, with the emergence of new countries in the global food market, such as Brazil and China. The 2008 crisis showed the limits of this food regime and triggered many debates on agri-food models and the possible ways in which the international food regime might evolve: i) the pursuit of the *corporate and environmental* regime strengthening the private management of commodities (partially framed by public safeguards set up on a global scale), the development of 'global value chains' targeted at various quality standards, and a new agrarian capitalism; ii) the emergence of a *civic food regime* around international negotiations between governments, firms, the scientific community and social movements, that recognises the diversity of agri-food models, their possible contributions to public goods, the legitimacy of national or regional food sovereignty, and the importance of agroecology and local know-how; ii) the 'return' to more fragmented international regulation that gives back a role to the State (bilateral agreements between States, negotiation between State and multinational firms, protectionism, etc.), with a diversity of trajectories and combinations of agri-food models, according to the national policies (and perhaps the hegemony of China).

The analysis of agri-food models and their place in different international food regimes also points to the need to reconsider the question of the particularities of the agri-food sector, with regard to the global trends of capitalism and other sectoral dynamics. It is necessary to identify those trends that express a global convergence and those that seem to renew the bases of sectoral particularities. This issue arises from both recent debates on food regimes (Bernstein, 2016; Friedmann, 2016) and discussion of agri-food models (Touzard et al., 2014). The general transformations of the 'work relationship' or 'activity relationship' in agriculture, the mediatisation and accountability of food supply chains, the weakening of national agricultural policies, the financialisation of activities and development of territorial governance, *inter alia*, are all features that are generally shared with the other sectors. At the same time, the basic characteristics of agricultural and agri-food activities (dependence on bioclimatic resources and risks; inherited family productive structures; importance of local knowledge; cultural or symbolic dimension of foods; etc.) maintain a series of specific problems and challenges (Touzard et al., 2014) and are an incentive for the maintenance or emergence of particular institutional arrangements, such as the construction of 'alternative agri-food systems' or negotiations on the 'environmental services' of agriculture (Aznar et al., 2007; Zuindeau, 2007). It is in the combination of these structural characteristics (which are never entirely exclusive to the sector) that the notion of agri-food particularity can be renewed, one that is affirmed to a greater or lesser degree, depending on the agri-model considered.

Conclusion

Studies in the French Regulation Theory tradition have contributed substantially to stabilising the notion of 'economic regime' on a sectoral scale. A period in the early 1980s of fruitful Regulation Theory-based research on agriculture led to an attempt in the mid-1990s to synthesise it (Allaire and Boyer, 1995). This made it possible to characterise the economic regime of agriculture in industrialised countries during the Fordist period, and then the different dimensions of the crisis into which this sectoral regime was entering. Studies of agriculture then increasingly focused on these different dimensions during the 2000s. These were characterised by a proliferation of interaction with other currents of thought, not only in institutionalist economics (Convention Theory, evolutionary economics, etc.) but also within other social sciences (sociology, political science, etc.), on a diversity of research trajectories. Yet in this period no new collective attempt was made to synthesise this work and characterise, within the Regulation Theory framework, the evolution of economic regimes occurring in the diversity of national agricultural and agri-food sectors and in a context of globalisation and liberalisation of trade. In this chapter, we argue that the current context of agriculture and food affords opportunities precisely to revive that effort to produce a collective synthesis that articulates research on several scales, taking into account a diversity of agri-food models and critically examining the mechanisms of their transitions towards an integration of the environmental challenges.

In this chapter, we have proposed two possible and complementary routes for the integration of Regulation Theory into efforts at international level to synthesise the diversity of economic regimes in agricultural and food sectors. The first route consists of a better insertion of Regulation Theory in multidisciplinary debates on ecological transitions. These debates afford an interesting possibility for multi-level integration of the analysis of institutional forms that support economic regimes. The second route consists of dialogue between research in Regulation Theory and that which characterises the diversity of agri-food models and the evolution of food regimes on a global scale. In both cases, the conceptual underpinnings of Regulation Theory retain a high heuristic value to contribute to a better understanding of economic regimes and their transformations, by emphasising certain institutional forms and the dynamics of institutionalised compromises between various groups of actors. In return, the research communities working on transition studies or food regimes, with the richness of their debates, networks and projects, largely focused on comparative and historical analyses, encourage the integration of analyses at the micro-, meso-, and macroeconomic level. These communities could prove to be fruitful and reciprocal allies in the further development of Regulation Theory.

Notes

1 *Crédit Agricole.*
2 *Centre d'Economie Rurales*, farmer-based organisations specialised in bookkeeping and economic advisory services.
3 *Sociétés d'aménagement foncier et d'établissement rural.*
4 Olivier-Salvagnac and Legagneux (2012) characterise these farms by their large economic scale, with specific legal forms that allow for investments from outside the farm household. These farms tend to rely on farm employees and/or on contract farming.

References

Albaladejo C., 2012. Les transformations de l'espace rural pampéen face à la mondialisation. *Annales de géographie*, 4, 387–409.

Allaire G., 1988a. Itinéraires et identités professionnels des travailleurs de l'agriculture. *Actes et Communications de l'INRA*, 3, 175–211.

Allaire G., 1988b. Le modèle de développement agricole des années 1960. *Économie rurale*, 184, 171–181.

Allaire G. 1991. Développement et formes de travail: les formes sociales du travail agricole, communication présentée au *septième congrès mondial de sociologie rurale*, Bologne (Italie).

Allaire G., 1996. Emergence d'un nouveau système productif en agriculture. *Canadian Journal of Agricultural Economics/Revue canadienne d'agroéconomie*, 44(4), 461–479.

Allaire G., 2002. L'économie de la qualité, en ses secteurs, ses territoires et ses mythes. *Géographie économie société*, 4(2), 155–180.

Allaire G., 2010. Applying Economic Sociology to Understand the Meaning of "Quality" in Food Markets. *Agricultural Economics*, 41(s1), 167–180.

Allaire G., 2013. Les communs comme infrastructure institutionnelle de l'économie march-ande. *Revue de la régulation*, 14. http://regulation.revues.org/10546.

Allaire G., Boyer R., 1995. *La grande transformation de l'agriculture: lectures convention-nalistes et régulationnistes*. Paris, Quae.

Allaire G., Daviron B., 2008. Régime d'institutionnalisation et d'intégration des marchés. In Chiffoleau Y., Dreyfus F. and Touzard J.-M. (Eds.), *Nouvelles figures des marchés agroalimentaires: apports croisés de la sociologie, de l'économie, et de la gestion.* Versailles, Éditions Quæ.

Allaire G., Sylvander B., 1997. Qualité spécifique et innovation territoriale. *Cahiers d'économie et sociologie rurales*, 24, 29–59.

Anseeuw W., 2004. *Reconversion professionnelle vers l'agriculture marchande et politique publique en Afrique du Sud*. Thèse de Doctorat, Université de Grenoble IIGrenoble, France.

Anseeuw W., Lay J., Messerli P., Giger M., Taylor M., 2013. Creating a Public Tool to Assess and Promote Transparency in Global Land Deals: The Experience of the Land Matrix. *Journal of Peasant Studies*, 40(3), 521–530.

Aznar O., Guérin M., Perrier-Cornet P., 2007. Agriculture de services, services environne-mentaux et politiques publiques: éléments d'analyse économique. *Revue d'Économie Régionale & Urbaine*, 2007(4), 573–587.

Barthelemy D., Nieddu M., 2003. Multifonctionnalité agricole: biens non marchands ou biens identitaires? *Économie rurale*, 273, 103–119.

Bartoli P., 1985. *La politique viticole, un exemple de régulation'*, communication présentée au colloque franco-hongrois d'économie rurale. Paris, INRA ESR.

Bartoli P., Boulet D., 1989. *Dynamique et régulation de la sphère agroalimentaire: l'exemple de la sphère viticole*. Thèse d'Etat, Université de Montpellier IMontpellier, France.

Bartoli P., Boulet D., 1990. Conditions d'une approche en termes de régulation sectorielle: le cas de la sphère viticole. *Cahiers d'économie et de sociologie rurales*, 17, 7–38.

Becker J., Raza W., 2000. La theorie de la regulation et l'ecologie politique: une separation inevitable? *Economies et Sociétés*, 34(1), 55–70.

Belletti G., Marescotti A., Touzard J.-M., 2017. Geographical Indications, Public Goods, and Sustainable Development: The Roles of Actors' Strategies and Public Policies. *World Development*, 98, 45–57.

Berkhout F., Smith A., Stirling A., 2004. Socio-Technological Regimes and Transition Contexts. In Elzen B., Geels F.W. and Green K. (Eds.), *System Innovation and the Transition to Sustainability: Theory, Evidence and Policy*. Cheltenham, Edward Elgar, pp. 48–75.

Bernstein H., 2016. Agrarian Political Economy and Modern World Capitalism: The Contributions of Food Regime Analysis. *The Journal of Peasant Studies*, 43(3), 611–647.

Berriet-Solliec M., 1999. *Les interventions décentralisées en agriculture. Essai sur la composante territoriale de la politique agricole*. Paris, L'Harmattan.

Berriet-Solliec M., Déprés C., Trouvé A., 2008. La territorialisation de la politique agricole en France. Vers un renouvellement de l'intervention publique en agriculture. In Laurent C. and Du Tertre C., (Eds.), *Secteurs et territoires dans les régulations émergentes (Laurent Dutertre)*. Paris, L'Harmattan, pp. 121–136.

Berriet-Solliec M., Mouriaux M.-F., Delorme H., Mundler P., Laurent C., Perraud D., 2006. Régulation de l'agriculture: les Régions comme nouveau lieu de mise en cohérence territoriale des politiques agricoles? La région Rhône-Alpes dans le contexte européen. *Canadian Journal of Regional Science*, 29(1), 55–73.

Bonanno A., Busch L., 2015. The International Political Economy of Agriculture and Food: An Introduction. In Bonanno A. and Busch L. (Eds.), *Handbook of the International Political Economy of Agriculture and Food*. New York, Edward Elgar Publishing, pp. 1–15.

Bosc P.-M., Losch B., 2002. Les agricultures familiales africaines face à la mondialisation: le défi d'une autre transition. *Oléagineux, Corps Gras, Lipides*, 9(6), 402–408.

Boulet D., Touzard J.M. 1995, Filière, Territoire et construction sociale de la qualité: l'exemple du marché du vin à la production', communication présentée au colloque *Qualification des produits et des territoires*, 2 et 3 octobre 1995.

Boyer R., 1989. Histoire des techniques et théories économiques. Vers un nouveau programme de recherche. *Cahiers du CEPREMAP*. 8903.

Boyer R., 1990. Les problèmes de la régulation face aux spécificités sectorielles. *Cahiers d'économie et de sociologie rurales*, 17, 39–76.

Boyer R., Chavance B., Godard O., 1991. *Les figures de l'irréversibilité en économie*. Paris, Editions de l'ecole des hautes études en science sociales.

Boyer R., Freyssenet M., 2000. *Les modèles productifs*. Paris, La Découverte.

Boyer R., Saillard Y., 1995. *Théorie de la Régulation: l'état des savoirs*. Paris, La Découverte.

Bui S., 2015. *Pour une approche territoriale des transitions écologiques. Analyse de la transition vers l'agroécologie dans la Biovallée*. Thèse de doctorat, Agroparistech, Paris.

Bye P., 1997. Productive Inertia and Technical Change. *Science Technology & Society*, 2(1), 131–150.

Chastel J.-M., 1995. *Le rôle des institutions dans l'évolution de la filière canne à sucre à La Réunion*. Thèse de Doctorat. Montpellier, ENSAM.

Chaumet J.-M., Pouch T., 2012. La Chine au risque de la dépendance alimentaire. *Oléagineux, Corps gras, Lipides*, 19(5), 290–299.

Chevassus-Lozza E., Gallezot J., Galliano D., 1999. Les déterminants des échanges internationaux intra-firme : le cas de l'agro-alimentaire français. *Revue d'économie industrielle*, 87(1), 31–44.

Chiffoleau Y., Dreyfus F., Touzard J.-M., 2008. *Nouvelles figures des marchés agroalimentaires: apports croisés de la sociologie, de l'économie, et de la gestion*. Versailles, Editions Quae.

Colonna P., Fournier S., Touzard J.-M., 2013. Food Systems. In Esnouf C., Russel M. and Bricas N. (Eds.), *Food Systems Sustainability*. Cambridge, UK, Cambridge University Press, pp. 69–100.

Coulomb P., Delorme H., 1987. Crise agricole, crise de politique. *Sociologie du travail*, 29(4), 385–413.

Coulomb P., Delorme H., Hervieu B., Lacombe P., 1990. *Les agriculteurs et la politique*. Paris, Presses de sciences po.

Courleux F., Depeyrot J.-N. 2014, La Chine, le nouveau stockeur en dernier ressort? communication présentée au colloque *Renouveler les approches institutionnalistes sur l'agriculture et l'alimentation: La 'grande transformation' 20 ans après*, Montpellier, 16 et 17 Juin.

Debailleul G., 1990. *Evolution de la politique agricole américaine*. Thèse de Docteur Ingénieur. Paris, INAP-G.

Delord B., Lacombe P., Touzard J.-M., 2000. Agricultural Systems and Agricultural Policy. In Colin J.P. and Crawford E. (Eds.), *Research on Agricultural Systems: Accomplishments, Perspectives and Issues*. New York, Nova Science Publishers, pp. 178–201.

Delorme H., 2002. *Vers une politique agricole commune multifontionnelle?* Paris, Presses de sciences po.

Dosi G., Coriat B., 1995. Evolutionnisme et régulation. Divergences et convergences. In Boyer R. and Saillard Y. (Eds.), *Théorie de la régulation. L'état des savoirs.* Paris, La Découverte, pp. 500–510.

Favereau O., 1995. Conventions et régulation. In Boyer R. and Saillard Y. (Eds.), *Théorie de la régulation. L'état des savoirs.* Paris, La Découverte, pp. 511–520.

Fèvre C., Pouch T., 2013. L'affirmation des multinationales de l'agroalimentaire des pays émergents. Le cas des firmes brésiliennes de la viande. *Économie rurale,* 334, 85–98.

Fleck L., 1980. *Genèse et développement d'un fait scientifique* (traduction de 2008). Paris, Flammarion.

Fonte M., 2002. Food Systems, Consumption Models and Risk Perception in Late Modernity. *International Journal of Sociology of Agriculture and Food,* 10, 13–21.

Fouilleux E., Goulet F., 2012. Firmes et développement durable: le nouvel esprit du productivisme. *Études rurales,* 2012(2), 131–146.

Fournier S., Touzard J.-M., 2014. La complexité des systèmes alimentaires: un atout pour la sécurité alimentaire? *VertigO-la revue électronique en sciences de l'environnement,* 14(1).

Friedmann H., 2005. From Colonialism to Green Capitalism: Social Movements and Emergence of Food Regimes. In Buttel F.H. and McMichael P. (Eds.), *New Directions in the Sociology of Global Development.* Bingley, UK, Emerald Group Publishing Limited, Book series Research in Rural Sociology and Development, Vol. 11, pp. 227–264.

Friedmann H., 2016. Commentary: Food Regime Analysis and Agrarian Questions: Widening the Conversation. *The Journal of Peasant Studies,* 43(3), 671–692.

Friedmann H., McMichael P., 1989. Agriculture and the State system: The Rise and Decline of National Agricultures, 1870 to the Present. *Sociologica Ruralis,* 29(2), 93–117.

Geels F.W., 2002. Technological Transitions as Evolutionary Reconfiguration Processes: A Multi-Level Perspective and a Case-Study. *Research Policy,* 31(8), 1257–1274.

Geels F.W., 2004. From Sectoral Systems of Innovation to Socio-Technical Systems. *Research Policy,* 33(6–7), 897–920.

Geels F.W., 2010. Ontologies, Socio-Technical Transitions (to Sustainability), and the Multi-Level Perspective. *Research Policy,* 39(4), 495–510.

Genieys W., Smith A., 2001. Idées et intégration européenne:'la grande transformation'du midi viticole. *Politique européenne,* 1, 43–62.

Genus A., Coles A.-M., 2008. Rethinking the Multi-Level Perspective of Technological Transitions. *Research Policy,* 37(9), 1436–1445.

Gibbs D., 1996. Integrating Sustainable Development and Economic Restructuring: A Role for Regulation Theory? *Geoforum,* 27(1), 1–10.

Gilly J.-P., Wallet F., 2005. Enchevêtrement des espaces de régulation et gouvernance territoriale. Les processus d'innovation institutionnelle dans la politique des Pays en France. *Revue d'Économie Régionale & Urbaine,* 2005(5), 699–722.

Goodman D., Dupuis E.M., Goodman M.K., 2012. *Alternative Food Networks: Knowledge, Practice, and Politics.* London, Routledge.

Goodwin M., Cloke P., Milbourne P., 1995. Regulation Theory and Rural Research: Theorising Contemporary Rural Change. *Environment and Planning A,* 27(8), 1245–1260.

Griffon M., 1994. *Rural Economy, Institutional Economics and Agriculture.* Rome, Cirad/FAO.

Grouiez P., 2010. *Les stratégies des communautés et la régulation sectorielle et territoriale des configurations productives: le cas de l'agroalimentaire russe.* Thèse de doctorat, Université de ReimsReims.

Grouiez P., 2013. Des kolkhozes à l'agrobusiness en Russie. *Etudes Rurales*, 2013(2), 49–62.

Holtz G., Brugnach M., Pahl-Wostl C., 2008. Specifying "Regime": A Framework for Defining and Describing Regimes in Transition Research. *Technological Forecasting and Social Change*, 75(5), 623–643.

Hugon P., 1993. Les trois temps de la pensée francophone en économie du développement. In Choquet C., Dollfus O., Le Roy E. and Vernières M. (Eds.), *Etat des savoirs sur le développement.* Paris, Karthala.

Karpik L., 2007. *L'économie des singularités.* Paris, Gallimard.

Labarthe P., 2005. Trajectoires d'innovation des services et inertie institutionnelle: Dynamique du conseil dans trois agricultures européennes. *Géographie, économie, société,* 7 (3), 289–311.

Labarthe P., 2006. *La privatisation du conseil agricole en question: Évolutions institutionnelles et performances des services de conseil dans trois pays européens (Allemagne, France, Pays-Bas).* Thèse de Doctorat, Université de Marne-la-Vallée, Marne-la-Vallée.

Labarthe P., 2009. Extension Services and Multifunctional Agriculture: Lessons Learnt from the French and Dutch Contexts and Approaches. *Journal of Environmental Management*, 90, S193–S202.

Labarthe P., 2010. Services immatériels et verrouillage technologique. Le cas du conseil technique aux agriculteurs. *Économies et sociétés*, 44(2), 173–196.

Lacombe P., 1998. Les agriculteurs dans la société: Quelles fonctions? Quels métiers? In Miclet G., Siriex L. and Thoyer S. (Eds.), *Agriculture et alimentation en quête de nouvelles légitimités.* Paris, INRA-Economica, pp. 282–288.

Lacroix A., Mollard A., 1990. Pourquoi les agriculteurs travaillent-ils tant? In Coulomb P., Delorme H., Hervieu B., Jollivet M. and Lacombe P. (Eds.), *Les agriculteurs et la politique.* Paris, Presses de sciences po, pp. 282–288.

Lacroix A., Mollard A., Bel F., 1995. L'approche sectorielle de la régulation: Une problématique à partir de l'agriculture. In Allaire G. and Boyer R. (Eds.), *La grande transformation de l'agriculture.* Paris, Economica.

Lamarche T., Nieddu M., Grouiez P., Chanteau J.-P., Labrousse A., Michel S., Vercueil J. 2015. A Regulationist Method of Meso-Analysis, communication présentée au forum *La théorie de la régulation à l'épreuve des crises,* Paris, 10–12 juin.

Landel P., 2015. *Participation et verrouillage technologique dans la transition écologique en agriculture – Le cas de l'Agriculture de Conservation en France et au Brésil.* Thèse de Doctorat, AgroParisTech, Paris, France.

Lataste F., 2014. *Place et enjeux des biens publics dans la Politique agricole commune : les apports d'une lecture institutionnaliste.* Thèse de Doctorat, Université de DijonDijon, France.

Lataste F., Berriet-Solliec M., Trouvé A., Lépicier D., 2012. Le second pilier de la Politique Agricole Commune: une politique à la carte. *Revue d'Économie Régionale & Urbaine*, 2012(3), 327–351.

Laurent C., 1992. *L'agriculture et son territoire dans la crise: analyse et démenti des prévisions sur la déprise des terres agricoles a partir d'observations réalisées dans le Pays d'Auge.* Thèse de Doctorat. Paris, Université Paris VII.

Laurent C., 1995. La fin de l'hégémonie de l'agriculture professionnelle sur le territoire. In Allaire G. and Boyer R. (Eds.), *La grande transformation de l'agriculture.* Paris, INRA-Economica, pp. 323–344.

Laurent C., 2013. The Ambiguities of French Mediterranean Agriculture: Images of the Multifunctional Agriculture to Mask Social Dumping? In Ortiz-Miranda D., Moragues-Faus A. and Alegre E.A. (Eds.), *Agriculture in Mediterranean Europe: Between Old and New Paradigms*. Bingley, UK, Emerald Group Publishing, Series Research in rural sociology and development, Vol. 19, pp. 315–332.

Laurent C., 2015. L'agriculture méditerranéenne française entre multifonctionnalité et dumping social. *Le Courrier de l'environnement de l'INRA*, 65, 123–134.

Laurent C., Berriet-Solliec M., Kirsch M., Labarthe P., Trouvé A., 2010. Multifunctionality of Agriculture, Public Policies and Scientific Evidences: Some Critical Issues of Contemporary Controversies. *Applied Studies in Agribusiness and Commerce–APSTRACT*, 2010(1–2), 53–58.

Laurent C., Berriet-Solliec M., Labarthe P., Trouvé A., 2012. Evidence-based policy: de la médecine aux politiques agricoles? *Notes et études socio-économiques*, 36, 79–101.

Laurent C., Cartier S., Fabre C., Mundler P., Ponchelet D., Rémy J., 1998. L'activité agricole des ménages ruraux et la cohésion économique et sociale. *Économie rurale*, 244, 12–21.

Laurent C., Du Tertre C., 2008. *Secteurs et territoires dans les régulations émergentes*. Paris, L'harmattan.

Laurent C., Mouriaux M.F., 2008. Secteurs, territoires, rapport social d'activités. In Laurent C. and Du Tertre C. (Eds.), *Secteurs et territoires dans les régulations émergentes*. Paris, L'Harmattan.

Le Roch C. 1993. Environnement et théorie de la régulation: Une approche à partir de l'agriculture, communication à l'école chercheurs INRA-ESR: Economie des institutions, organisée à Hyères (France), 27–30 September.

Lipietz A., 1995. Ecologie politique régulationniste ou économie de l'environnement? In Boyer R. and Saillard Y. (Eds.), *Théorie de la régulation: l'état des savoirs*. Paris, La Découverte.

Loconto A., Fouilleux E., 2014. Politics of private regulation: ISEAL and the Shaping of Transnational Sustainability Governance. *Regulation & Governance*, 8(2), 166–185.

Lopez M. 2006, Regulation Redux, Regional Studies and the Sociology of Agriculture, in *Proceedings of the conference of the Rural Sociology Society*, Kentucky, 10 août 2006.

Losch B., 2000. La Côte d'Ivoire en quête d'un nouveau projet national. *Politique Africaine*, 2000(2), 5–25.

Madelrieux S., Nettier B., Dobremez L. 2010, L'exploitation agricole, la famille et le travail: Nouvelles formes, nouvelles régulations? communication présentée aux *Journées d'étude INRA-Cirad: le travail en agriculture dans les sciences pour l'action*.

Malassis L., 1979. *Economie Agroalimentaire*. Paris, France, Cujas.

Malerba F., 2002. Sectoral Systems of Innovation and Production. *Research Policy*, 31(2), 247–264.

Markard J., Truffer B., 2008. Technological Innovation Systems and the Multi-Level Perspective: Towards an Integrated Framework. *Research Policy*, 37(4), 596–615.

Marsden T., 1992. Exploring a Rural Sociology for the Fordist Transition. *Sociologia Ruralis*, 32(2–3), 209–230.

Marzin J., 2006. *L'impact de la microfinance sur une communauté villageoise. Le cas de Gandaogo au Ganzourgou*. Thèse de Doctorat. Strasbourg, Université Louis Pasteur.

McMichael P., 2009. A Food Regime Analysis of the "World Food Crisis". *Agriculture and Human Values*, 26, 281–295.

Muchnik J., Requier-Desjardins D., Sautier D., Touzard J.M., 2007. Systèmes agroalimentaires localisés. *Economies et Sociétés*, 29, 1465–1484.

Mundler P., Laurent C., 2003. Flexibilité du travail en agriculture: méthodes d'observation et évolutions en cours. *Ruralia. sciences sociales et mondes ruraux contemporains*, 12/13.

Nefussi J., 1987. *Les industries agro-alimentaires en france croissance et financement 1950-1985. Essai sur l'intégration financière et la dynamique industrielle*. Thèse de Docteur Ingénieur. Paris, INAP-G.

Nguyen G., Purseigle F., 2012. Les exploitations agricoles à l'épreuve de la firme: L'exemple de la Camargue. *Etudes rurales*, 190, 99–118.

Nieddu M., Gaignette A., 2000. L'agriculture française entre logiques sectorielles et territoriales (1960-1985). *Cahiers d'Economie et de Sociologie rurales*, 54, 48–87.

Nieddu M., Garnier E., Bliard C., 2010. L'émergence d'une chimie doublement verte. *Revue d'économie industrielle*, 132, 53–84.

Nieddu M., Vivien F.-D., 2015. La chimie verte, une fausse rupture? Les trajectoires de la transition écologique. *Revue française de socio-economie*, 2015(2), 139–153.

Nieddu M., Vivien F.-D., 2016. La bioraffinerie comme objet transitionnel de la bioéconomie. *Économie Rurale*, 349–350, 7–11.

Olivier-Salvagnac V., Legagneux B., 2012. l'agriculture de firme: Un fait émergent dans le contexte agricole français. *Etudes rurales*, 190, 77–97.

Pecqueur B., 2001. Gouvernance et régulation: Un retour sur la nature du territoire. *Géographie, économie, société*, 3(2), 229–245.

Pecqueur B., 2007. L'économie territoriale: Une autre analyse de la globalisation. *L'Économie Politique*, 2007(1), 41–52.

Pecqueur B., Zimmermann J.-B., 2004. *Economie de proximités*. Paris, Lavoisier.

Perraud D. 1985, Crise laitière et modes de régulation, communication présentée au *Colloque franco-québecois de Rimouski*, 7–10 octobre, pp. 43–49.

Petit P., 1998. Formes structurelles et régimes de croissance de l'après fordisme. Recherches & Régulation Working Papers. n° K 1998-1.

Plumecocq G., 2014. The Second Generation of Ecological Economics: How Far has the Apple Fallen from the Tree? *Ecological Economics*, 107, 457–468.

Ponte S., 2016. Convention Theory in the Anglophone Agro-Food Literature: Past, Present and Future. *Journal of Rural Studies*, 44, 12–23.

Quemia M., 2001. Théorie de la régulation et développement: Trajectoires latino-américaines. *L'année de la régulation*, 5, 57–103.

Rastoin J.-L., Ghersi G., 2010. *Le système alimentaire mondial*. Versailles, Éditions Quae.

Requier-Desjardins D., Guibert M., Bühler E.A., 2014. La diversité des formes d'agricultures d'entreprise au prisme des réalités latino-américaines. *Économie rurale*, 344, 45–60.

Saulais L., 2015. Foodservice, Health and Nutrition: Responsibility, Strategies and Perspectives. In Sloan P., Legrand W. and Hindley C. (Eds.), *The Routledge Handbook of Sustainable Food and Gastronomy*. New York and London, Routledge, pp. 256–263.

Servolin C., 1989. *L'agriculture moderne*. Paris, éditions Seuil.

Shove E., Walker G., 2007. Transitions Ahead: Politics, Practice, and Sustainable Transition Management. *Environment and Planning A*, 39(4), 763–770.

Shucksmith M., 2010. Disintegrated Rural Development? Neo-Endogenous Rural Development, Planning and Place-Shaping in Diffused Power Contexts. *Sociologia Ruralis*, 50(1), 1–14.

Smith A., Costa O., Maillard J., 2007. *Vin et politique. Bordeaux, la France, la mondialisation*. Paris, Les Presses de Sciences Po.

Smith A., Stirling A., Berkhout F., 2005. The Governance of Sustainable Socio-Technical Transitions. *Research Policy*, 34(10), 1491–1510.

Tordjman H., 2008. La construction d'une marchandise: Le cas des semences. *Annales. Histoire, sciences sociales*, 63(6), 1341–1368.

Torre A., Filippi M., 2005. *Proximités et changements socio-économiques dans les mondes ruraux*. Paris, Inra Editions.

Touzard J.-M., 1995. Régulation sectorielle, dynamique régionale et transformation d'un système productif localisé: L'exemple viticole languedocien. In Allaire G. and Boyer R. (Eds.), *La grande transformation de l'agriculture*. Paris, INRA-Economica, pp. 293–322.

Touzard J.-M., 2000. Coordination locale, innovation et Régulation, l'exemple de la transition vin de masse – vin de qualité. *Revue d'Economie Régionalee et Urbaine*, 3, 589–605.

Touzard J.-M., 2009. Quels apports de la Théorie de la Régulation à l'analyse des transformations agroalimentaires actuelles ? *Economies et Sociétés*, série AG, 31, 1923–1934.

Touzard J.-M., Temple L., Faure G., Triomphe B., 2014. Systèmes d'innovation et communautés de connaissances dans le secteur agricole et agroalimentaire. *Innovations, Cahiers d'économie et de management de l'innovation*, 43, 13–38.

Trouvé A., 2007. *Le rôle des Régions européennes dans la redéfinition des politiques agricoles*. Thèse de Doctorat, Université de DijonDijon.

Trouvé A., 2009. Les régions, porteuses de nouveaux compromis pour l'agriculture? *Revue de la régulation*, 5.

Trouvé A., Berriet-Solliec M., 2010. Regionalization in European Agricultural Policy: Institutional Actualities, Issues and Prospects. *Regional Studies*, 44(8), 1005–1017.

Vanloqueren G., Baret P.V., 2009. How Agricultural Research Systems Shape a Technological Regime that Develops Genetic Engineering but Locks out Agroecological Innovations. *Research Policy*, 38(6), 971–983.

Wilkinson J., 1997. A New Paradigm for Economic Analysis? Recent Convergences in French Social Science and an Exploration of the Convention Theory Approach with a Consideration of Its Application to the Analysis of the Agrofood System. *International Journal of Human Resource Management*, 6(3), 335–339.

Wilkinson J., 2002. The Final Foods Industry and the Changing Face of the Global Agro-Food System. *Sociologia Ruralis*, 42(4), 329–346.

Zuindeau B., 2007. Régulation School and Environment: Theoretical Proposals and Avenues of Research. *Ecological Economics*, 62(2), 281–290.

4 Convention Theory in Anglophone agri-food studies

French legacies, circulation and new perspectives

Emmanuelle Cheyns and Stefano Ponte[1]

Introduction

The French school of Convention Theory (CT) has influenced various branches of agri-food studies in the past two decades, as part of a wider trend in the Anglophone social sciences (Wilkinson, 1997; Raikes et al., 2000; Biggart and Beamish, 2003; Stark, 2011; Diaz-Bone, 2012; du Gay and Morgan, 2013; Barnett, 2014). Convention Theory in Anglophone agri-food studies has attracted attention particularly in sociology, economic geography, regional studies, and international political economy, partly inspiring the analysis of the 'quality turn', alternative food networks, and various forms of coordination and governance in global value chains. The diffusion of Convention Theory in this literature took place particularly after the publication in English of *Worlds of Production* (Storper and Salais, 1997), *The New Spirit of Capitalism* (Boltanski and Chiapello, 1999 [2005]) and *On Justification* (Boltanski and Thévenot, 1991 [2006]). More recently, the related 'regimes of engagement' approach, drawing from French pragmatic sociology (Thévenot, 2006, 2007), has also started attracting attention in this field.

In this chapter, we examine how Convention Theory (also known as *l'économie des conventions*) was introduced in Anglophone[2] agri-food studies, what results and patterns of diffusion have ensued, what originalities, divergences or similarities have developed in comparison to the French school, and perspectives for further developments. In the first section we explain the origins and filiations with the original French school of Convention Theory. In the second section we examine a collection of 51 articles to explain how Convention Theory has spread to Anglophone agri-food studies, and what issues have been addressed. In the third section we discuss similarities and divergences that characterise the Anglophone literature in comparison to the French school, and highlight new issues and approaches it introduced. Finally, we discuss the extension of this programme towards the related 'regimes of engagement' approach, which evolved from French pragmatic sociology.

The French school of Convention Theory: origins and early diffusion

Convention Theory (CT) was first adopted in the Anglophone literature in the mid-1990s, mainly in reference to the approach known in France as '*l'économie des conventions*.[3] This approach emerged in the mid-1980s and initially encompassed the concerns of institutionalist economists, sociologists and historians. It became formalised as a current of 'heterodox economics' with the publication of a 'manifesto' issue of *Revue économique*, with a joint introduction by the founders of the movement (Dupuy et al., 1989). This school was concerned with solving problems of coordination in situations of uncertainty, in particular when the determination of the quality of goods and labour was involved. It called into question the postulates of neoclassical economics relating to substantive rationality of agents and of completeness of contracts and rules, and argued in favour of procedural rationality and incompleteness of information or 'radical uncertainty' (Orléan, 1989). With this cognitivist orientation, Convention Theory, as well as the Regulation Theory (RT) with whom it had close connections (see Chapter 2), became an alternative paradigm in economics, at least in France (Favereau, 1989).

The form of coordination between individuals postulated in neoclassical models of the economy is purely market-based: coordination takes place through the market, under the exclusive influence of 'an invisible hand'. In contrast, Convention Theory argues 'that agreement between individuals, even when it is limited to the contract of a market exchange, is not possible without a common framework and a constitutive convention'[4] (Dupuy et al., 1989: 142; see also Orléan, 1994). Thus, rules are not only 'constraints' or 'contracts', but also 'collective cognitive devices' (Favereau, 1989). Going beyond the opposition between 'holism' and 'individualism', a convention is defined as 'a regularity that has its source in social interactions but presents itself to the actors in an objectified form' (Dupuy et al., 1989: 145).

Boltanski and Thévenot, in their seminal work *On Justification* (1991 [2006]), contribute to this theory by characterising and modelling a diversity of forms of coordination in addition to 'market'. They provide a framework for analysing the forms of legitimacy on which individuals can rely to justify their agreements in practice, referring to particular views of the common good. To deal with critique, justifications need to rely on reality tests. Boltanski and Thévenot (ibid.) argue that these tests involve both people and objects and are related to modes of coordination based on six historical 'orders of worth' (also translated as 'worlds'). These worlds are organised around different principles of qualification.

In the *inspired* world, the common principle is spiritual or creative enhancement through passion. The objects and arrangements that equip worth stem from a mind and a body that are prepared for the break of creation. The test, which eludes measure, is an adventurous lived experience or is attached to a unique set of practices that cannot be certified through formal audits. Firms draw on the concept of creativity.

The *domestic* world is related to the family and the common principle of traditional benevolence, care provision and trustworthiness; it extends the chains of hierarchical and personal dependency through generations. The test is based on oral evidence (firms draw on the concept of loyalty).

In the world of *fame*, the common principle is celebrity or renown in public opinion. A higher state of worthiness is reached when one becomes famous and visible. Qualified objects are recognition signs, and the test is to be known and visible. Firms use the concept of reputation. In the *civic* world, the common principle is collective and anonymous solidarity and the state of worthiness depends on one's public agency as representing the collective. The subjects in this world are delegates, representatives, and members. The objects support the representation of the collective—e.g. slogans, policies, rules and legal forms. The test is reached through meetings and assemblies, which produce representation. Firms refer to the idea of responsibility.

In the *market* world, the common principle is market competition. The test is constructed via deal-making, and evidence is provided by the price of market goods and services, including labour. Firms organise themselves around the concept of competitiveness.

In the *industrial* world, the common principle is efficiency. The test is based on technical efficiency and scalability, plus proper functionality. Evidence for testing is technical and objectively measurable. Firms evoke the concept of productivity.

The first Anglophone paper on Convention Theory in the field of agri-food studies appeared in 1997 in *Economy and Society* by John Wilkinson, a sociologist at the Federal Rural University of Rio de Janeiro.[5] Wilkinson examines two French currents in heterodox economics – Convention Theory and Regulation Theory which he sees as a promising alternative paradigm in economics. He introduces Convention Theory through the work of Salais and Thévenot (1986), underscoring the initial focus of the French school of Convention Theory on the wage relation and on processes by which labour is 'qualified'. He then analyses the cross-fertilisation of work by Boltanski and Thévenot (1991), Favereau (1994), Eymard-Duvernay (1994) and Salais and Storper (1992).

Wilkinson underlines three fundamental contributions by Boltanski and Thévenot (1991):

- A key focus on the 'situation' in which action takes place, related to a specific relationship between persons and things. This calls into question the intrinsic notion of social class.
- Attention to collective action, seen as constituted by processes of justification and testing. Understanding people's engagement in justifiable action makes it possible to relativise the concept of 'interests', which are not permanent and univocal attributes of groups.
- A plurality of meaning of what is 'just', which they develop through the model of multiple 'worlds' that exist without a determined hierarchy between different forms of legitimacy.

Wilkinson underscores the heuristic value of Convention Theory in work by Eymard-Duvernay (1994) on firm typologies and conventions of quality, and in the 'worlds of production' approach of Storper and Salais (1997). Finally, he highlights emerging applications of this approach in agri-food studies, such as the various contributions included in *La grande transformation* (Allaire and Boyer, 1995), thus helping raise awareness of existing empirical studies.

The contributions that followed Wilkinson's article in 1997 to 2003 all took inspiration from the group of French 'founding authors' mentioned above (but also including Nicolas and Valceschini, 1995; and Sylvander, 1995). These contributions are authored by researchers at the University of Cardiff (Murdoch and Miele, 1999; Marsden et al., 2000; Murdoch et al., 2000; Parrott et al., 2002), a few American universities (Busch, 2000; Barham, 2002, 2003; Raynolds, 2002; Biggart and Beamish, 2003; Freidberg, 2003), a Mexican university (Renard, 2003), and a Danish research centre (Raikes et al., 2000). Most of these early contributors had close connections or collaborations with the research community in France.[6] Subsequently, Anglophone agri-food studies started to rely more heavily on English translations of key works, thus relatively side-lining other important contributions in French (notably those by Eymard-Duvernay, 1989, 2006a, 2006b).

Convention Theory in the Anglophone agri-food literature

General observations

In this section, we examine the Anglophone literature on agri-food studies published in 1997–2014, which engaged with Convention Theory.[7] The review includes 51 articles, books and book chapters that were selected first via searches in Google Scholar and Scopus on 'convention(s)', 'worth', 'Convention Theory', and 'pragmatic sociology', then filtered to match whether they apply to the agri-food sector[8] and qualify as 'Anglophone' (see Endnote above). These entries were then complemented by other relevant contributions that were referenced in the articles already in the list.

The contributors to this literature are institutionally affiliated in three main geographical locations: the US (17 entries), Nordic countries (14), and the UK (11), with others based in Spain (4), Italy (1), New Zealand (3), Portugal (1), Germany (1), Brazil (1), and Mexico (1). Contributors are mostly focused on geography, planning and environmental studies (20), other multidisciplinary environments (15), and sociology (14). Publications have mainly appeared in journals (45 entries), the most important being *Journal of Rural Studies* (9) and *Sociologia Ruralis* (6), followed by *Economy and Society* (4), *Agriculture and Human Values* (2), *Geoforum* (2), *Journal of Agrarian Change* (2) and *Regional Studies* (2).

From a thematic standpoint, the earliest and most popular area of application of Convention Theory related to discussions on the 'quality turn' and the emergence of 'alternative agri-food networks' (AAFNs) (23 entries). Between 1997 and 2004, almost all publications applied Convention Theory in these areas, mostly by scholars based in the US and the UK.[9] But from 2005, two further

thematic areas emerged in the application of Convention Theory to (i) coordination and governance in global value chains (10 entries), mainly published by
researchers based in Denmark (e.g., Gibbon and Ponte, 2005; Ponte and Gibbon,
2005) and the UK (Gwynne, 2006; Tallontire, 2007), and (ii) innovation and
institutional change (7 entries), published by researchers in the US (Cidell and
Alberts, 2006; Guthey, 2008), Spain (Lindkvist and Sánchez, 2008; Sánchez-
Hernández, 2011), Norway (Straete, 2004) and Italy (Barbera and Audifredi,
2012). From 2009 onwards, other thematic areas started to appear, dealing with
environmental and land management (Ekbia and Evans, 2009; Blok, 2013;
Nyberg and Wright, 2013), consumption and household food provisioning
(Andersen, 2011; Evans, 2011; Truninger, 2011), and farm labour management
(Gibbon and Riisgaard, 2014; Riisgaard and Gibbon, 2014).

From an empirical perspective, 38 papers focus on specific sectors, regions
and case studies. They cover a wide array of sectors and products, from niche
and local/regional foods to global commodities. The wine industry has been the
most popular target (10 entries), followed by coffee (4) and cut flowers, fresh
fruit and vegetables (4). Organic, fair trade and other sustainability certifications
(8) and geographic indications (6) have also attracted substantial empirical
attention. The majority of these contributions focus on case studies in Europe,
North America and Oceania (24).

Analytically, the literature has developed along two distinct (but sometimes
overlapping) approaches: a first that engages with an agri-food adaptation of the
'worlds of production' framework (Salais and Storper, 1992; Storper and Salais,
1997); and a second that applies the 'orders of worth' approach of Boltanski and
Thévenot (1991[2006]) and further elaborations of 'quality conventions'
(Eymard-Duvernay, 1989; Sylvander, 1995; Thévenot, 1995).

Worlds of production approaches

In agri-food studies, the 'worlds of production' approach of Salais and Storper
was adapted to the special features of the sector and translated into four 'worlds
of food', characterised by distinctive local/regional cultural, ecological and
political/institutional logics (Morgan et al., 2006; see also Murdoch and Miele,
1999; Murdoch et al., 2000): (i) a world of mass industrial food production,
where standardised technologies are applied to produce large volumes of generic
foods sold to mass markets; (ii) a world of niche production, where standardised
technologies are used to produce high-quality and differentiated products for
niche markets; (iii) a world of local production, where artisanal/traditional
techniques are used to produce specialist foods sold to clients through close
relationships; and (iv) a world of high technology production, where specialised
processes are used to deliver new, special or functional foods to mass markets
(Sánchez-Hernández et al., 2010: 470).

In some cases, the 'worlds of food' approach has been used in the agri-food
literature to explain the strategic positioning of individual firms and their movements between worlds (Murdoch and Miele, 1999, 2004) and/or to highlight how

different worlds of production coexist within individual firms (e.g. Trabalzi, 2007). More common has been the tendency to highlight collective approaches that strategically position specific clusters, localities and regions through trajectories of learning, innovation, clustering, and institutional change (Cidell and Alberts, 2006; Guthey, 2008; Lindkvist and Sánchez, 2008; Sánchez-Hernández et al., 2010; Sánchez-Hernández, 2011), or to specify the innovation systems and regulatory interventions that allow individual firms to move between worlds (Stræte, 2004; Barbera and Audifredi, 2012). For example, Cidell and Alberts (2006) examine quality conventions in the chocolate industry, and in particular the negotiations of quality that are related to the location of chocolate manufacturing rather than where cocoa is grown. They highlight that agri-food quality can be linked to the geographies of manufacturing and innovation, in particular where these innovations were first introduced. Guthey (2008), in his analysis of the wine industry in northern California, argues that changes in production practices, quality and performance can originate from local collective social processes, including the shaping of conventions. He concludes that practice and local specificity are as important as 'nature' and global forces in the making of quality.

Similarly, Sanchez-Hernandez and colleagues (Lindkvist and Sánchez, 2008; Sánchez-Hernández et al., 2010; Sánchez-Hernández, 2011) through the study of wine innovation pathways in Castilla y Leon (Spain) (also in comparison to the salt fish industry in Norway), show that innovation can arise from explicit attempts by value chain players to 'adapt their conventions and to attach new qualities to their core product... [and] improve their market performance' (Sánchez-Hernández, 2011: 105; see also Stræte, 2004). Lindkvist and Sánchez (2008) challenge the established Southern/Northern Europe binomial characterisation of dominant conventions (Parrott et al., 2002; Morgan et al., 2006), actually showing that wine producers adapted their production systems to new market demands in Spain, while salt fish producers in Norway continued with their traditional conventions and lacked innovation. This possible counter-trend is strengthened by the case study of wine in Piedmont (Italy), where the 'methanol scandal' of the mid-1980s triggered a change in quality conventions that led to a major and successful change in the institutional configuration of wine production in the region (Barbera and Audifredi, 2012). Collectively, these authors argue that innovation can be embedded in the passage from one world of production to another through changing quality conventions, a process that can be facilitated by appropriate organisational structures in local production systems.

Orders of 'worth'/'worlds' approaches

Quality turn and alternative agri-food networks

A popular application of Convention Theory in the Anglophone agri-food literature has been attributing one or another type of convention to the disclosure of quality to facilitate coordination efforts. Much discussion has focused on examining the content of civic/ecological and domestic conventions, and whether these are being folded within a compromise of market and industrial conventions that allows 'alternative' quality traits to be mainstreamed and standardised (Murdoch et al., 2000; Barham, 2002, 2003; Raynolds, 2002, 2004, 2012a, 2012b; Freidberg, 2003, 2004; Renard, 2003; Kirwan, 2006; Raynolds et al., 2007; Rosin, 2008; Andersen, 2011).

The 'quality turn' literature has relied almost exclusively on case studies based in industrialised countries (and especially in Europe). It argues that consumers have turned away from industrial agri-food products and towards 'high-quality' products, including those characterised by organic or low external input practices, specific locations or regions, and those supplied through farmers' markets, short/local food supply chains, and agro-tourism or other kinds of multifunctional agricultural enterprises. This 'quality turn' is explained in part by consumers' heightened reflexivity (both in relation to 'intrinsic' quality and to production and process methods) and in part by reactions to repeated food scares in the 1990s (BSE, e-coli, salmonella). The combined result led to the increasing importance of 'transparency' in agri-food networks – embedded in practices of quality assurance, traceability, geographic origin, sustainable agro-ecological practices and direct marketing schemes (Goodman, 2004).

In Convention Theory contributions, these trends have been framed as part of a general movement from industrial conventions (and the logic of mass production) to domestic conventions based on trust, tradition and place (Murdoch and Miele, 1999; Murdoch et al., 2000). Murdoch et al. (2000) in particular argue that quality is coming to be seen as inherent in more 'local' and 'natural' foods, and thus 'quality food production systems are being re-embedded in local ecologies' (ibid.: 103). Although they recognise that 'alternative conventions' are not sufficient to fundamentally transform the global food system, they suggest that 'domestic and ecological criteria can be used by local producers to secure their own positions in the networks on favourable terms' (ibid.: 119).

Work with more explicit North–South framings has been more critical of the potential of civic and domestic conventions to transform agri-food systems. For example, Freidberg (2003; see also Freidberg, 2004) is critical of both interpersonal trust as the basis of 'quality', and of the ability of alternative agri-food networks to promote meaningful socio-economic change in developing countries. In her historical and comparative analysis of conventions in Anglophone (Zambia to the UK) and Francophone (Burkina Faso to France) horticultural trade, Freidberg finds that 'relationships based on trust ... are often just situations where one or all parties has no choice but to hope for luck or mercy ... Economies of quality ... are not necessarily less exploitative than others' (2003: 98). She also notes that, in North–South trade, 'situations of exchange' are such that actors do not necessarily share ethical or behavioural norms. Rather, producers comply with retailers' demands because they have no choice (see also Busch and Tanaka, 1996), even though the ecological and socio-cultural conditions of production are different from those predominant in the country of consumption.

Likewise, Raynolds (2002, 2004, 2012a, 2012b; Raynolds et al., 2007), in her work on organics, fair trade and other sustainability certification systems, has employed Convention Theory to understand how quality contestations arise and are resolved in global commodity networks, and how 'contestations over divergent qualifications and how collective enrolment in particular conventions permit forms of control at a distance' (Raynolds, 2002: 409). In particular, she sees fair trade as enacting a 'mode of ordering of connectivity',[10] where discursive and material relations are based on renewed investment in 'trust' (domestic convention), while they take the form of 'partnerships' and 'alliances'. Furthermore, she highlights how fair trade refers to civic norms and qualifications that are based on collective responsibility for (and evaluation of) societal benefits – thus extending domestic conventions to socially and spatially distant peoples and spaces. At the same time, she argues that fair trade is rooted in important ways in market conventions as it deals with mainstream distributors and retailers, and in industrial conventions rooted in formal standards, inspections and certifications (on similar lines, see Renard, 2003). She recognises that 'certifiers may align with profit-driven corporations in reasserting industrial standards and commercial price competition, with fair trade principles converted into auditable attributes and certification practices enabling new forms of control at a distance' (Raynolds, 2012b: 286). Along with much of the agri-food literature on this subject, on the one hand Raynolds sees the dangers of 'mainstreaming' in alternative food networks, and on the other hand she argues that fair trade and organic certifications are better able to deliver progressive social change than voluntary initiatives that are based on industrial-type certifications.

Global value chain governance and labour management

One of the other ways Convention Theory has been applied in the agri-food literature has been through the analysis of how different quality conventions shape certain forms of coordination or organisation, and the specific dynamics of governance in global value chains (Ponte and Gibbon, 2005; Tallontire, 2007; Ponte, 2009; Sánchez-Hernández, 2011; Coq-Huelva et al., 2012; Ponte and Sturgeon, 2014). This literature examines how different orders of worth and related organisational principles can lead to different foci of justification once they are challenged; how these challenges are based on different sets of testing questions and measures of product quality; and how they have different transmission potential along value chains (Daviron and Ponte, 2005; Gibbon and Ponte, 2005; Ponte, 2009; Ponte and Daviron, 2011). This literature shows that while quality conventions typically overlap, one or a specific combination (for example, market and industrial, or domestic and fame) often form a dominant underpinning for linkages in a value chain node at a particular time. Furthermore, it examines how dominant conventions may 'travel' along a chain, explains the factors that makes them travel, and identifies which actors have the normative power to impose one convention over another beyond a single node in the value chain.

Furthermore, Ponte and Sturgeon (2014) examine the possible role of different quality conventions in facilitating specific kinds of linkages at individual nodes in the value chain (market, modular, relational, captive and hierarchical; as in Gereffi et al., 2005). They argue that market linkages are facilitated by market conventions in simple transactions where there is a straightforward association of product quality with price. Modular linkages are typically enabled by industrial conventions and the standards that underlie them, while spatially embedded domestic conventions go hand-in-hand with relational linkages and long-term, trust-based inter-firm relationships. Yet, they caution that these associations are loose and not deterministic, given that different types of conventions can coexist in each node of the value chain, firms can apply different conventions with different suppliers seeking exchange in the same node of the value chain, and different quality conventions can be applied simultaneously with the same supplier when negotiating product portfolios or long-term contracts.

In an empirical application of this approach, Ponte (2009) examined quality conventions in the wine value chain between South Africa and the UK, showing how certain instruments of verification that 'test' specific quality conventions translate into supply relations and divisions of labour that are employed to govern value chains in particular ways. Such instruments of verification make 'visible' and 'translate' the complex and hidden negotiations, interactions and representations that lay behind the crystallisation of a convention (or the overlap of conventions) at a particular time and position in the value chain. He argues that the moment of testing is important because it renders justifications explicit, 'pulls in' specific socio-technical devices, and recruits different types of knowledge and expertise. Finally, he shows how wine value chain operators may apply different conventions with different clients at the same time.

More recently, Convention Theory has also been employed to analyse labour management systems in large-scale farming in Africa (Gibbon and Riisgaard, 2014; Riisgaard and Gibbon, 2014). While the previous agri-food literature on this topic (from different analytical perspectives) had argued that farm labour management in Africa had undergone a transition from a 'domestic' to a 'market' system in the 1980s and early 1990s, Gibbon and Riisgaard (2014) argue that the present configuration can be explained as a combination of industrial and civic conventions. They use Convention Theory to critically interrogate established views on organisational dynamics in labour-intensive farming of cut flowers in Kenya and extend their argument to high-value crops in Africa more generally. They find that industrial and civic conventions are dominant and that market and domestic approaches are present but residual. In a related contribution, Riisgaard and Gibbon (2014) make a call to 'go beyond the traditional Convention Theory tradition of describing how conventions manifest themselves to asking *why certain conventions prevail* in a specific context at a specific point in time' (ibid.: 261). They examine the conditions under which certain conventions and specific combinations can emerge and become dominant in relation to labour management in cut flower farms in Kenya. In their case study, they find that while compromises between market and industrial conventions are necessary to operate a business, compromises involving civic conventions can only materialise as a result of political imperatives caused by popular struggles. Riisgaard and Gibbon (2014) also suggest that the stabilisation of an industrial convention may be a necessary precondition for civic elements to emerge, and that adequate and continuous political pressure needs to accompany this process.

Convergence and originality of the Anglophone literature

Similarities and divergences with the French school of Convention Theory

The Anglophone studies reviewed so far have strong similarities with the French school of Convention Theory. Like the French contributions, they make visible a diversity of forms of coordination based on a 'plurality of conventions' (Eymard-Duvernay, 1989; Thévenot, 1989; Favereau and Thévenot, 1991). Collectively, they formalise a 'pluralism' of coordination forms, as opposed to a reductive vision of economic coordination. The Anglophone literature has paid much attention not only to historical transitions from one dominant convention to another, but also from one combination of conventions to another. Recent work has begun to explore the importance of pluralistic quality conventions firms can rely upon – depending on the end-market they target or the different situations in which they find themselves in. In general, many of these studies introduce the formulation of pluralism of conventions, which makes it possible to handle greater complexity in empirical situations. Thus, the dual characterisation of a North–South division of Europe, based on a prevalence of conventions specific to each of these two major regions, is repositioned in a more nuanced approach (Lindkvist and Sánchez, 2008; Barbera and Audifredi, 2012). Similarly, the trajectories of one-dimensional change between 'worlds of production' or 'worlds of food' are enriched by composite dynamics. Rosin and Campbell (2009) in particular rely on this complexity to challenge the linear and dual view of other approaches to organic farming in New Zealand in terms of 'conventionalisation' and 'bifurcation'. Some studies have also introduced a temporal dimension, thus explaining innovation through the dynamic development of various equilibria of conventions. For example, Sánchez-Hernández et al. (2010) explain innovation in the regional production of wine in Spain during major changes in conventions, and in particular in the transition from an interpersonal to an industrial world.

The dissatisfaction with characterising empirical facts by one or another convention has led to examinations of how multiple justifications are used by actors simultaneously, as opposed to selective engagement in a single world (see, Busch, 2000; Rosin, 2008; Truninger, 2011). The relationship to 'situated action', central to the French school, gradually found place in some of these studies, even though they did not necessarily explicitly reference it. Engagement in specific conventions is associated with pluralities of markets, audiences and clients. Murdoch and Miele (2004), for example, compare a fast food company and the slow food movement to show how companies use different justifications to adjust to different audiences. Ponte (2009) shows how the same operators mobilise different conventions of quality depending upon the end-market they target, activating a portfolio of parallel conventions rather than a stabilised compromise (see also Trabalzi, 2007). Andersen (2011) and Truninger (2011) analyse the conventions associated with different purchasing or supply situations.

Other key concepts developed by the French school of Convention Theory have been less utilised in the Anglophone literature, although they can be found in some studies. Examining different forms of exercise of power has been the subject of much discussion in the French academic debate. Although critics have argued that the notion of power virtually disappeared in the application of Convention Theory to coordination mechanisms (Bourdieu, 1997), Convention Theory scholars have counter-argued that power is at the heart of Convention Theory (Thévenot, 2012, 2015a, 2015b, 2016). They argue that the understanding of power has moved towards the study of oppression, linked to the imposition of certain forms of normality (associated with worths) or of 'legitimised conventions' (Orléan, 2004). Every order of worth specifies a way in which people and things are qualified for a certain characterisation of the common good. The conventional forms that equip life and the capacities that are linked to them inevitably generate power asymmetries (Thévenot, 2012). In the early 2000s, Murdoch et al. (2000) and Raynolds (2002) emphasised the potential of Convention Theory to analyse different forms of exercise of power. However, this call was not answered in the Anglophone literature on agri-food studies until quite recently, when Gibbon and Riisgaard (2014, see also Cheyns and Riisgaard, 2014; Riisgaard and Gibbon, 2014) started to examine the normative power of conventions, in a constructive dialogue between Convention Theory and political economy. These authors challenge the vision of power that is only linked to strategic capabilities and rationality. They move from an exclusive attention to the authority of the state and the economic power of firms, towards a greater variety of forms of legitimate authority, which is also at the heart of the French school of Convention Theory (Orléan, 2004; Thévenot, 2015a). Other work has also focused on specifying the different formats of knowledge, know-how and expertise that are embedded in conventions and what they mean in terms of power dynamics (Ponte and Cheyns, 2013; Ponte and Sturgeon, 2014).

In Convention Theory theory, 'qualified objects' and 'tests' are valuable elements for analysing conventions. One of the characteristics of the French school is to consider objects in terms that are broader than their materiality of goods for 'production' or 'consumption', i.e., also as points of reference or guides for action and bases for coordination (Norman, 1993). The 'test', defined as a device (composed of qualified objects and persons) allowing the attribution of a state of worth (Boltanski and Thévenot, 1991) makes it possible to grasp the justifications at play and their critical tensions. Its empirical identification is also essential to avoid reducing all human action as relating to justification, in the sense understood by the theory. Convention Theory applications have sometimes strayed away from these major elements and some studies (including in France) have tended to reify 'worths' in organisational devices. In the Anglophone literature, some studies have paid attention to tests and conventional objects, although they are limited in number. Ponte (2009), Nyberg and Wright (2013), and Ponte and Sturgeon (2014) focus on tests as an analytical category and highlight the critical tensions generated by the use of different quality verification instruments or by the marketing of new green products. By referring to different worths of the common good, these moments are valuable for understanding how people justify their actions in practice and how they choose what is appropriate in a given situation. For their part, Ekbia and Evans (2009) pay particular attention to the qualification of objects related to the transmission of information.

Finally, much of the Anglophone literature that applies the 'order of worth' approach of Boltanski and Thévenot has focused on 'quality conventions'. Work on 'labour conventions' (for exceptions, see Gibbon and Riisgaard, 2014; Riisgaard and Gibbon, 2014) or conventions applied to the internal organisation of firms is still at its infancy in agri-food studies. And the interdependence between labour conventions (or firm typologies) and quality conventions (Eymard-Duvernay, 1989, 1994) has been overlooked. An exception is Rosin's (2008) work, which reveals the effects of audits on the reconstruction of the identities of milk and kiwi producers in New Zealand. He shows that an audit, because it 'alters the spirit of farming' (p. 46), modifies the quality conventions between producers and processors. Product quality moves from being determined by the producer as a professional figure to being determined by the individual capacity to comply with the audit.

Innovations in the Anglophone literature

The Anglophone literature has also proposed new analytical and theoretical directions. Here, we focus on three contributions that are particularly original. First, most Convention Theory studies consider compromises as relatively stabilised agreements. On the contrary, Nyberg and Wright (2013) show the importance of looking at 'unstable and temporary compromises' that include market conventions. They explain how companies work towards compromises that include market conventions to resolve disputes arising from different perceptions of what it means to be 'good for the environment'. They show that these compromises are unstable and contested, but also constantly renewed. Companies engaged in environmental management respond to criticism through the framing of goods that can be 'compatible with plural worlds ... for example, both the market and the green world ... such as "green" cars or carbon offsets' (Nyberg and Wright, 2013: 409). They also show that, in several cases, these goods are again called into question. The companies subjected to criticism then undergo further adjustments and frame new 'environmentally friendly' compromises; for example, by proposing 'second-generation' biofuels as a response to criticism of 'first generation' biofuel. Overall, these are examples of formatting of action oriented towards a search for consensus and a rejection of antagonistic positions. Stakeholders do not seek to clarify their agreement by referring to one or another worth, thus avoiding the possibly expensive and risky work of unveiling criticism. This depoliticisation process facilitates greater 'market domination' by validating market mechanisms (ibid.: 420). Nyberg and Wright (2013) highlight how the 'business world' is gradually taking on the role of steward of the environment through the partnerships it forges with environmental conservation NGOs and government authorities. This production of temporary compromises, by removing criticism, contributes to the degradation of the environment in favour of the market. These arguments can also be found in the French school of Convention Theory (e.g. Boltanski, 2011; Boltanski and Chiapello, 1999; Thévenot, 2009), which reported the hegemony of the 'market world' and its subjugation of the other worlds. In general, however, this approach has been relatively less developed in the French school of Convention Theory and thus suggests new venues for research.

Second, Ekbia and Evans (2009) provide novel directions by focusing on formats of information. They examine what formats guide producers in the US Midwest in their land use decisions and argue that disseminated information received from different sources acquires different values, in a similar way to 'objects', which acquire different values in each order of worth. Ebkia and Evans introduce the concept of 'information regime' to examine 'the situated practices of daily life involved in the creation and enactment of information' (p. 341), each of which assigns a value to a specific information format. They find clear linkages between the owners' land use decisions and different information regimes. This study is in line with the origin of the differentiation of worlds in Convention Theory – which links to the different types of 'investment of forms', knowledge and 'forms of the probable' attached to different orders of worth (Thévenot, 2009) – but also provides further analytical directions.

Third, Kirwan (2006) studies the manner in which producers and consumers coordinate their reciprocal expectations about product quality by face-to-face and personalised exchanges at farmers' markets. Drawing from Offer (1997), Kirwan develops the analytical category of 'regard convention', which reflects the attention accorded to the other and the reciprocal consideration of the parties in an exchange. This convention makes it possible to reveal quality attributes that are not inherent to the product (and do not determine its quality), but which, like domestic conventions, are constructed through personal interactions. Regard is thus the guarantor of the relationship's authenticity. Kirwan notes that this convention brings a non-market benefit into the overall assessment of the experience of exchange. Consumers refer to these exchanges as more 'intimate, profound and pleasurable' and note their different temporality (they 'have time to talk' p. 309), as opposed to the functional and mechanical relationship of market exchange. Proximity, trust, reciprocity and social connection are important elements of this convention, Although the 'regard convention' could be seen as a combination of traits usually covered within a combination of domestic convention (confidence gained in personalised interactions, anchoring to a place, and personal ties) and civic convention (collective and social well-being), Kirwan argues that it needs to be understood as a distinctive convention.

Regimes of engagement

Beyond justification

In the Anglophone literature on agri-food studies, some authors have run up against the limits of Convention Theory in explaining specific situations and empirical observations. Some modalities of action cannot be explained through the mechanisms of justification relying on general orders of worth. In Kirwan's study (2006), for example, other modalities of action appear in 'regard'. These are close to care and an 'engagement in familiarity' (Thévenot, 2006) that arises from proximity, rather than from seeking public legitimacy. In a different way, Nyberg and Wright (2013) show that stakeholders focus on strategic negotiation of interests based on individual preferences and avoid relying on the common good. These modalities of action converge on a 'liberal grammar' of individuals, different from that of justification and of qualification of common goods (Thévenot, 2014, 2015c). Similarly, Raynolds' work on new fair trade conventions shows that personalised interactions can also rely on the negotiation of interests through alliances and partnerships, rather than on 'justifiable action'. We thus see in these studies the necessity to consider a more general theory of action.

The extension of the theoretical programme of the French school of Convention Theory first appeared in the early 2000s, with a collective publication at the *Conventions et institutions* symposium in 2003, which highlighted a second 'pluralism of coordination' (Eymard-Duvernay et al., 2006). This second pluralism sought to study a very different form of coordination, based on engagements of proximity, which falls short of the requirements of public justification. This extension is based on the work of Laurent Thévenot and the sociology he has developed around different 'regimes of engagement' (Thévenot, 2006). Alongside 'justifiable action' (Boltanski and Thévenot, 1991), Thévenot developed a program of research aimed at examining other modalities of action: those of 'action in a plan' and of 'familiar engagement' (Thévenot, 2007). He draws on grammars of commonality to analyse the different way people voice and resolve differences. In addition to a grammar of plural orders of worth, he elaborated two others: the 'liberal grammar' of individuals engaging in a plan, and the 'grammar of personal affinities to common places' (Thévenot, 2014, 2015c).

This extension to plural regimes of engagement and modalities of coordination, from the most public to the most intimate (Thévenot, 2007), made the rounds of the sociological Anglophone literature through a different disciplinary entry point, that of 'French pragmatic sociology' (see in particular, Silber, 2003; Jagd, 2011; Luhtakallio, 2012; Blok, 2013, 2015; Friedland et al., 2014; Luhtakallio and Eliasoph, 2014; Scott and Pasqualoni, 2014; Hansen, 2016). But so far, scholars in the field of agri-food, environmental and rural studies have made limited use of this approach. In the next sub-sections, we examine two areas where this is starting to happen[11]: (i) the literature examining governance of natural resources through standard-setting and multi-stakeholders initiatives; and (ii) the literature on 'modes of valuations' of the environment.

Governing through standards and multi-stakeholder initiatives

Several studies have analysed the participatory mechanisms that are used by multi-stakeholder initiatives to claim better governance of natural resources and to gain legitimacy. Some have examined the processes of depoliticisation or technicisation entailed in setting and applying sustainability standards and certifications. They highlight the oppression that multi-stakeholder governance exerts on the qualification of the common good – the expression of what is considered 'just' and 'unjust' in practice (Cheyns, 2011; Nyberg and Wright, 2013). These studies examine various alliances of firms and professionalised transnational NGOs (Cheyns, 2011; Blok, 2013; Nyberg and Wright, 2013; Blok and Meilvang, 2015), which turn ecological issues into opportunities for building 'green markets'. These alliances frame coordination towards plan-oriented action formats and a common technical language (Cheyns, 2011; Ponte and Cheyns, 2013; Blok and Meilvang, 2015). In these initiatives, the tensions, divergences and conflicts imposed by these framings are often addressed without public debate. By analysing these modes of action and oppression, these studies have the power to 're-politicise' contemporary governance mechanisms, making them contestable again (Hansen, 2016).

Other studies show how governance through 'multi-stakeholderism' undermines the familiar attachments of certain participants. For example, Blok and Meilvang (2015) show how some activists, invested in the process of sustainable development of a former harbour in Copenhagen, used images to express, share and make public various familiar attachments to the city and its ecology. At the same time, sustainability management emerged at the heart of strategic governance by industry and NGOs, which ascribed value to technical and detached expertise. In this 'regime of engagement in a plan' (Thévenot, 2007), opponents are forced to mobilise references to technical standards (e.g. around air pollution) rather than familiar attachments. The analysis of the visuals used by activists reveals alternative ecologies that are undermined by the power of the plan. These images reflect site-specific sensitivities and embodied attachments that are difficult to verbalise and which require other means of expression. They remind participants of the existence of an alternative political valuation that is more hospitable to a wider diversity of familiar attachments (see also Richard-Ferroudji and Barreteau, 2012).

A related direction taken in this literature is one linking the exercise of power to the legitimacy of certain forms of what is probable, according to what is considered to be probative or proof (Thévenot, 2009). In transnational multi-stakeholder fora, the statistical form of the probable is often used to discredit monographic information, since the latter is supposed to concern only one singular case, opposed to the generalisation of a statistical fact. For example, Cheyns (2011), Silva-Castañeda (2012) and Ponte and Cheyns (2013) have shown how the dominance of macro and statistical variables led to the disqualification of the voices of 'local communities' in transnational multi-stakeholder initiatives for the standardisation of sustainability in palm oil production. More than representing a conflict between different orders of worth to establish legitimacy, it is a form of knowledge resting on 'familiar engagements' (Thévenot, 2006) that is disqualified in the process. Thus, accounts of monographic experiences, based on personal accounts or on personal or ancient marks on nature (old trees planted by ancestors, rivers), which constitute evidence for those who live and share a same link with the environment, are side-lined because they are not shareable with 'stakeholders' who are foreign to the location concerned (Cheyns, 2011; Silva-Castañeda, 2012). Similarly, 'local communities' affected by the expansion of oil palm cultivation are excluded from the debate when they attempt to report the damage done to their daily lives and when they recount, with lively emotion, the loss of customary lands, sacred trees or the destruction of their environment. They communicate a strong attachment that is embarrassing and foreign to other participants. The latter prefer a liberal grammar, which favours the expression of negotiable interests in a balancing of interests and engagement in a plan (Cheyns, 2014; Silva-Castañeda, 2015; Thévenot, 2015c).

Valuations of the environment

The new attention on the regime of familiar engagement has also led to work examining the environment as dwelled-in space (Centemeri, 2015; Centemeri and Renou, 2015). Centemeri (2015) invites us to go beyond the issue of (in)commensurability in environmental conflicts linked to orders of worth. She introduces another form of incommensurability that is more 'radical'. It is linked to languages and practices of valuation based on personal and intimate attachments to the environment resulting from familiarisation over time. In familiar engagement, the dwelled-in environment is 'a place that a person values because he/she moves and feels "at ease" in it, and because memories are deposited there'. Indeed, the 'person is "distributed" in his/her dwelled-in environment, which becomes a constitutive part of the person so that, if affected, the consequences rebound directly on him' (Centemeri, 2015: 312). These attachments are valuable in ways that exclude commensuration. Valuations can be understood and shared, but do not constitute legitimate arguments in a public debate (see also Blok and Meilvang, 2015).

This approach is key to understanding the tensions and suffering created by the transformation of a complaint around the loss of land into metrics of shared 'interests', a process usually proposed in so-called 'win-win' models and/or through monetary compensation mechanisms. The composition of different voices in contemporary mechanisms for the multi-stakeholder management of resources rests on the expression of a plurality of interests prepared for a negotiation (Thévenot, 2014). In this mode of composition, people are looking for win-win solutions, starting from a belief that they can reach a shared interest (Richard-Ferroudji and Barreteau, 2012; Silva-Castañeda, 2015). For example,

Silva-Castañeda (2015) examines the 'options' offered to Indonesian communities when their land has been lost to the benefit of large-scale plantations. With monetary compensation, land is ascribed a monetary value that is supposed to compensate for its loss. In 'win-win partnerships', the plantation promises to plant a new plot of oil palm, with part of the income earmarked for the community. This plot is usually located in a different place from where the communities lay claim to land. Land thus becomes a productive resource, without consideration of its dwelled-in value (Silva-Castañeda, 2015, 2016). These schemes are very often rejected by local communities, or accepted reluctantly over time (Centemeri, 2015). Local communities expect different responses from their grievances: recognition of the suffering caused by their loss, compassion, and the establishment of fault and legal redress (Silva-Castañeda, 2015). The processes of negotiation often entail the translation of rights and attachments into 'mutually beneficial options' for companies and local communities. These shared interests, however, fall wholly within the economic domain. Therefore, the results of these negotiations do not meet community expectations, nor fulfil promises to protect land rights (ibid.).

Conclusion

Convention Theory has had important impacts in various strands of the Anglophone agri-food literature in the past two decades. Two of these impacts are worth reiterating: (i) it guided discussions of how different conceptions, operationalisations and valuations of 'quality' shape organisation, coordination and exchange; and (ii) it steered attention away from a predominant preoccupation with transaction costs in explaining the dynamics of production, exchange and consumption, and towards more pluralistic and sociological understandings. These two developments in turn extended the palette of, and provided new interactions between, critical approaches in agri-food studies dealing with structured inequalities, distribution of value added, and the potential of alternative organisational systems. More recently, a better understanding of these dynamics has also been made possible by analytical instruments that go beyond the framework of 'justification' to also examine other regimes of engagement, such as 'action in plan' and 'familiar engagement', and the specific forms of oppression that a particular regime may exert over others.

At the same time, at least three further analytical, methodological and empirical developments are needed to carry this research agenda forward. First, it is evident that despite increasing interest in combining Convention Theory and political economy approaches in view of permeating structural/macro analyses of agri-food industries with micro and meso dimensions of power, empirical case studies of this kind are still rare. Second, recent work has highlighted the importance of unstable, multiple and parallel orders of worth, and of temporary settlements, but methodological discussions of how different types of orders come to be characterised as such have been lacking. Finally, the work of pragmatic sociology has highlighted the necessity to move towards a more general theory of action.[12] Although not yet widely circulated in the Anglophone literature on agri-food studies, the regimes of engagement approach provides a useful venue. Work that has explicitly or implicitly drawn from it has provided key insights on the relations between land use and the environment, the plurality of valuations of human relationships and their dependence on the environment, as well as the practices that weaken these existential bases of human life.

Notes

1   The authors have contributed in equal manner.
2   By 'Anglophone literature', we mean work published in English by scholars who are not institutionally based in France and who have not been trained there.
3   For more details, see Dupuy et al. (1989), Orléan (1994), Batifoulier (2001), Eymard-Duvernay (2006a, 2006b), Bessis (2008), Diaz-Bone and Thévenot (2010), Diaz-Bone (2011), and Diaz-Bone and Salais (2011). For the purposes of our discussion in this chapter, hereafter we use the term 'Convention Theory' to also include 'l'économie des conventions' approach.
4   French quotations have been translated by the authors.
5   Two papers by Busch, Busch (1992), Busch and Tanaka (1996), were published before Wilkinson's article (1997). While they cite the work of Boltanski and Thévenot (1991), they do not explicitly refer to Convention Theory.
6   Lawrence Busch discovered the work of Boltanski and Thévenot during a sabbatical year at IRD (the French National Research Institute for Sustainable Development, formerly ORSTOM) in 1987, when he also met Bruno Latour. Just after the publication of *La grande transformation*, John Wilkinson spent a sabbatical year in Paris where he was exposed to Convention theory. Elizabeth Barham, who was conducting an inquiry into social movements in France, attended a presentation of *La grande transformation* in Toulouse. Gilles Allaire and Michael Watts organized a seminar in Berkeley in 1998, which included John Wilkinson, Terry Marsden, Lawrence Busch, Michael Storper, Raymond Jussaume and other researchers of the RC40 of the International Association of Rural Sociology, and several authors of chapters in *La grande transformation*, including Laurent Thévenot. Benoit Daviron was invited to Copenhagen in 1999 to introduce Convention theory to a group of researchers working on global value chains, including Phil Raikes, Peter Gibbon and Stefano Ponte.
7   The literature review included in this section was first published in Ponte (2016). The version included in this chapter is much more condensed. Additional information and details on the 51 entries analysed here are available in the original article.
8   We consider 'agri-food' studies contributions dealing with the production, processing, trade, distribution, cooking, use and re-use/re-cycle of agricultural, fishery and forestry products for food, feed and industrial purposes, and their environmental interactions.
9   See authors mentioned in the previous section (1997–2003 period), and Goodman (2004), Raynolds (2004) and Murdoch and Miele (2004).
10 In the mode of ordering of connectivity 'stories are told of partnership, alliance, responsibility and fairness' (Whatmore and Thorne, 1997: 294–295).
11 In this section, we rely on the relevant literature published in English between 2011 and 2016, also covering authors who are institutionally affiliated in France. This is because the number of articles on these issues is still small.
12 Although Thévenot's (2006) book on regimes of engagement has not yet been translated into English, several articles and book chapter in English are available (Thévenot, 2002, 2007, 2009, 2014, 2015a).

References

Allaire G., Boyer R. 1995. *La grande transformation de l'agriculture: Lectures convention-nalistes et regulationnistes*. Paris, INRA-Economica.

Andersen A.H. 2011. Organic Food and the Plural Moralities of Food Provisioning. *Journal of Rural Studies*, 27(4), 440–450.

Barbera F., Audifredi S. 2012. In Pursuit of Quality: The Institutional Change of Wine Production Market in Piedmont. *Sociologia Ruralis*, 52(3), 311–331.

Barham E. 2002. Towards A Theory of Values-Based Labeling. *Agriculture and Human Values*, 19, 349–360.

Barham E. 2003. Translating Terroir: The Global Challenge of French AOC Labeling. *Journal of Rural Studies*, 19, 127–138.

Barnett C. 2014. Geography and Ethics III from Moral Geographies to Geographies of Worth. *Progress in Human Geography*, 38(1), 151–160.

Batifoulier P. (dir.), 2001. *Théorie des conventions*. Paris: Economica.

Bessis F. 2008. Quelques convergences remarquables entre l'Économie des Conventions et la Théorie de la Régulation. *Revue Française de Socio-Économie*, 1(1), 9–25.

Biggart N.W., Beamish T.D. 2003. The Economic Sociology of Conventions: Habit, Custom, Practice, and Routine in Market Order. *Annual Review of Sociology*, 29(1), 443–464.

Blok A. 2013. Pragmatic Sociology as Political Ecology on the Many Worths of Nature(s). *European Journal of Social Theory*, 16(4), 492–510.

Blok A. 2015. Attachment to the Common-Place: Pragmatic Sociology and the Aesthetic Cosmopolitics of Eco-House Design in Kyoto, Japan. *European Journal of Cultural and Political Sociology*, 2(2), 122–145.

Blok A., Meilvang M.L. 2015. Picturing Urban Green Attachments: Civic Activists Moving Between Familiar and Public Engagements in the City. *Sociology*, 49(1), 19–37.

Boltanski L. 2011. *On Critique: A Sociology of Emancipation*. Cambridge, Polity Press.

Boltanski L., Chiapello E. 1999. *Le nouvel esprit du capitalisme*. Paris, Gallimard. [English translation (2005) *The New Spirit of Capitalism*. London: Verso].

Boltanski O., Thévenot L. 1991. *De la justification. Les économies de la grandeur*. Paris, Gallimard. [English translation (2006) *On Justification: Economies of Worth*. Princeton: Princeton University Press].

Bourdieu P. 1997. *Méditations pascaliennes*. Paris, Le Seuil.

Busch L. 1992. Metatheories and Better Theories: A Reply to Ruttan. *International Journal of the Sociology of Agriculture and Food*, 2, 44–49.

Busch L. 2000. The Moral Economy of Grades and Standards. *Journal of Rural Studies*, 16, 273–283.

Busch L., Tanaka K. 1996. Rites of Passage: Constructing Quality in a Commodity Subsector. *Science, Technology and Human Values*, 21(1), 3–27.

Centemeri L. 2015. Reframing Problems of Incommensurability in Environmental Conflicts through Pragmatic Sociology. From Value Pluralism to the Plurality of Modes of Engagement with the Environment. *Environmental Values*, 24(3), 299–320.

Centemeri L., Renou G., 2015. Métabolisme social et langages de valuation. Apports et limites de l'économie écologique de Joan Martinez-Alier à la compréhension des inégalités environnementales. <hal-01163219>

Cheyns E. 2011. Multi-Stakeholder Initiatives for Sustainable Agriculture: Limits of the "Inclusiveness" Paradigm. In Ponte S., Gibbon P. and Vestergaard J. (Eds.), *Governing through Standards: Origins, Drivers and Limitations*, Basingstoke (Hampshire, Royaume-Uni), Palgrave Macmillan, pp. 318–354.

Cheyns E. 2014. Making "minority voices" Heard in Transnational Roundtables: The Role of Local NGOs in Reintroducing Justice and Attachments. *Agriculture and Human Values*, 31(3), 409–423.

Cheyns E., Riisgaard L. 2014. Introduction to the Symposium: The Exercise of Power through Multi-Stakeholder Initiatives for Sustainable Agriculture and its Inclusion and Exclusion Outcomes. *Agriculture and Human Values*, 31(3), 439–453.

Cidell J.L., Alberts H.C. 2006. Constructing Quality: The Multinational Histories of Chocolate. *Geoforum*, 37(6), 999–1007.

Coq-Huelva D., García-Brenes M.D., Sabuco-i-Cantó A. 2012. Commodity Chains, Quality Conventions and the Transformation of Agro-Ecosystems: Olive Groves and Olive Oil Production in Two Andalusian Case Studies. *European Urban and Regional Studies*, 19 (1), 77–91.

Daviron B., Ponte S. 2005. *The Coffee Paradox: Global Markets, Commodity Trade and the Elusive Promise of Development*. London, Zed Books.

Diaz-Bone R. 2011. The Methodological Standpoint of the "économie des conventions". *Historical Social Research/Historische Sozialforschung*, 36(4), 43–63doi: 10.12759/hsr.36.2011.4.43-63.

Diaz-Bone R. 2012. Elaborating the Conceptual Difference between Conventions and Institutions. *Historical Social Research*, 37(4), 64–75.

Diaz-Bone R., Salais R. 2011. Economics of Convention and the History of Economies: Towards a Transdisciplinary Approach in Economic History. *Historical Social Research/ Historische Sozialforschung*, 36(4), 7–39. doi: 10.12759/hsr.36.2011.4.7-39.

Diaz-Bone R., Thévenot L., 2010. 'La sociologie des conventions. La théorie des conventions, élément central des nouvelles sciences sociales françaises', *Trivium* [En ligne], 5 | 2010, consulté le 28 septembre 2017. URL: http://trivium.revues.org/3626.

Du Gay P., Morgan G. (Eds.), 2013. *New Spirits of Capitalism? Crises, Justifications, and Dynamics*. Oxford, Oxford University Press.

Dupuy J.P., Eymard-Duvernay F., Favereau O., Orléan A., Salais R., Thévenot L. 1989. Introduction au numéro spécial de la revue économique sur l'économie des conventions. *Revue Economique*, 40(2), 141–145.

Ekbia H.R., Evans T.P. 2009. Regimes of Information: Land Use, Management, and Policy. *The Information Society*, 25(5), 328–343.

Evans D. 2011. Consuming Conventions: Sustainable Consumption, Ecological Citizenship and the Worlds of Worth. *Journal of Rural Studies*, 27(2), 109–115.

Eymard-Duvernay F. 1989. Conventions de qualité et formes de coordination. *Revue Economique*, 40(2), 329–359.

Eymard-Duvernay F. 1994. Coordination des échanges de l'entreprise et qualité des biens. In Orléan A. (Ed.), *Analyse économique des conventions, Paris, PUF*.

Eymard-Duvernay F. (Ed.), 2006a. *L'économie des conventions, méthodes et résultats. Tome 1: Débats*. Paris: La Découverte.

Eymard-Duvernay F. (Ed.), 2006b. *L'économie des conventions, méthodes et résultats. Tome 2: Développements*. Paris: La Découverte.

Eymard-Duvernay F., Favereau O., Salais R., Thévenot L., Orléan A. 2006. 1. Valeurs, coordination et rationalité: trois thèmes mis en relation par l'économie des conventions. In François Eymard-Duvernay (Ed.), *L'économie des conventions, méthodes et résultats* (coll. Recherches), Paris, La Découverte, pp. 23–44.

Favereau O. 1989. Marchés internes, marches externes. *Revue économique*, 40(2), 273–328.

Favereau O. 1994. Règle, organisation et apprentissage collectif: un paradigme non standard pour trois théories hétérodoxes. In Orlean A. (Ed.), *Analyse économique des conventions*, Paris, Presses Universitaires de France.

Favereau O., Thévenot L. 1991. *réflexion sur une notion d'équilibre utilisable dans une économie de marchés et d'organisations*. Ronéo, ERMES/ EHESS, pp. 40.

Freidberg S. 2003. Culture, Conventions and Colonial Constructs of Rurality in South-North Horticultural Trades. *Journal of Rural Studies*, 19, 97–109.

Freidberg S. 2004. *French Beans and Food Scares: Culture and Commerce in an Anxious Age*. New York, Oxford University Press.

Friedland R., Mohr J., Roose H., Gardinali P. 2014. The Institutional Logics of Love: Measuring Intimate Life. *Theory and Society*, 43(3–4), 333–370.

Gereffi G., Humphrey J., Sturgeon T., 2005. The Governance of Global Value Chains. *Review of International Political Economy*, 12(1), 78–104.

Gibbon P., Ponte S. 2005. *Trading Down: Africa, Value Chains and the Global Economy*. Philadelphia, Temple University Press.

Gibbon P., Riisgaard L. 2014. A New System of Labour Management in African Large-Scale Agriculture? *Journal of Agrarian Change*, 14(1), 94–128.

Goodman D. 2004. Rural Europe Redux? Reflections on Alternative Agri-food Networks and Paradigm Change. *Sociologia ruralis*, 44(1), 3–16.

Guthey G.T. 2008. Agro-Industrial Conventions: Some Evidence from Northern California's Wine Industry. *The Geographical Journal*, 174, 138–148.

Gwynne R.N. 2006. Governance and the Wine Commodity Chain: Upstream and Downstream Strategies in New Zealand and Chilean Wine Firms. *Asia Pacific Viewpoint*, 47, 381–395.

Hansen M.P. 2016. Non-Normative Critique Foucault and Pragmatic Sociology as Tactical Re-Politicization. *European Journal of Social Theory*, 19(1), 127–145.

Jagd S. 2011. Pragmatic Sociology and Competing Orders of Worth in Organizations. *European Journal of Social Theory*, 14(3), 343–359.

Kirwan J. 2006. The Interpersonal World of Direct Marketing: Examining Conventions of Quality at UK Farmers' Markets. *Journal of Rural Studies*, 22, 301–312.

Lindkvist K.B., Sánchez J.L. 2008. Conventions and Innovation: A Comparison of Two Localized Natural Resource-Based Industries. *Regional Studies*, 42, 343–354.

Luhtakallio E. 2012. *Practicing Democracy. Activism and Politics in France and Finland*. Basingstoke, Palgrave Macmillan.

Luhtakallio E., Eliasoph N. 2014. Ethnography of Politics and Political Communication: Studies in Sociology and Political Science. In Hall Jamieson K. and Kenski K. (dirs.), *Oxford Handbook of Political Communication*, Oxford, Oxford University Press, Online Publication Date: Sep. 2014. doi: 10.1093/oxfordhb/9780199793471.013.28.

Marsden T., Banks J., Bristow G. 2000. Food Supply Chain Approaches: Exploring their Role in Rural Development. *Sociologia Ruralis*, 40, 424–438.

Morgan K., Marsden T., Murdoch J. 2006. *Worlds of Food: Place, Power, and Provenance in the Food Chain*. Oxford, Oxford University Press.

Murdoch J., Marsden T., Banks J. 2000. Quality, Nature and Embeddedness: Some Theoretical Considerations in the Context of the Food Sector. *Economic Geography*, 76, 107–125.

Murdoch J., Miele M. 1999. 'Back to Nature': Changing 'Worlds of Production' in the Food Sector. *Sociologia Ruralis*, 39, 465–483.

Murdoch J., Miele M. 2004. Culinary Networks and Cultural Connections: A Conventions Perspective. In Hughes A. and Reimer S. (Eds.), *Geographies of Commodity Chains*, London, Routledge, pp. 102–119.

Nicolas F., Valceschini E. (Eds.), 1995. *Agro-alimentaire: une économie de la qualité*. Paris: INRA/Economica.

Norman D.A. 1993. Les artefacts cognitifs. In Conein B., Dodier N. and Thévenot L. (Eds.), *Les objets dans l'action, de la maison au laboratoire*, Raisons pratiques, 4 Paris, EHESS, pp. 15–34.

Nyberg D., Wright C. 2013. Corporate Corruption of the Environment: Sustainability as a Process of Compromise. *The British Journal of Sociology*, 64(3), 405–424.

Orléan A. 1989. Pour une approche cognitive des conventions économiques. *Revue Economique*, 40(2), 241–272.

Orléan A. 1994. *Analyse économique des conventions*. Paris, PUF.

Orléan A. 2004. L'économie des conventions: Définitions et résultats, préface to the 4th édition of A. Orléan (nouvelle éd.). In *Analyse économique des conventions*, Paris, Presses Universitaires de France.

Parrott N., Wilson N., Murdoch J. 2002. Spatializing Quality: Regional Protection and the Alternative Geography of Food. *European Urban and Regional Studies*, 9, 241–261.

Ponte S. 2009. Governing through Quality: Conventions and Supply Relations in the Value Chain for South African Wine. *Sociologia Ruralis*, 49(3), 236–257.

Ponte S. 2016. Convention Theory in the Anglophone Agri-food Literature: Past, Present and Future. *Journal of Rural Studies*, 44(SupplementC), 12–23.

Ponte S., Cheyns E. 2013. Voluntary Standards and the Governance of Sustainability Networks. *Global Networks*, 13(4), 459–477.

Ponte S., Daviron B. 2011. Creating and Controlling Symbolic Value. In Bandelj N. and Wherry F. (Eds.), *The Cultural Wealth of Nations*, Standford, Stanford University Press.

Ponte S., Gibbon P. 2005. Quality Standards, Conventions and the Governance of Global Value Chains. *Economy and Society*, 34(1), 1–31.

Ponte S., Sturgeon T. 2014. Explaining Governance in Global Value Chains: A Modular Theory-Building Effort. *Review of International Political Economy*, 21(1), 195–223.

Raikes P., Jensen M.F., Ponte S. 2000. Global Commodity Chain Analysis and the French Filiére Approach: Comparison and Critique. *Economy and Society*, 29(3), 390–417.

Raynolds L.T. 2002. Consumer/Producer Links in Fair Trade Coffee Networks. *Sociologia Ruralis*, 42(4), 404–424.

Raynolds L.T. 2004. The Globalization of Organic Agri-food Networks. *World Development*, 32(5), 725–743.

Raynolds L.T. 2012a. Fair Trade: Social Regulation in Global Food Markets. *Journal of Rural Studies*, 28(3), 276–287.

Raynolds L.T. 2012b. Fair Trade Flowers: Global Certification, Environmental Sustainability, and Labor Standards. *Rural Sociology*, 77(4), 493–519.

Raynolds L.T., Murray D., Heller A. 2007. Regulating Sustainability in the Coffee Sector: A Comparative Analysis of Third-Party Environmental and Social Certification Initiatives. *Agriculture and Human Values*, 24(2), 147–163.

Renard M.C. 2003. Fair Trade: Quality, Market and Conventions. *Journal of Rural Studies*, 19, 87–96.

Richard-Ferroudji A., Barreteau O. 2012. Assembling Different Forms of Knowledge for Participative Water Management: Insights from the Concert'eau Game. In Claeys C. and Jacqué M. (Eds.), *Environmental Democracy Facing Uncertainty*, pp. 19. <hal-00777847>http://hal.ird.fr/hal-00777847/document (consulté le 13 octobre 2016).

Riisgaard L., Gibbon P. 2014. Labour Management on Contemporary Kenyan Cut Flower Farms: Foundations of an Industrial–Civic Compromise. *Journal of Agrarian Change*, 14 (2), 260–285.

Rosin C. 2008. The Conventions of Agri-Environmental Practice in New Zealand: Farmers, Retail Driven Audit Schemes and a New Spirit of Farming. *GeoJournal*, 73, 45–54.

Rosin C., Campbell H. 2009. Beyond Bifurcation: Examining the Conventions of Organic Agriculture in New Zealand. *Journal of Rural Studies*, 25, 35–47.

Salais R., Storper M. 1992. The Four 'worlds' of Contemporary Industry. *Cambridge Journal of Economics*, 16, 169–193.

Salais R., Thévenot L. (Eds.), 1986. *Le travail, marché, règles, conventions*. Paris: INSEE-Economica.

Sánchez-Hernández J.L. 2011. The Food Value Chain as a Locus for (dis)agreement: Conventions and Qualities in the Spanish Wine and Norwegian Salted Cod Industries. *Geografiska Annaler: Series B, Human Geography*, 93(2), 105–119.

Sánchez-Hernández J.L., Aparicio-Amador J., Alonso-Santos J.L. 2010. The Shift between Worlds of Production as an Innovative Process in the Wine Industry in Castile and Leon (Spain). *Geoforum*, 41(3), 469–478.

Scott A., Pasqualoni P.P. 2014. The Making of a Paradigm: Exploring the Potential of Economy of Conventions and Pragmatic Sociology. In Adler P.S., du Gay P., Morgan G. and Reed M. (Eds.), *Oxford Handbook of Sociology, Social Theory and Organization Studies*, Oxford, Oxford University Press, pp. 64–87.

Silber I.F. 2003. Pragmatic Sociology as Cultural Sociology. *European Journal of Social Theory*, 6(4), 427–449.

Silva-Castañeda L. 2012. A Forest of Evidence: Third-Party Certification and Multiple Forms of Proof—A Case Study of Oil Palm Plantations in Indonesia. *Agriculture and Human Values*, 29(3), 361–370.

Silva-Castañeda L. 2015. What Kind of Space? Multi-Stakeholder Initiatives and the Protection of Land Rights. *International Journal of Sociology of Agriculture and Food*, 22(2), 67–83.

Silva-Castañeda L. 2016. In the Shadow of Benchmarks: Normative and Ontological Issues in the Governance of Land. *Environment and Planning A*, 48(4), 681–698.

Stark D. 2011. *The Sense of Dissonance: Accounts of Worth in Economic Life*. Princeton and Oxford, Princeton University Press.

Storper M., Salais R. 1997. *Worlds of Production: The Action Frameworks of the Economy*. Cambridge, Harvard University Press.

Stræte E.P. 2004. Innovation and Changing 'Worlds of Production': Case Studies from Norwegian Dairies. *European Urban and Regional Studies*, 11, 227–241.

Sylvander B. 1995. Convention de qualité, concurrence et coopération: Cas du label rouge dans la filière volaille. In Allaire G. and Boyer R. (Eds.), *La grande transformation de l'agriculture. Lectures conventionnalistes et régulationnistes*, Paris, INRA/Economica, pp. 73–96.

Tallontire A. 2007. CSR and Regulation: Towards a Framework for Understanding Private Standards Initiatives in the Agri-Food Chain. *Third World Quarterly*, 28(4), 775–791.

Thévenot L. 2012. Convening the Company of Historians to Go into Conventions, Powers, Critiques and Engagements. *Historical Social Research*, 37(4), 22–35.

Thévenot L. 1989. Economie et politique de l'entreprise; économies de l'efficacité et de la confiance. In Boltanski L. and Thévenot L. (Eds.), *Justesse et justice dans le travail*, Paris, Cahiers de Centre d'Etudes de l'Emploi, PUF, pp. 135–207.

Thévenot L. 1995. Des marchés aux normes. In: Allaire G., Boyer R. (dir.), *La grande transformation de l'agriculture: lectures conventionnalistes et régulationnistes*. Versailles/Paris, Inra éditions/Economica, 33–51.

Thévenot L. 2002. Conventions of Co-Ordination and the Framing of Uncertainty. In Fullbrook E. (Ed.), *Intersubjectivity in Economics*, London and New York, Routledge.

Thévenot L. 2006. *L'action au pluriel: sociologie des régimes d'engagement*. Paris, Découverte.

Thévenot L. 2007. The Plurality of Cognitive Formats and Engagements: Moving between the Familiar and the Public. *European Journal of Social Theory*, 10(3), 413–427.

Thévenot L. 2009. Governing Life by Standards: A View from Engagements. *Social Studies of Science*, 39(5), 793–813.

Thévenot L. 2014. Voicing Concern and Difference: From Public Spaces to Common-Places. *European Journal of Cultural and Political Sociology*, 1(1), 7–34.

Thévenot L. 2015a. Certifying the World: Power Infrastructures and Practices in Economies of Conventional Forms. In Aspers Patrik and Dodd Nigel (Eds.), *Re-Imagining Economic Sociology*. Oxford: Oxford University Press, 195–223.

Thévenot L. 2015b. Vous avez dit 'capital'? Extension de la notion et mise en question d'inégalités et de pouvoirs de domination. *Annales Histoire Sciences Sociales*, 70(1), 69–80.

Thévenot L. 2015c. Making Commonality in the Plural, on the Basis of Binding Engagements. In Dumouchel P. and Gotoh R. (Eds.), *Social Bonds as Freedom: Revising the Dichotomy of the Universal and the Particular*, New York, Berghahn, pp. 82–108.

Thévenot L. 2016. Le pouvoir des conventions. In Batifoulier P. et alii (Eds.), *Dictionnaire des conventions*, Villeneuve d'Ascq, Presses Universitaires du Septentrion, pp. 203–207.

Trabalzi F. 2007. Crossing Conventions in Localized Food Networks: Insights from Southern Italy. *Environment and Planning A*, 39, 283–300.

Truninger M. 2011. Cooking with Bimby in a Moment of Recruitment: Exploring Conventions and Practice Perspectives. *Journal of Consumer Culture*, 11 (1), 37–59.

Whatmore S., Thorne L. 1997. Nourishing Networks: Alternative Geographies of Food. In Goodman D. and Watts M. (Eds.), *Globalizing Food, Agrarian Questions and Global Restructuring*, London, Routledge, pp. 287–304

Wilkinson J. 1997. A New Paradigm for Economic Analysis. *Economy and Society*, 26, 305–339.

5 The new autocracy in food and agriculture

Lawrence Busch

Introduction

Sixty years ago, nearly all agri-food standards were enforced through some form of government regulation, i.e., there were *de jure* standards required to participate in a given market. Of course, even as early as the late 1920s, processors adhered to de facto firm-based standards, as these were necessary for mass production of food products (National Industrial Conference Board, 1929). However, such de facto standards were demanded by particular processors in order to facilitate the internal operations of their particular firms. Moreover, they were only enforced through visual inspection in the purchasing of farm products; and, since different processors desired different qualities of producers, they played a relatively minor role in commerce.

However, what processed, packaged food did was to permit the development of supermarkets. Grocery stores were designed to handle bulk products, with employees filling customers' orders by selecting items, and weighing and packaging the goods for sale. In contrast, supermarkets replaced bulk products with packaged ones and shifted the job of selecting the goods to the customers themselves. Hence, shoppers could walk through the store and collect the (already packaged) goods they desired. This saved considerably on labour costs, allowing lower prices so as to attract more customers (Cochoy, 2011). That said, as long as supermarkets were individually owned or merely chains of a few stores, processors could and did set prices and formats.

Moreover, for industrialised nations that were self-sufficient in food production, most of the agri-food chain was characterised by *national* production, processing and retailing. The only significant exceptions to this rule were a few tropical commodities (mostly legal stimulants such as coffee, sugar, cocoa and tea, but also including bananas) and certain luxury products such as foie gras. In addition, most producers and processors produced 'for the market,' bringing their goods to wholesale markets where most grocers engaged in a 'cash and carry' approach, purchasing whatever was available in the wholesale markets to resell to final consumers.

In contrast, several transformations to food and agriculture have occurred over the last 60 years. First, we have witnessed the rise of large supermarket chains that have significant national market shares and therefore can dictate prices and qualities to processors and producers. For example, in France in 2009 the five largest firms had a 65% share of the market. Among other nations, concentration runs as high as 90% for three firms in Portugal (2011) and as low as 35% for four firms in the US (2006) (Nicholson and Young, 2012). However, the US figures are misleading as concentration levels within a given metropolitan area are much higher (Cotterill, 1999).

Second, over recent decades the larger supermarket chains have begun to operate internationally. The collapse of the Eastern Bloc opened the door to the relatively well-off Eastern European nations (Dries et al., 2007). But most of the growth has been in poor nations, where supermarkets are able to use their higher standards, greater purchasing power and sophisticated purchasing systems to attract middle-income consumers. Hence, supermarket growth in Asia (Reardon et al., 2014), Latin America (Reardon et al., 2007) and Africa (Weatherspoon and Reardon, 2003) has been extraordinarily rapid. And, much of this involves the expansion of chains headquartered in Western nations.

Finally, we now live in a world characterised by 'supply chains.' The notion of a supply chain has been traced to two rather different origins. One perspective links it to the emergence of systems theory in the 1950s (New, 1997), while the other argues that it emerged from the successes of the Japanese automotive industry in the 1970s (Cox, 1999). Likely both played a role in their development as the advent of supply chain management is simultaneously a new branch of economic thought and a new way of organising industries (Busch, 2007). It is performative in the sense understood by Callon (1998, Callon et al., 2007) and his colleagues; as economists have defined the formats and advantages of supply chain management, so practitioners have transformed their operations into supply chains.

A key ingredient in supply chain management is that every actor in the supply chain is expected to conform to a wide and ever-growing variety of (increasingly international) de facto standards. Conformity is enforced through systems of audit that extend from the behaviour of CEOs to janitors, from farm supply companies to farmers to processors to retailers. One can distinguish four components to this Brave New World in which we now live. First, there is the de facto tripartite standards regime (tsr) as noted in other chapters. Second, there is the extension of assembly line technologies perfected by Ford a century ago to much of the agri-food chain. Third, there is a New Taylorism, in which something akin to the time and motion studies developed by Taylor has been extended to all professions. Fourth, there is the rise of Big Data, made possible by advances in information technologies, that permits all of this to become real. In this chapter, I will examine each of these four institutional transformations as they relate to food and agriculture, asking how they are responding to the crises facing us today, of climate change, environmental pollution, financial instability and obesity, among others. In conclusion, I ask what the consequences might be for democratic governance. I begin with the TSR.

The tripartite standards regime

While the widespread creation of formal, written standards for products dates back at least a century, it is only in the last few decades that specific institutions have been developed to certify that particular products, people or processes meet those standards and that certifiers are themselves accredited to certify others. My colleague, John Stone, proposed that this triple transformation – linking (i) standards, (ii) certifications and (iii) accreditations – be called the 'tripartite standards regime' (Loconto et al., 2012). What the TSR does is to establish a private global system of governance that extends far beyond that of individual firms. Let me explain.

The first agri-food standards established were those of individual firms engaged in the mass production of goods such as farm machinery and packaged food products.[1] For machinery, each part would be standardised so as make assembly a fairly straightforward operation as well as to facilitate repair. For packaged goods, standards would define the qualities of the goods as well as the size and shape of packaging and labels. These firm standards replaced the differentiation of craft production. But standards for individual firms led to a proliferation of products and processes, the components of which were not interchangeable. For example, screws employed in creating farm machinery, cans used for processed foods, cases for fresh fruit and other intermediate products varied by company.

The 'Great War' of 1914–18 showed the world the importance of standards. Not only munitions but uniforms, vehicles, aircraft and, of course, rations that were not standardized failed miserably: Ammunition that did not fit all weapons of the same kind could not be fired at the enemy. Spare parts from one manufacturer would not fit another machine. Unstandardized rations failed to deliver adequate nutrition to soldiers and took more space in vehicles. (Frontard, 1994).

By the early 1920s, the lessons of the War had been learned. It became apparent that a lack of industry-wide standards was wasteful of materials, labour and time. As Herbert Hoover (quoted in Cotton, 1922: 144), then the US Secretary of Commerce, put it, 'There is one thing that stands out about American industry that comes up daily to the Department, and that is the remarkable efficiency of the individual industry and the very considerable inefficiency of collective industry.' Hence, in the US and in other industrial nations, the hugely complex process of developing industry-wide standards began.

Such standards served several purposes. First, they made contracting easier as buyers could specify more precisely just what was wanted by referring to a standard. Second, they reduced costs considerably as parts, packaging and labelling suppliers could reduce unnecessary variation. Third, many goods could be purchased sight unseen, since the relevant characteristics were specified in the standards.[2] Fourth, suppliers within a given nation could be put in direct competition with each other to the benefit of buyers, since each would be producing 'the same' thing. Finally, once such standards were widely used, they permitted well-capitalised firms to increase their market share by using assembly line or continuous process production to reduce costs while enhancing profits. Moreover, since most trade was limited by national boundaries, standard contracts, warehouse receipts and related documents made it easier to resolve conflicts between buyers and sellers in national courts.

While some international trade in food and agricultural products has taken place for millennia, the advent of a truly global food regime was made possible through the various meetings of the General Agreement on Tariffs and Trade (GATT) leading to the creation of the World Trade Organization and other international agreements. The WTO (1994) had the effect of taking the Codex Alimentarius – until then a rather backwater agency jointly responsible to the Food and Agriculture Organization and the World Health Organization – into the centre of debates. Referred to in the Sanitary and Phytosanitary Agreement, its decisions became essential to international trade in food products. However, since the Codex's mandate was rather narrow, it was insufficient for the needs of those firms able and desiring to engage in large-scale international trade (e.g., Carrefour, Heinz, International Harvester) as well as for non-governmental organisations (NGOs) wishing to put pressure on large companies for a wide range of reasons, including enhanced animal welfare, fair trade and environmental sustainability. While action by official international agencies was possible in principle, it became clear to both large companies and NGOs that getting such action would take years if not decades to complete. Instead, there was rising pressure to create *global de facto* standards to certify to. Hence, organisations such as GlobalGAP and The Consumer Goods Forum were organised by the largest companies so as to harmonise standards, while ISEAL and Fairtrade International were developed by international proponents of sustainability and fair trade.

However, this, in turn, created a new problem. Since buyers and sellers in global markets often operated under different national legal regimes, disputes about the quality of delivered products – whether fresh tomatoes or parts for tractor assembly – required complex and costly court cases. Moreover, processors and especially retailers were concerned that unsafe or poor-quality products would result in a loss of sales. The solution was to be found in the creation of certifying agencies. While a few food certifiers had been around for a century or more (e.g. the American Institute of Baking), most came into being in just the last few decades. The European Commission recently identified 441 different schemes within its Member States (European Commission, 2015). In recent years, food certification proved itself to be a very lucrative business, attracting the attention of certifiers in other industries. For example, Det Norske Veritas, an established certifier of the seaworthiness of ships, expanded its role to include food products.

However, since anyone could hang a sign outside his office proclaiming to be a certifier, it became necessary to find means to accredit certifiers. And, since the larger certifying companies had branched out into virtually all industries, accreditation could not be limited merely to certifiers engaged with food and agricultural products and processes. Ultimately, the International Accreditation

Forum (2015) and the International Laboratory Accreditation Cooperation (2015) were formed in 1993 and 1996, respectively.[3] The former accredits national programmes of accreditation for all products and processes, while the latter focuses on ensuring standard laboratory tests are used by accrediting national bodies charged with the oversight of lab testing. They are international, non-governmental or quasi-governmental organisations that now accredit national accreditors who accredit certifiers who certify that particular standards are adhered to by particular firms and farms across all industries. Soon after their establishment, considerable efforts were made by both the US and the European Union to promote the establishment of National Accreditation Bodies in those nations that did not have them (Donaldson, 2005; Loconto and Busch, 2010).

At the same time, NGOs desirous of promoting, for example animal welfare or fair trade standards found that, with increased international trade, this task could no longer be pursued solely within national boundaries. However, since the firms involved were well-known to consumers in all the countries where the products were sold, they began to pressure those firms into supporting their efforts. In addition, they have formed alliances with both large producers (e.g. Unilever and Rainforest Alliance), processors (e.g. Coca-Cola and the Nature Conservancy) and with retailers (e.g. McDonald's and the Environmental Defense Fund). They have also participated in so-called multi-stakeholder initiatives (MSIs) in many poor nations (e.g. the Roundtable for Sustainable Palm Oil (RSPO)), and retailers have formed alliances to head off NGO actions (e.g. the Ethical Trading Initiative).

In short, the TSR emerged in order to satisfy the needs of large agri-food (and other) corporations desirous of operating in newly opened global markets and of NGOs desirous of subordinating that trade to goals pursued by those NGOs. Importantly, while obviously large corporations are not run according to democratic principles – indeed, they resemble medieval landed aristocracies in that it is stock ownership that provides voice – the same is true for NGOs. While many might find their actions just and justifiable, they are part of what has been called the NGO-industrial complex (Gereffi et al., 2001). But the development of the TSR is only one facet of the new autocracy. Let us also consider assembly line production.

The new assembly lines

Assembly line production was first employed in the processing of food. In the United Kingdom, biscuits were manufactured by assembly lines as early as 1833. In the United States at about the same time, hogs were slaughtered and butchered as part of a disassembly line process (Giedion, 1975 [1948]). But, as is well-known, it was Ford who popularised assembly line production in the early 20th century, making his Model T car available to the masses while paying wages significantly higher than his competitors. By the mid-20th century, nearly all food processing plants employed some sort of assembly line technologies, some partly automated, most involving a minute division of labour. By the 1970s assembly lines had been extended to tomato harvesting, with machines on which workers could stand removing blemished tomatoes from a conveyor belt while they moved through the fields (Friedland and Barton, 1975), and to fast food restaurants where sandwiches could be made in assembly line fashion (Reiter, 1991). However, as assembly line technologies were expanded, Ford's high wages gradually disappeared. In many other nations considerable resistance was to be found, including in France. New technologies that reduce the cost of small-scale production units, a renewed interest in local foods and the limits of biological processes suggest that such technologies may be fast approaching their limits. Yet, slowly these technologies have spread such that they are found today in nearly every nation on the planet.

In contemporary food processing plants as well as in food service and retail establishments, assembly lines are commonplace. In a processing plant, preparing the raw product is often accomplished by disassembling the parts. Fresh pine-apples, for example, will have their tops removed, their skin removed and their core removed in three distinct processes in succession. Similarly, in the US and many other nations, nearly all slaughtering and butchering of animals is done in central facilities where workers perform the same operations again and again, often to such a degree as to cause carpal tunnel syndrome. One recent study found that 8.7% of poultry processing workers had carpal tunnel syndrome, more than double the rate of other manual workers (Cartwright et al., 2012; see also Lloyd and James, 2008).

In food service operations, individual portions of salads are assembled in much the same way. And, one only need to observe at McDonald's to watch the assembly of hamburger, chicken, fish and other sandwiches. All of this allows the employment of low-skill, part-time and sometimes undocumented workers at whatever the minimum wage might be. But assembly lines only work at peak efficiency when the actions of workers can be well-controlled. Hence, various forms of Taylorism have spread.

The New Taylorism

A century ago, engineer Frederick Winslow Taylor (1911) developed the practice of what he called 'scientific management.' Taylor believed that management had to move beyond 'rule of thumb' practices by bringing science to bear on job organisation. Such an approach, he believed, would also reduce 'soldiering' and other activities that would slow the production line. Moreover, Taylor was clear: worker judgement was to be curtailed and decisions were to be placed squarely in the hands of managers. Using stopwatches and other equipment, the repetitive actions of manual labourers were measured in an attempt to determine the 'most efficient' means by which to accomplish a given task. His approach was so successful that it became a founding pillar for the discipline of ergonomics. It was not only employed in Western nations, but was also put into practice by Alexei Gastev of the Soviet Central Labour Institute in the 1920s. Gastev believed that he could promote Taylorism as a means to the creation of a 'Soviet Americanism' (Bailes, 1977).

However, Taylor and his supporters saw their approach as inherently limited to the kind of minute division of labour and repetitive work found in assembly line production or in certain mining operations. Only manual labour could be subject to this sort of industrial discipline. Managers and professional workers would be exempt.

But he has been proven wrong. His approach has been applied of late to virtually all productive endeavours. A New Taylorism is now applied to craft and intellectual labour including virtually all professions. This has been accomplished through the creation of what Michael Power (1997) has called 'the audit society.'

Here we can see how the New Taylorism has taken two new forms: certifications and performance monitoring. Certifications, a part of the TSR as noted above, are now required of all participants in agri-food supply chains, from input suppliers to farmers, processors and even retailers. As I have argued elsewhere (Busch, 2011), certifications are nearly always based on 'best practices.' This is particularly true of ISO 9000 and 14000 standards, but it is also to be found in many other certification schemes. In each instance, 'best practices' are defined by technical experts, usually with little or no input from practitioners. Consider also that certifications involve what is known in the jargon as 'conformity assessment.' The focus is as much or more on the adherence to best practices as it is on the final product. For example, both Hazard Analysis and Critical Control Points (HACCP) standards and organic standards focus on such best practices and not on the final product. Indeed, in the worst cases, producers may follow best practices and produce an inferior product because other human and non-human actors intervene. Thus, Hatanaka (2010) has shown how Indonesian shrimp fishers were required to adhere to best practices designed elsewhere that failed to consider local conditions, including significant pollution of estuaries from industrial sources.

But certifications are only one part of the New Taylorism. Performance monitoring involves audits of particular individuals often not engaged in assembly line production. Everyone from janitors to CEOs can be monitored by a combination of methods. First, employees engaged in routine work are often subject to regular audits based on a set of predefined measures. Moreover, incentives can be given to those who do very well according to the measures, while sanctions are provided for those who do poorly. Among the activities now frequently monitored are the rapidity at which food orders are entered into a computer database, the time required by a truck driver to deliver food to a given retailer as well as the driver's adherence to a predefined fixed route (measured by Global Positioning Systems) or the promptness of handling of telephone orders at a fresh produce market.

In addition, managers are often audited based on the performance of their store relative to others in similar places. For example, a Walmart manager might be audited based on a comparison of sales in her store as compared to those of stores of a similar size located in neighbourhoods with similar demographic characteristics. A manager of a processing plant might be evaluated based on the volume of throughput per day. And, managers of fast food restaurants may be evaluated based on the volume of meals sold as compared to other stores with similar customer demographics. And, of course, CEOs of larger firms are now evaluated in large part based on quarterly profits of their company.

In addition to these 'internal' audits, one can have purchasers evaluate employee performance. Many full-service restaurants provide survey forms to customers where they can rate the quality of the service; those forms are then used to evaluate, reward or punish servers. Similarly, the larger supermarket chains and food processors often have consumer complaint phones and invite comments on their web sites, and fast food restaurants often encourage consumers to answer on-line questionnaires by offering a discount on a future purchase or a chance in a lottery. All are designed to help in auditing the behaviour of employees as well as the quality of the product.

But, given the enormous amount of data that must be collected to engage in frequent audits, certifications and other measures, rapid information technologies have become an essential part of these transformations. Indeed, without such technologies, much of the auditing would be too difficult or too costly. I now turn to those technologies.

Agri-food in an information age

Essential to the spread of assembly line technologies within various supply chains and throughout the agri-food system, as well as of Taylorism to all occupations, has been the rise of what is often referred to as Big Data. That is to say, the ability to collect and analyse large-scale data sets has made possible (i) the creation of a minute division of labour in countless industries where previously craft or professional production was the norm, and (ii) the application of Taylorist principles to more complex forms of work. A few examples should suffice:

- Fast food restaurants now commonly use computer software to schedule employees. This allows restaurants to arrange employee hours to conform as closely as possible to business volume (Love and Hoey, 1990). This also allows the decomposition of business volume into sales of specific food items, allowing greater use of just-in-time stocking of restaurants. In addition, it promotes the substitutability of workers as most jobs require easily learned skills. Indeed, fast food restaurants have become sufficiently standardised as to allow the use of simulation modelling and operations research to enhance profits (Swart and Donno, 1981).
- Walmart specifies that its suppliers deliver during certain fixed time frames. Failure to do so results in a penalty for the supplier (Bianco and Zellner, 2003). (This allows Walmart to spread the unloading work over longer periods of time and to minimise the space and labour needed for unloading.) In addition, since 1983 it has maintained a large private satellite

communication system that links its many stores together by voice, video and data (White, 2004). This allows Walmart to have unprecedented control over its supply chains as well as what goes on inside its stores.

- It has now become commonplace among food processors and others for tractor-trailers to be fitted with Global Positioning Systems (GPS) such that drivers can be tracked during their entire voyage.
- Bar codes, initially introduced as a means to reduce wait times at super-market check-outs, have taken much of the mental work out of cashiering, replacing it with merely sliding items over a scanner (Brown, 1997). This has also permitted supermarkets to monitor minutely the actions of cashiers and to remove most judgements from the process.
- Point of sale (POS) systems are now commonplace in supermarkets and even smaller food stores. They allow managers to monitor cashiers' behaviour more carefully. As one software vendor explains,

Cashiers using our grocery store POS software will sign in under their own names to use the register, which leads to increased employee accountability for errors and missing money. Also, any manager can access the transaction history of a register to quickly find just about anything, or even print up a report. Cashier Live even includes a time clock, which will make it easy for you to do payroll at the end of a pay-period.

(CashierLive, 2015)

In addition, POS software can be used by large chains to manage their suppliers. Knowing how much of a particular store keeping unit (SKU) has been sold allows those retailers to demand just-in-time delivery from their suppliers. Moreover, large chains can use multi-store data to discipline managers who do not produce at expected levels.

But data is never raw. Data must be created and this can only be done by standardising, certifying, accrediting, measuring and recording. When used largely to monitor the behaviour of machines, as is the case in HACCP, it is often unproblematic. As long as the right hazards are identified and the right measures are employed at Critical Control Points, then the collection of Big Data can be used to ensure, for example, the safety of canned food. However, Big Data can also be – and increasingly is – employed to monitor behaviour of workers at all levels.

However, Big Data can pose significant problems when data may/must be shared across organisational boundaries. As Allaire and Wolf (2004) describe with respect to pig breeding, software developed by the Pig Improvement Company (PIC) allows its owners to share data from growers to evaluate their proprietary genetic lines. However, some growers are uncomfortable with this approach and use other software so as to avoid providing those data to PIC. Similarly, many American cattle growers have resisted proposed nanotechnol-ogy-enabled ear tags for cattle that would collect real-time disease information. They are concerned that such data could be used by the meat packing industry to minimise cattle prices (Whyte et al., 2014). In both instances, it is clear that sources of resistance have been and likely will continue to be found which will put limits on the use of Big Data.

That said, at the limit, Big Data can be used to eliminate much of the manual *and* mental work involved in the entire agri-food supply system. For example, radio frequency identification devices (RFID) are already being used to monitor inventory and place reorders. If the large supermarket chains overcome resistance, RFID will also be used to tally customers' orders instantly and to eliminate most cashier and inventory control jobs. Similarly, the use of sensors to create driverless farm equipment as well as to apply seeds, fertilisers and pesticides based on sensor readings offers the possibility of an agriculture that no longer requires anyone in the field. And sensors attached to farm animals can be used to monitor for diseases as well as ensure that feed quality is linked to the nutritional needs of specific animals, in both instances dramatically reducing the need for most labour (e.g. Zhang and Pierce, 2013). Automated dairies and precision agriculture are already commonplace in much of the industrial world. Similarly, Tetra Pak® food packaging is nearly fully automated, requiring only a few mechanics who wait for the occasional alarm when a part of the automated process breaks down. How far this trend will continue remains to be seen, but the long-term consequences are worthy of considerable research.

Markets and bureaucracies

It is also important to emphasise that there is great irony in this. Neoclassical economists have insisted that markets have the great virtue of avoiding bureaucracy. Similarly, even neoliberal economists, who believe that the State must be mobilised to promote markets, insist that markets have this desirable character; it is what makes it possible for freedom to be equated with markets. Indeed, we have been told that free markets are virtually the antithesis of organisations with their complex bureaucracies.

Moreover, 40 years ago Oliver Williamson (1975) told us how organisations chose between markets and hierarchies based on minimising transaction costs. In so doing, he provided a means for understanding the decisions as to what to keep inside the firm and what to buy from others. While Williamson was rightly criticised for forgetting that production had to take place within a hierarchy (Dietrich, 1994), he eventually modified his views to permit that. But what Williamson and his followers appear to have been unaware of is that modern markets *require* bureaucracies.

There is little doubt that virtually no bureaucracy was required in (largely mythic) small village markets where, for example, surplus food might have been exchanged for pottery. There, goods were simple and easily examined for quality. In addition, anyone cheating would be aware that future sales were jeopardised. However, even by medieval times, markets required a considerable bureaucracy, often enforced by guilds (Steel, 2008): the standard size of a loaf of bread carved into the wall of the Strasbourg cathedral provided a measure, but inspectors were required to ensure that vendors conformed to that measure. In contrast, as noted above, modern markets, and especially the global markets constructed over the last several decades, demand the creation of multiple massive bureaucracies. These bureaucracies, as highlighted in Table 5.1, set the rules, design the measures, audit market participants, settle disputes, and otherwise format each market as well as create competitions and quasi-markets.

As such, a huge international bureaucracy consisting of standards organisations, certifying firms, accreditors, developers of measures, data collectors and analysers has been established in order to create global markets (cf., Garcia-Parpet, 2007). Table 5.1 provides a list of some of the bureaucracies necessary for global trade in standardised food and agricultural products. As the reader will note, some are specific to agri-food products, while others extend to virtually all aspects of society. Together, they form an invisible system of governance that is largely lacking in accountability to any particular government and is far removed from any connection to those not part of national and international elites. To emphasise the size and scope of this bureaucracy, consider that according to its website, one certifier alone, SGS, has 80,000 employees and 1,650 offices engaging in certification of a vast range of materials and processes (SGS 2015). One unintended consequence of this is the suppression of innovation. After all, as more and more people are subjected to the New Taylorism, to 'best practices,' to incessant audits, innovation becomes more and more likely to result in punishment. Indeed, even democratic governance is at risk.

Conclusions: can democracy survive?

In his classic work, *Exit, Voice and Loyalty*, Albert O. Hirschman (1970) noted that only two possible types of response are possible when we are faced with a situation

Table 5.1 Some of the bureaucracies needed to make global agri-food markets function

(Inter)governmental	World Trade Organization
	International Monetary Fund
	World Intellectual Property Organization
	Regional trade agreements, e.g., NAFTA, Mercosur
	Commission on Phytosanitary Measures (IPPC)
	Office International des Epizooties
	International and National Bureaus of Weights and Measures
	Codex Alimentarius
	Convention on Biological Diversity
Private	Consumer Goods Forum
	International Accreditation Forum
	International Laboratory Accreditation Cooperation
	International Organization for Standardization (ISO)
	Certifying agencies
Public/Private	National standards bodies (e.g., AFNOR)
	National accreditation bodies

not to our liking: leave the situation or object. Both, in turn, are linked to loyalty. We may object because we are loyal to the organisation, i.e., we believe that it will change as a result of our voice. Conversely, we may exit because we believe that changes resulting from voice are unlikely to be forthcoming. Clearly, without voice democracy is impossible. Yet, the four transformations described above are largely designed to transform society by restricting voice.

The neoliberal utopia designed by Friedman (2002 [1962]), Hayek (1973–1979, 2007 [1944]), Becker (1964), Buchanan (1968) and others is all about turning all institutions into markets and competitions where, we are told, freedom will be optimised. But that kind of society – in which everything is standardised, certified and accredited, in which assembly line production prevails, in which everyone is audited incessantly, all made possible by the advent of high speed information technologies – is one in which all choices are binary: we either buy or don't buy, we either stay or leave. There is little space for voice. Freedom requires more than merely binary decisions.

We might object to being constantly audited, but the market society we have built gives us few other options. Indeed, as Power (1997) suggests, it is nearly impossible to be against audits, whether it is that of a CEO audited based on quarterly profits or of a truck driver audited for timeliness, sticking to the prescribed route and even fuel usage. Yet, the audits themselves create the very people whose behaviour the proponents insist must be monitored. They encourage us to consider ourselves as isolated individuals out to maximise our human capital and to avoid any unnecessary work. Put differently, the approach of methodological individualism common to mainstream economics becomes reality. Moreover, this transformation is not limited in any way to the agri-food sector; it has become ubiquitous in much of the world. Yet, this new autocracy undermines democracy by minimising or eliminating all opportunities for voice.

Notes

1 As Cochoy (2002) has suggested, packaging of food products simultaneously makes possible such things as knowing the ingredients used and the net weight. It also prevents the consumer from seeing what is inside the package. Hence, certain products could be packaged successfully while others remained unpackaged.
2 Raymond (2013) notes that wholesale buyers of fruits and vegetables call such products 'beton', alluding to their lack of ripeness.
3 These dates should be taken as indicative, as in both cases considerable effort was necessary prior to international recognition by the key trading nations.

References

Allaire G., Wolf S., 2004. Cognitive Representations and Institutional Hybridity in Agro-food Innovation. *Science, Technology & Human Values*, 29(4), 431–458.
Bailes K.E., 1977. Alexei Gastev and the Soviet Controversy over Taylorism. *Soviet Studies*, 29(3), 373–394.

Becker G., 1964. *Human Capital: A Theoretical and Empirical Analysis*. Chicago, University of Chicago Press.

Bianco A., Zellner W., 2003. Is Wal-Mart Too Powerful? *Business Week*, 6 octobre, pp. 100–104, 106, 108, 110.

Brown S.A., 1997. *Revolution at the Checkout Counter*. Cambridge (MA), Harvard University Press.

Buchanan J.M., 1968. *The Demand and Supply of Public Goods*. Chicago, Rand McNally.

Busch L., 2007. Performing the Economy, Performing Science: From Neoclassical to Supply Chain Models in the Agrifood Sector. *Economy and Society*, 36(3), 439–468.

Busch L., 2011. *Standards: Recipes for Reality*. Cambridge (MA), MIT Press.

Callon M., (dir.), 1998. *The Laws of the Markets*. Oxford, Basil Blackwell.

Callon M., Millo Y., Muniesa F., (dir.), 2007. *Market Devices*. Oxford, Blackwell.

Cartwright M.S., Walker F.O., Blocker J.N., Schulz M.R., Arcury T.A., Grzywacz J.G., Mora D., Chen H., Marín A.J., Quandt S.A., 2012. The Prevalence of Carpal Tunnel Syndrome in Latino Poultry Processing Workers and Other Latino Manual Workers. *Journal of Occupational and Environmental Medicine*, 54(2), 198–201.

CashierLive, 2015. *Grocery Store Point-of-Sale*. Chicago, CashierLive, www.cashierlive. com/ (consulté le 9 mai 2015).

Cochoy F., 2002. *Une sociologie du packaging ou l'âne de Buridan face au marché*. Paris, Presses universitaires de France.

Cochoy F., 2011. "Market-things inside": Insights from Progressive Grocer (United States, 1929–1959). In Cayla J. and Zwick D. (dir.), *Inside Marketing*. Oxford, Oxford University Press, pp. 58–86.

Cotterill R.W., 1999. *Continuing Concentration in Food Industries Globally: Strategic Challenges to an Unstable Status Quo*. Storrs, Food Marketing Policy Center, Department of Agricultural and Resource Economics, Storrs, University of Connecticut.

Cotton, 1922. Sixty-Six to Eleven. *Cotton*, janvier, 144, 150.

Cox A., 1999. Power, Value and Supply Chain Management. *Supply Chain Management: An International Journal*, 4(4), 167–175.

De Raymond A.B., 2013. *En toute saison: le marché des fruits et légumes en France*. Rennes, Presses universitaires de Rennes.

Dietrich M., 1994. *Transaction Cost Economics and Beyond: Toward a New Economics of the Firm*. Londres, Routledge.

Donaldson J.L., 2005. *Directory of National Accreditation Bodies*. Gaithersburg (MD), National Institute of Standards and Technology.

Dries L., Reardon T., van Kerckhove E., 2007. The Impact of Retail Investments in the Czech Republic, Slovakia, Poland and the Russian Federation. In Swinnen J.F.M. (dir.), *Global Supply Chains, Standards and the Poor: How the Globalization of Food Systems and Standards Affects Rural Development and Poverty*. Cambridge (MA), Cabi, pp. 228–240.

European Commission, 2015. *Systèmes de certification de la qualité des denrées alimentaires*. Bruxelles (Brussel), Agriculture et développement rural, Bruxelles, European Commission. http://ec.europa.eu/agriculture/quality/certification/index_fr.htm (consulté le 13 juillet 2015).

Friedland W.H., Barton A., 1975. *Destalking the Wily Tomato*. Research Monograph No. 15, Department of Applied Behavioral Sciences, Davis, University of California.

Friedman M., 2002 [1962]. *Capitalism and Freedom*. Chicago, University of Chicago Press.

Frontard R., 1994. Histoire de la norme. *Culture Technique*, 29, 18–27.

Garcia-Parpet M.-F., 2007. The Social Construction of a Perfect Market: The Strawberry Auction at Fontaines-en-Sologne. In MacKenzie D., Muniesa F. and Siu L. (dir.), *Do Economists Make Markets? On the Performativity of Economics*. Princeton, Princeton University Press, pp. 20–53.

Gereffi G., Garcia-Johnson R., Sasser E., 2001. The NGO-Industrial Complex. *Foreign Relations*, 125, 56–65.

Giedion S., 1975 [1948]. *Mechanization Takes Command*. New York, W. W. Norton.

Hatanaka M., 2010. Certification, Partnership, and Morality in an Organic Shrimp Network: Rethinking Transnational Alternative Agrifood Networks. *World Development*, 38(5), 706–716.

Hayek F.A., 1973–1979. *Law, Legislation and Liberty*. Vol. 3, Chicago, University Of Chicago Press.

Hayek F.A., 2007 [1944]. *The Road to Serfdom*. Chicago, University Of Chicago Press.

Hirschman A.O., 1970. *Exit, Voice, and Loyalty: Responses to Decline in Firms, Organizations, and States*. Cambridge (MA), Harvard University Press.

Lloyd C., James S., 2008. Too Much Pressure? Retailer Power and Occupational Health and Safety in the Food Processing Industry. *Work, Employment and Society*, 22(4), 713–730.

Loconto A., Busch L., 2010. Standards, Techno-Economic Networks, and Playing Fields: Performing the Global Market Economy. *Review of International Political Economy*, 17, 507–536.

Loconto A., Stone J.V., Busch L., 2012. Tripartite Standards Regime. In Rtizer G. (dir.), *The Wiley-Blackwell Encyclopedia of Globalization*. Malden (MA), Blackwell Publishing Ltd., pp. 2044–2051.

Love R.R., Hoey J.M., 1990. Management Science Improves Fast-Food Operations. *Interfaces*, 20(2), 21–29.

National Industrial Conference Board, 1929. *Industrial Standardization*. New York, National Industrial Conference Board.

New S.J., 1997. The Scope of Supply Chain Management Research. *Supply Chain Management*, 2(1), 15–22.

Nicholson C., Young B., 2012. *The Relationship between Supermarkets and Suppliers: What are the Implications for Consumers?* Londres, Consumers International.

Power M., 1997. *The Audit Society: Rituals of Verification*. Oxford, Oxford University Press.

Reardon T., Berdegue J., Flores L., Balsevich F., Hernandez R., 2007. Supermarkets, Horticultural Supply Chains, and Small Farmers in Central America. *FAO Commodities and Trade Proceedings*, pp. 95–104.

Reardon T., Chen K.Z., Minten B., Adriano L., The Anh D., Wang J., Gupta S.D., 2014. The Quiet Revolution in Asia's Rice Value Chains. *Annals of the New York Academy of Sciences*, 1331, 106–118.

Reiter E., 1991. *Making Fast Food: From the Frying Pan into the Fryer*. Montréal, McGill-Queens University Press.

Steel C., 2008. *Hungry City: How Food Shapes Our Lives*. Londres, Vintage.

Swart W., Donno L., 1981. Simulation Modeling Improves Operations, Planning, and Productivity of Fast Food Restaurants. *Interfaces*, 11(6), 35–47.

Taylor F.W., 1911. *The Principles of Scientific Management*. New York, Harper.

Weatherspoon D.D., Reardon T., 2003. The Rise of Supermarkets in Africa: Implications for Agrifood Systems and the Rural Poor. *Development Policy Review*, 21(3), 333–356.

White C., 2004. *Strategic Management*. New York, Palgrave Macmillan.

Whyte K.P., List M., Stone J.V., Grooms D., Gasteyer S., Thompson P.B., Busch L., Buskirk D., Giorda E., Bouri H., 2014. Uberveillance, Standards, and Anticipation: A Case Study on Nanobiosensors in U.S. Cattle. In Michael M.G. and Michael K. (dir.), *Uberveillance and the Social Implications of Microchip Implants: Emerging Technologies*. Hershey (Pennsylvanie), IGI Global, pp. 251–269.

Williamson O.E., 1975. *Markets and Hierarchies*. New York, Free Press.

Zhang Q., Pierce F.J., 2013. *Agricultural Automation: Fundamentals and Practices*. Boca Raton (FL), CRC Press.

Part II

Ongoing transformations of the agri-economy

6 Energy, biomass and hegemony

A long history of transformations of agricultures

Benoit Daviron and Gilles Allaire

Introduction: three temporalities for analysing contemporary transformations in agriculture

To be able to see the future, one has to look backwards in history. To open a discussion on the future of agriculture and food, this chapter proposes re-examining the industrialisation of agriculture and food through a triple temporal perspective: that of the 'general economy' that places the history of societies within the history of nature and life as a holistic process; that of the sequence of socio-ecological (or metabolic) regimes in the history of humanity; and that of the sequence of hegemonic configurations of the world economy in the history of capitalism. We hope to show that this is not an overambitious endeavour.

Our first point of departure is Georges Bataille's concept of an economy of 'expenditure', which represents a total break from dominant utilitarian concepts. His analysis predated certain conclusions later established by thermodynamics – conclusions that permitted a more precise development of the 'general economy' perspective and its implications for human societies.

The analyses on the material base of human societies, the sources of materials and energy, constitute our second reference. These analyses are the basis for an entire current of ecological economics and have been illustrated by the works of the Vienna Institute of Social Ecology.

Lastly, in this chapter we subscribe to the abundant literature produced, following the foundational works of Fernand Braudel (1979), by Immanuel Wallerstein (2006) or Giovanni Arrighi (1994), which proposes an analysis of the world as a hierarchical collective with a centre and peripheries, and world history as a sequence of hegemonic configurations. We aim to demonstrate that specific conditions of production and consumption of biomass can be linked to the different hegemonic configurations which have appeared since the 16th century: the United Provinces, the United Kingdom, and lastly, the United States.

The first three sections of the chapter are organised around each one of the three perspectives studied. The fourth section looks at ongoing transformations in the world of agriculture and more widely in the sources and uses of biomass.

The general economy: the 'accursed share' and the dissipation of energy

Very few authors have gone beyond providing long historical perspectives and identifying historical metabolic regimes, to invite readers to consider a radically wider perspective by placing life and human societies within the dynamic of the cosmos. In 'The Accursed Share' (1967 [1949]), the writer and philosopher Georges Bataille takes up the idea, already developed in 1932 in 'The Notion of Expenditure', that living matter (over the course of ages) and human societies (over the course of centuries) appropriate an increasing flow of energy. The development of forms of life with the emergence of sexual reproduction, of plants, followed by herbivores, then carnivores, and later humanity, and the increasing complexity of the forms of life, allow living matter as a collective to mobilise and expend ever-increasing amounts of energy. It is not an analogy between the evolution of life on earth and that of humanity that leads Bataille to the notion of a 'general economy', but a continuity.

Human activity, regardless of the ends that man thinks are being pursued, can be viewed as an extension of the general activity of living matter, here defined as the appropriation of energy flows received by earth: '*Beyond our immediate ends, man's activity in fact pursues the useless and infinite fulfillment of the universe*' (Bataille, 1967 [1949]: 49). Apparently unwittingly, Bataille's thinking was in line with the characterisation of living matter that was progressively being established in the 1920s by the works of authors such as Lotka, who argued that natural selection favoured species that were more efficient in capturing the energy necessary for their preservation (Lotka, 1922). Bataille extended this vision of living matter to human matters.[1]

And thus the passage of societies from the hunter-gatherer stage, to the agrarian stage followed by the industrial stage can be interpreted along this vision. The mastery of fire was the first large innovation to substantially increase energy expenditure by human societies (Crosby, 2006). The advent of agriculture (in the Neolithic period) and techniques of food preservation later further increased energy expenditure by human societies, although the consequences of this transition on the social level were more ambivalent.[2]

With the revolutions in industrial techniques – the use of coal and later oil – the expenditure of ever-increasing amounts of energy took on the form of the accumulation of physical capital. But as Guillaume (1988: 1005) points out,

> this response can only be provisional: when 'equipped', human society becomes a collective capable of capturing ever larger quantities of energy, which it must then waste (in massive wars, for example) or reinvest. Such a system is destined to rapidly reach its limits, especially as its different components do not always develop in a coherent way.
>
> (p. 105)

For Guillaume, what is lacking in Bataille's analysis is the idea of entropy. Two elements from the now generally accepted theory of thermodynamics provide a better understanding and expansion of the work of Bataille. The first is the notion of 'energy dissipating structures', according to the expression used by Ilya Prigogine (1969), which accounts for structures whose existence is linked to a permanent exchange of energy and of matter, and sometimes information, with their environment. The second element is the third principle of thermodynamics, according to which the evolution of dissipative structures is directed towards maximising the flows of energies that pass through them: in other words, an ever greater production of entropy or ever greater dissipation of energy until disequilibrium followed by collapse occurs.

Thus for Guillaume (1988: 106):

> It is also because he did not consider entropy, that Bataille underestimated the capacity of the industrial world to push further the boundaries of its development. What do post-industrial societies do in effect? They organise practices (bureaucracy) and signs (information and communication). This organisation—and this is the second principle of thermodynamics—absorbs a lot of energy and contributes to the waste of excess energy. The notion of expenditure thus needs to be expanded to that of bureaucratic and communicational expenditure.
>
> (p. 106)

This form of expenditure did not emerge with 'post-industrial societies'. Earlier occurrences can be seen for example in the bureaucracy of the Middle Empire, which enabled centralised management of agriculture. Nonetheless, the capacity to dissipate energy under the form of normalisation and bureaucracy spiked significantly during the course of the last century (as described in Chapters 1 and 4).

Bataille goes further than biologists or physicists by affirming that living organisms widely dispose of an excess of energy and that this excess must be spent in one way or another:

> The living organism, in a situation determined by the play of energy on the surface of the globe, receives more energy than is necessary for maintaining life; the excess energy (wealth) can be used for the growth of the system (e.g. an organism); if the system can no longer grow, or if the excess cannot be completely absorbed in its growth, it must necessarily be lost without profit; it must be spent, willingly or not, gloriously or catastrophically.
>
> (op. cit. 49)

The issue for human societies is thus how to use this excess. This is what leads Bataille to label the excess energy as the 'accursed share' that cannot be exchanged quid pro quo according to the laws of conventional economics, and must be 'sacrificed' in order for the organism (the dissipative structure) to find a functional equilibrium, and this throughout the organism's life. If it is not stored, for a certain time (in particular in the form of productive capital), the excess energy must be spent, 'consumed in pure loss'. Games, feasts, sacrifices and eroticism which accord value to futilities are some means for this.

The theory of Bataille – that of a living world condemned to find ways of dissipating excess energy, and societies faced with the same fate – does not mean that men live in abundance. Abundance, moreover, as demonstrated by Sahlins (1976), is a cultural or civilisational concept. The 'dissipative structures' in question are not human beings as biological and spiritual organisms, but human societies viewed with their territories and their metabolisms,[3] or in other words, in terms of ecosystems. A society disposes of resources that are limited by the controlled territory (biomass, human or animal labour and technology) and at the same time produces new resources and a new population through its dynamism, its metabolism. Thus, human societies are constantly confronted with the question of balance between population and resources. As anthropologists teach us, the control of resources occurs through the control of demography through, among other elements, very frequent use of infanticide (see Wilkinson, 1973; Sahlins, 1976; Clark, 2007) as a form of sacrifice of the 'accursed share', just like wars and migrations.

In line with the third principle of thermodynamics, the size and the complexity of the energy dissipating structures that are societies, has grown continuously and today covers a globalised economy and society. It is precisely this point that impassions Bataille when he wrote 'The Accursed Share' in 1949, following the outbreak of the Second World War. He was pleading for a 'peaceful coexistence' in order to work for the end of poverty through development, in opposition to the prospect of the Cold War (the accursed share going into the arms race) which would emerge, and which by maintaining under-development would not avoid new explosions.

For our discussion, it is striking to note that Bataille talks in his writings of the 'play of energy on the surface of the globe' at a moment when society around him and his own material life increasingly depended on energy extracted from the depths of the globe! This exploitation of fossil resources, by offering humans – or more precisely, some humans – abundant and cheap energy, was to create a turning point in human history. This turning point is at the heart of the ecological economics work on the socio-ecological metabolism of industrial nations that we will now address.

Ecological economics and socio-ecological metabolism

Ecological economics has its origins in 19th century Russian thinkers (Martínez-Alier, 1997) but its development is linked to concerns about the limits to growth at the end of the 1960s. Economic ecology, at least in its most original form, is differentiated from mainstream environmental economics through its concern to link natural sciences with social sciences (Spash, 1999). It represents another perspective of general economics, albeit less general than that propounded by Bataille. The approach focuses on:

- The impossibility of infinite growth due to the physical limits of the planet, which is in line with movements in favour of de-growth.
- The incommensurability of values, that is, the impossibility of converting everything into monetary value and defining in a purely objective manner an 'ecologically correct' price or financial compensation (Martínez-Alier and Muradian, 2015).
- The link between issues of distribution and ecology created by the ability that dominant actors have to push resource extraction and its 'negative externalities' on to the weakest social groups (Martínez-Alier, 2014).
- The distinction between *oikonomia* and *chremastistika* as those terms were defined by Aristotle: *oikonomia* means supply of material resources to the *oikos*, the wider family, while the latter term refers to the art of studying markets to make money (Daly et al., 1994).

Biomass in past and present metabolic regimes

Among the sources of matter and energy that mankind uses to feed and clothe itself, to move around or for shelter (and also to build palaces or wage wars), biomass which includes all types of organic matter, holds an essential and very specific place.

As explained by Wrigley (2010), for most of human history, humans have depended on biomass not only for providing food but also as the almost unique source of raw materials and energy. Biomass provides combustibles, fibres and skins for clothing, a good part of the materials necessary for shelter, and even, through animals and humans themselves, the essential part of mechanical energy. It also plays an essential role in maintaining soil fertility. Lastly, biomass provides much of the raw materials and thermal energy (charcoal) needed for most trades: carpentry, glass-making, iron making, shoemaking, brewing, hat-making, etc.

The role of biomass in human endeavours was radically transformed by the growth in the use of fossil energies from the 18th century on. European societies, and later the rest of the world, moved from an economy (or metabolism) that could be characterised as 'solar' to a 'mineral' economy.[4] The particularity of such a mineral economy, characteristic of the Industrial Revolution and subsequent periods, is that it essentially obtained its resources by exploiting what lay underground. The transformation was most evident in the field of energy. Within a few decades, coal and then oil and natural gas (and less significantly uranium) were established as almost the sole sources of energy (Kander et al., 2014). The supply of materials was also transformed, as products made from biomass were substituted with products synthesised or derived from minerals whose extraction and processing was made possible by the abundance of energy. Note that fossil resources – coal, oil, or gas – are considered biomass. Sieferle uses the highly colourful term of 'subterranean forest' to refer to these resources (Sieferle, 2001). However, fossil fuels are a type of biomass that does not renew itself (at least not at a speed relevant for human history). For humans, therefore, the stock of mineral resources is finite and not linked to the flow of solar energy.

Fisher-Kowalski, Krausmann and the other researchers of the Institute of Social Ecology in Vienna have developed a an impressive conceptual framework and quantitative analyses covering a long period (see for example Fischer-Kowalski and Haberl, 2007b; Krausmann et al., 2008a; Krausmann and Fischer-Kowalski, 2017). The notion of social metabolism plays a central role in their work; originally formulated by Karl Marx, it denotes 'the need that humans have to obtain their means of subsistence through an exchange with nature, in a process that is socially organised and connected with labour' (Krausmann and Fischer-Kowalski, 2013:340). Social metabolism has been further elaborated to characterise specific forms of societal production and consumption with different uses of energy and matter, different productions of waste and then different relations with ecosystems that are called metabolic regimes: 'a specific funda-mental pattern of interaction between (human) society and natural systems' (Fischer-Kowalski and Haberl, 2007a: 8). Hunter-gatherer societies, agrarian societies and industrial societies are thus supported by different metabolic regimes with very different ways to get and use biomass.

The agrarian metabolism is based upon the controlled transformation of ecosystems with the aim of increasing the utilisable yield of biomass. Labour is invested in redesigning ecosystems and increasing the yield of utilisable biomass that can be harvested per unit area. The basic precondition for this form of subsistence is that significantly more energy in the form of biomass must be produced than is expended in the form of human (and animal) labour. The higher the surplus, the more complex the societal structures become. However, this surplus is never particularly high, as a system must be very well organised for the work of ten farm families to be able to sustain more than one or two other households. Although agrarian societies have the potential to be ecologically sustainable in energetic terms, because they make use of renewable flows and do not consume exhaustible resources, the reliance of the agrarian regime upon a massive transformation of nature is associated with risks and leads to a range of specific environmental problems. In most regions, deforestation was a precondi-tion for the spread of agriculture. The transformation of the ecosystem leads to changes in fauna and flora and the human-induced transfer of plants, livestock and parasites has many unwanted side-effects (for example, see Crosby, 1986). The close contact with livestock encourages the spread of parasites and infec-tious diseases and in cities, water and air become polluted.

The massive use of fossil fuels, coal and oil, relying on the large-scale exploitation of non-renewable stocks, is at the very core of industrial society. Combined with a series of technological and social innovations it gradually extends the inherent growth limits by abrogating negative feedbacks operating in the agrarian regime:

> The availability of an area-independent source of energy and the fossil fuel–powered transformation of agriculture from an energy-providing activity to a

sink of useful energy are the two main factors that made it possible to almost completely decouple energy provision from land use and the control of territory. All of the sociometabolic constraints stemming from the controlled solar energy system are abolished: Energy turns from a scarce to an abundant resource, labour productivity in agriculture and industry can be increased by orders of magnitude, the energy cost of long-distance trans- port declines, and the number of people who can be nourished from one unit of land multiplies, allowing for an unprecedented growth of urban agglomerations.

(Krausmann et al., 2008a: 643)

Table 6.1 shows a comparison of energy used in agrarian versus industrial metabolic regimes.

These analyses provide a fascinating new vision of the material basis of what is commonly call 'progress', 'development' or 'modernisation', but remain quite limited with regard to the process provoking or enabling the succession and the global diffusion of metabolic regime. The notion of 'socio-ecological transition' is mostly descriptive when it is used for the past and becomes normative, a motto, when dealing with the present situation.

Accumulation and unequal exchange of matter and energy

Hornborg reminds us that all accumulation implies unequal exchange and uses this idea to designate the asymmetrical transfer of matter and/or energy that enables the increase of the productive capacity of a group at the expense of another (Hornborg, 2003: 8). Defending the need to maintain a distinction between the material/biophysical dimension and cultural/semiotic dimension of exchange, Hornborg notes:

Any history of consumption will make it abundantly clear that the first condition for accumulation is that there is a cultural demand for the commodity

Table 6.1 Metabolic profile of the agrarian and industrial metabolic regimes

	Unit	Agrarian	Industrial
Energy use per capita	GJ/cap	40–70	150–400
Material use per capita	t/cap	3–6	15–25
Population density	cap/km^2	<40	<400
Agricultural population share of total population	%	>80	<10
Energy use per area	GJ/ha	<30	<600
Material use per area	t/ha.	<2	<50
Biomass share of energy use	%	>95	10–30

Source: Krausmann et al. (2008a: 643)

in question ... But contrary to mainstream economists, we must recognize that a second condition for accumulation is the material organization of production. It is this biophysical dimension of economic processes that the mainstream economists' preoccupation with utility neglects and that has been the common denominator of the many materialist challenges of this preoccupation from Karl Marx to ecological economics [not to forget Georges Bataille]. A crucial task is to offer such a challenge, which acknowledges the biophysical dimension but without equating it with value.

(Hornborg, 2012: 13)

Hornborg is drawing on the works of Bunker who proposes analysing unequal exchange in terms of extraction (Bunker, 1985), and to add to the concept of 'mode of production' that of 'mode of extraction' to cover the situation of so-called 'developing' countries. For Bunker, all production takes place through the transformation of matter and energy that has been extracted from a specific place:

This matter and energy which circulate through the productive system are partially and temporarily stored in a useful form which encourages increasingly complex social organization and a stronger productive capacity of the physical environment. Conversely, the loss of energy and matter in the territory from which they are extracted and the disturbance of the social system and of natural living systems this involves, increasingly simplifies social organization and its natural environment by reducing both the flow of energy and its use.

(Bunker, 1985: 13)

We are confronted again with the idea of dissipative structures whose survival implies a permanent supply of energy from their environment but also continuous exportation of entropy towards this same 'environment', as the supply occurs through the transfer of what Bunker calls 'extractive commodities'. The unequal exchange enables both the centre and the periphery to exist as dissipative structures: the transfer of wealth towards the centre allows the former to accumulate capital and a level of consumption that dissipates entropy, while the dissipative peripheral structure is able to survive but only through the impoverishment of its territory (and in the end, the migration of its populace).

Frontiers

The idea that accumulation is impossible without extraction (transfer of matter and energy from one region to another) underscores the importance of frontiers in the long history of capitalism. 'Pioneer fronts' generally refers to spaces characterised either by the cultivation of land previously uncultivated or by the exploitation of 'natural' (non-anthropogenic) biomass.

In the first case, frontiers do not just represent a process of spatial expansion, but also the intensification of the exploitation (colonisation) of nature. They involve the process of colonising nature as defined by Fischer-Kowalski and Haberl:

> To feed their metabolisms, societies transform natural systems in such a way as to maximise their social utility. Natural ecosystems are replaced by agricultural ecosystems (meadows, fields) designed to produce as much useful biomass as possible or they are converted into built spaces. Animals are domesticated; genetic codes of species are modified so as to increase their resistance to disease, to pesticides, or to pharmaceutical products. These interactions between the social system and the natural system cannot be interpreted in terms of metabolic exchange of matter and energy. They are of a different nature. With reference to the word '*colonus*', which in Latin designates the farmer, we will call this mode of intervention on natural systems 'colonisation' and designate through this term all human activities which deliberately change key parameters of natural systems and actively maintain them in a state different from that which would prevail in absence of such interventions.
>
> (Fischer-Kowalski and Haberl, 1998: 575)

The second frontier situation involves above all extractive resources as studied by Bunker ('natural' forests, 'wild' animals, sea products, etc.). To the present biomass, we can add past biomass that has been fossilised, i.e. coal, oil and natural gas.

The history of capitalism from the 16th century[5] onwards may thus be presented as the history of a frontier. Its point of departure is Western Europe from where it spread initially both westwards and eastwards (Poland, Ukraine, Romania, Russia, etc.) The advancement of this great frontier accelerated further from 1750 onwards, pushed by population growth and the development of transport and communications networks (McNeill, 1992). And this time the whole world was involved. During the 18th century, with the development of coal usage, England opened a new frontier for Europe – that of fossil energy use. This was no longer a horizontal frontier but a vertical one using a new type of distant biomass, one that came from *elsewhen*, rather than *elsewhere* to use Catton's terms (Catton, 1982: 41).

The notion of pioneer front has been frequently used and much discussed by historians, in particular in the works on the United States using the term 'frontier'. Following the foundational work by Turner (1893), most discussion has concerned the role of the frontier in the formation of political institutions or the economic trajectory of the countries involved (see Barbier, 2011, for a recent summary). But the existence of a global pioneer front, starting from around the 16th century and its role in the prosperity of Europe, is at the heart of Webb's book entitled The Great Frontier (Webb, 1964):

> What was the essential character of the frontier? It was inherently a vast body of wealth without proprietors … This sudden, continuing and ever-increasing flood of wealth precipitated on the Metropolis [the term Webb

uses to designate Europe] a business boom such as the world has never before and probably never can know again ... This boom began when Columbus returned from his first voyage, rose slowly and continued at an ever-accelerating pace until the frontier which fed it was no more. Assuming that the frontier closed in 1890 or 1900, it may be said that the boom lasted about four hundred years.

(Webb, 1964: 13)

Our analysis is an extension of Webb's analysis with two new developments however. First, we consider that the 'great frontier' did not just involve neo-Europe.[6] Our analysis covers a much larger space and included various territories from tropical regions but also Eurasia (Siberia or the Central Asian steppes, for example). Second, by including fossil biomass, the analysis accounts for the existence of an internal frontier in Europe (and later in several other parts of the world). This double spatial expansion also leads us to consider that the frontier logic did not halt in 1900 and that it is actually still very much alive.

Hegemonies

Like Bonneuil and Fressoz, we want to articulate the history of the world systems with the history of the earth system (Bonneuil and Fressoz, 2016, 2017). These authors have realised a very relevant and useful criticism of the notion of Anthropocene as it has been emerging and been employed in the social and scientific debate since the beginning of the 2000s (see in particular Steffen et al., 2007). Against a representation of the current environmental crisis explained in relation to the actuation of an undifferentiated and abstract homo sapiens (or human being), they underline the role and responsibility of specific social groups and processes. One of these is capitalism and for this reason they speak of Capitalocene[7] and they consider, as an historical actor of the Anthropocene, the successive metabolic regimes generated by the world systems over the past 250 years. With this perspective we can now analyse the transformations of the earth system, not in terms of a vague Anthropos but as a historical system of domination that organises in distinct manners the global flows of matter, energy, commodity and capital.

This chapter is also in line with the analyses that have been undertaken by Anglophone scholars on food regimes. These analyses, which aim to 'link international relations of food production and consumption to the forms of accumulation that can be distinguished according to various periods of capitalist transformations since 1870' (Friedmann and McMichael, 1989: 95), use a periodisation of history that is based on the sequence of hegemonic configurations.

The following regimes have thus been described:

- The first regime (1870 to 1930) centred on the United Kingdom and characterised by exports of tropical products from the colonies and staple grains and animal products from British dominions.

- The second regime (1950 to 1970) in which the United States plays a central role thanks to their market share, the massive use of food aid and the exporting of an intensive agro-industrial model (factors that are relativised in Chapter 7).

Authors in this current of thought also talk of the existence in the present-day situation of a third new regime. In his recent work, McMichael propounds the idea of a 'corporate regime' characterised by the power of firms (McMichael, 2009, 2013). Friedmann (2005) prefers to speak of a 'corporate-environmental food regime' to underscore the importance of environmental issues and sustainability standards.

Our analysis however differs from the work under the food regime approach in three ways. (i) It focuses on biomass, and its sources and uses. The specialisation of agriculture, and of rural spaces, solely on food production is a unique feature of the 20th century that is closely tied to the extensive use of fossil energy. It is highly likely that the future will call into question this specialisation of agriculture on food (see Chapter 8). (ii) It is interested in the role of biomass in the socio-ecological metabolism of hegemonies. We are not only interested in the contribution of agriculture and biomass to the accumulation of capital, but also in the exchanges of matter and energy on which all of human society depends. This resonates with the works of Moore who proposes substituting the concept of world economy with that of world ecology (Moore, 2003, 2010). (iii) It is based on a longer historical perspective. Going back to the hegemony of the United Provinces, prior to the 'Industrial Revolution', makes it possible to understand the conditions for hegemony in a world without fossil energy and also to better understand the particularity and precariousness of the situation we have been in for the past two centuries and the current challenges for the agricultural sector.

United Provinces

At the end of the 16th and beginning of the 17th centuries, the United Provinces, who had freed themselves from Habsburg domination, were an unrivalled economic and military power. One of their distinct features was an abundant urban population, a precocious occurrence in Europe with the exception of Flanders and the Mediterranean coast. In 1675, 42% of the population of the United Provinces lived in towns, and this rose to 60% in the province of Holland, with 200,000 inhabitants in the town of Amsterdam alone. At the same period, the urbanisation rate of England stood at around 15% and that of France at 12%.

On the local level, the United Provinces had two original resources at their disposal on which their socio-ecological metabolism was partially based:

- An agriculture specialised around animal production (dairy products especially), textile fibres (flax) and tobacco. A good part of the techniques that were used at the time, the use of legumes for example, came from nearby

Flanders where they had been developed from the Middle Ages, the golden
age for Ghent, Bruges and Antwerp (Slicher Van Bath, 1963: 71).

• The use of peat. In this forest-free territory, peat replaced wood for the
supply of thermal energy and enabled the development of several energy-
consuming activities: ironworks, breweries, brick-making, refineries, dyeing,
and so on (De Decker, 2015).

Above all, however, the United Provinces founded their metabolism and their
wealth on the mobilisation of distant biomass, through the creation of markets.
According to Moore, the Provinces represented a real breaking point in the long
history of Eurasian pioneer fronts:

> The pioneer fronts in search of merchandise replaced the pioneer fronts in
> search of resources. Global expansion substituted regional expansion as a
> response to socio-ecological pressures. And where, expansion had once
> reduced tensions created by demographic pressure, fast-growing capitalism
> reversed this logic.

(Moore, 2010: 35)

This capacity to mobilise distant resources was based on a remarkable mastery of
the combination of wood, water and wind, which characterise what Mumford
calls the 'neotechnic phase' ('the dawn of modern techniques') (Mumford, 1934:
110), and which was materialised by mills, canals and sailboats. It is estimated
that the fleet of the United Provinces in the middle of the 17th century was equal
to the entire fleet of the other European countries. The United Provinces not only
had an impressive transportation capacity but also very low costs and this
allowed them to establish quasi-monopolies for trade of heavy goods, that is,
bulk trade (wood, seeds, salt, fish, flax, hemp, tar, etc.).

Exports of grain from the Baltic reached a peak in the middle of the 17th
century and the United Provinces controlled 80–90% of these exports (van
Tielhof, 2002: 73). Amsterdam stood at the centre of this market, its stock
exchange allowing the centralisation and dissemination of information on quantities and prices. While half of the grain from the Baltic delivered to Amsterdam
was re-exported towards Southern Europe, the other half served to feed all the
towns and countryside of the United Provinces. In the middle of the 17th century,
grain imports fed more than half of the one million inhabitants living in the
provinces of Holland, Utrecht, Friesland and Groningen (de Vries, 1974: 172).[8]

But grains were not the only product the Dutch were trading. Caught in the
Baltic Sea, in the North Sea and later in the North Atlantic, fish (herring) was
abundantly re-exported (80% according to de Vries and Van Der Woude (1997:
251)) as far as Poland. Wood and forest products (pitch and tar, ashes and
potash), indispensable for construction and especially for the highly successful
naval industry, were also imported from the Baltic, as well as from Scandinavia.
There was also wool imported from England or Spain, and flax from Russia which
supplied an exporting textile industry. To this we must add furs – omnipresent in

the dress of the European elite of the era – from Finnish or Russian forests. And last but not least, from the point of view of capitalist accumulation, there were spices. The Dutch succeeded in the middle of the 17th century in displacing the Portuguese from their stronghold on the Indian Ocean.

The hegemonic position of the United Provinces was challenged by competition from France and England from the mid-17th century. The Dutch played a founding role in the development of sugar cane plantations in northern Brazil, which they controlled at the time, but they were later ousted from this activity. The full-scale development of slave plantations in the Americas became an affair for the French and the English, and was part of their mercantilist policies. This radically new mode of long-distance trade of biomass – new in the sense that it came with the direct control of the supplying territory and the *ex nihilo* creation of plantations based on imported labour of African slaves – cannot be linked to a hegemonic configuration. It is rather the product of one of the phases of the fragmentation of trade that are part of hegemonic transitions. Sugar, despite what Wallerstein suggests (1974: 44), was not, even at the end of the 18th century, an essential component of European diets.[9] It was still a spice; essential for capital accumulation certainly, but not particularly significant for the socio-ecological metabolism of European countries. Things would change in the course of the next century.

United Kingdom

As we already mentioned, the Industrial Revolution, which is the basis of the hegemonic status of the United Kingdom, was accompanied by a radical shift in the country's socio-ecologic metabolism (Sieferle, 2001; Krausmann et al., 2008b; Wrigley, 2010). Coal, a fossil biomass, progressively became the dominant energy source over the course of the 17th and 18th centuries. Its abundance enabled a quadrupling of per capita energy consumption between 1650 and 1850. By the latter date, coal accounted for 90% of energy consumption of the country (Warde, 2007).

But, during the English hegemonic phase, the use of fossil energy was accompanied by an increased use of distant biomass. Until 1850, non-food biomass was imported in very rapidly rising quantities. The main imports were textile fibres, first of which was cotton, and dyes. In 1845, raw materials made up, in terms of value, two-thirds of biomass imports, fibres alone accounting for 36% and dyes for 7% (Davis, 1979). It is this large flow of forest and agricultural products that makes possible the conquest of England by the machines and the steam engines (Bonneuil and Fressoz, 2016).

But the abolition of the Corn Laws in 1846 changed the situation. In the decades that followed, the United Kingdom increased its imports of food products considerably. On the eve of the First World War, the country was in a position of extreme 'external dependence' for its biomass, at levels that have not been seen since. In addition to raw materials intended for industry (cotton, wool, flax, oils and fats, rubber, etc.), wheat, meat, butter and fruits were also massively imported. In 1913, imports represented 58% of food calories consumed in the country (Board of Trade, 1917) and implied the use of land surfaces equal to all of the cultivated land in the United Kingdom (Krausmann et al., 2008b).

English agriculture at that moment then refocused on animal production, developing the genetic selection of animals (sheep, cows, and horses) following the works of Robert Blakewell (1725–95), considered the first animal farmer to have used rational selection methods for cattle (Vissac, 2002). At the same time, agriculture started to adopt practices of buying inputs for animal feed and fertilisation. The intensification of English agriculture was thus based as much on advances in the 'colonisation of nature' with the rationalisation of animal selection, as on the purchase of inputs (which were in part imported). English agriculture thus initiated the disintegration of agriculture and animal farming challenging the original, and historically deeply founded, characteristic of the West European rural world.

The United Kingdom was also able to import so much because it had an abundant supply to offer. The introduction of the steam engine as a source of mechanical energy for transport played a key role in this. For the first time in the history of humanity, a source of energy other than muscle (animal or human) was available for land transportation. It suddenly became possible to move great quantities of distant biomass, including within continents. The radical transformation of transport conditions did not only affect products. The cost of migration was also significantly reduced and this led to massive movements of European and Asian populations accompanied by profound changes in the methods of governing work. The European indentured workers of the 16th and 17th centuries, whose journey had been paid in exchange for several years of obligatory work, were replaced by African slaves in the 18th century and then by Asian, mainly Indian, indentured labour. The 19th century was exceptional because for the time in the history of capitalism, there was a mass migration of 'free' workers (McNeill, 1992: 55).

The development of the railroad opened up what can be considered the golden age of continental pioneer fronts. Colonisation by European migrants of territories previously occupied by hunter-gatherers or herders took place simultaneously in the great prairies of North America, the steppes east of Russia, the veldt of South Africa, the pampas of Argentina, the Australian bush, the Manchurian plain, the Atlantic Forest of Brazil and many other places.

Distant countries were not alone in supplying the United Kingdom. Several European countries were also suppliers: the Mediterranean countries for wine and oranges, and of greater consequence for the future of European agriculture, Denmark and the Netherlands specialised in butter and pork meat production by diverting (so to speak) part of the flow of cereals and oilseeds originating from distant countries. In these latter two countries, a radicalised version developed of the strategy of specialisation in animal production and disintegration of agriculture and animal farming as had already been conceived by English farmers during the 19th century.

United States

The United States was a product of English expansion and became the biggest neo-European country, to use Crosby's expression, during the course of the 19th century. Thanks to the railroad and immigration, it acquired a continental dimension and thus a population, a stock of resources and a market with no equivalent among European countries (with the exception of Russia). These characteristics laid the base for its hegemony in the 20th century.

But during the first decades of the 20th century, American agriculture was faced with a double crisis: a crisis of production and a crisis of markets. According to Turner, with the end of the frontier during the 1890s, the logic of depletive use of soil fertility, which became dominant when the reserves of 'virgin' land seemed inexhaustible, reached its limits. Agricultural per capita production dropped significantly from 1900. And from the same date, agricultural exports started to stagnate (and yet the economy of the country had been built on these) and imports to rise steeply. During the 1920s, the agricultural trade balance became negative (see Figure 6.1). Finally at the end of that decade, yields from the whole country started to fall as concurrently new competitors, from countries where pioneer fronts had opened later on, such as Argentina or Australia, were appearing.

At the same time, from the onset of the First World War, protective measures closed agricultural markets. War taught all European belligerents the dangers of

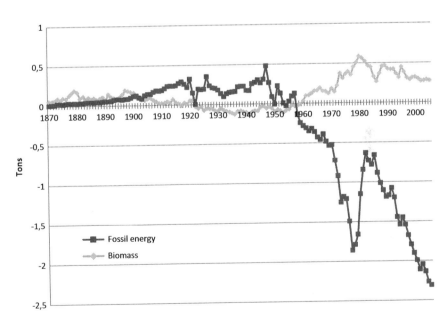

Figure 6.1 United States: net balance (export-import) of trade per capita in biomass and fossil energy, 1870–2005.

too great a reliance on distant supply markets. A second objective was added to that of national security following the crisis of the 1930s and the Second World War: that of fighting unemployment and overproduction with regulatory policies. This was to become the primary preoccupation in economic policy in Europe and North America. Participation in external exchange was clearly subordinated to the goals of full employment of factors and the stability of internal markets (Block, 1977; Ruggie, 1982). In the field of agriculture, self-sufficiency thus progressively became the norm, even if this goal was sometimes impossible to attain (as in Japan, Chapter 10) or was expressed in an imperial framework (such as in France between 1930 and 1957) or a regional framework (as in Europe after 1957).

Yet, very early on, in a context of increasing rivalry between hegemonic contenders (in particular Germany), coal mining was no longer considered a means to seek ever more distant biomass, but a means of substituting it with synthetic products. German chemistry, which developed from the production of dyes (Hohenberg, 1967), progressively branched out to different sectors of materials (plastics, fibres, rubber) and, of key significance, the production of nitrate fertilisers (Smil, 2001). The development of chemistry enabled simultaneously a reduction in the variety of uses of biomass and an increase in the production capacity of a given territory. The general move towards a self-centred national economy found its material base in this, at least from the point of view of biomass. If a country was able to control coal reserves, and later oil reserves (and a few other mineral reserves for obtaining phosphorus and potassium), it appeared possible to attain limitless growth in agricultural production. Biomass self-sufficiency thus became possible, all the more so as agriculture was now almost exclusively focused on food production for humans thanks to the development of synthetic products and the possibility of transforming thermal energy into mechanical energy.

In this context, the end of the crisis for American agriculture implied a radical change in modes of production and consumption. On the production side, the solution lay in the full integration of agriculture in the mineral economy. A soaring rise in work productivity and yields was enabled from 1940 onwards by the adoption of the tractor, which was particularly well-adapted to the high level of mechanisation that occurred in US agriculture from the mid-19th century in response to labour shortages, and the use of synthetic fertilisers, combined with the dissemination of improved and hybrid varieties, as well as the use of pesticides (see Figure 6.2). Contrary to animal selection in Europe, which had started under the control of herders, plant selection in the United States saw the establishment of seed businesses that were separate from farmers (Lewontin and Berlan, 1990). On the supply side, one also had the development of animal production based on the consumption of grains and soybeans. The agricultural model that established itself in the United States in this period should thus be viewed as a hybrid of the innovations that brought responses to the challenges facing American agriculture at the end of the 19th century and the at beginning of the 20th century (mechanisation and then motorisation in response to labour

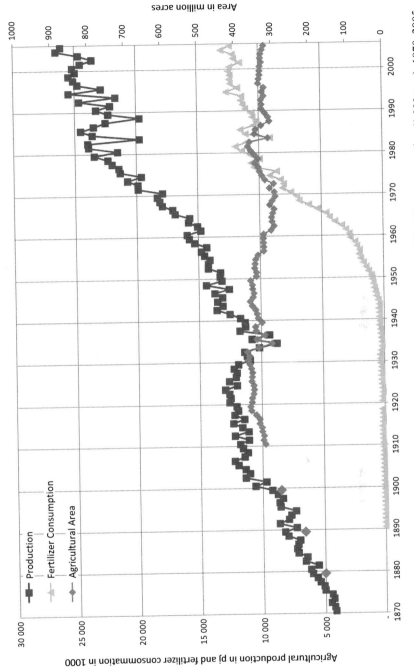

Figure 6.2 United States agricultural production (PJoules), agricultural area (million acres) and fertiliser consumption (1,000 tons), 1870–2005.

shortages), and those which had brought responses to challenges for West European agriculture (specialisation in animal production using, in the European case, purchased and imported grain). On the consumption side, the growth in the consumption of animal products and internal food aid initially resolved the risk of overproduction at a time when widespread adoption of national self-sufficiency strategies was limiting international markets (Daviron, 2008).

The end of the crisis also involved the adoption of a highly interventionist policy (border protection, stocks, direct aid, etc.) aimed at stabilising farm incomes and securing increasing markets.

This set of both technical and institutional 'solutions' was later adopted by European countries after the Second World War. The technical component and the self-sufficiency objective were also adopted by several developing countries under the so-called 'Green Revolution', even if some countries under their industrialisation strategies taxed agriculture more than they supported it.

The agriculture chapter of the GATT, the only mechanism for regulating trade policies put in place after the war, endorsed and authorised interventionist policies (Daviron and Voituriez, 2006). It gave very wide latitude for the use of non-tariff barriers and export subsidies. Waivers that were granted to the United States in 1955 confirmed this exceptional status of agriculture. In effect, until the conclusion of the Uruguay Round, virtually all instruments to guarantee the protection of national agriculture were 'legally' authorised (see Hopkins and Puchala, 1980; Cohn, 1993).

The international division of labour between biomass exporting countries and industrial goods exporting countries that prevailed during the English hegemony was well and truly over. Trade in agriculture grew very weakly – a lot less than trade in other products. Markets for agricultural products became residual markets where only surpluses and deficits were traded.

From the 1960s, the Unites States however, attained once again a surplus balance in its biomass trade. From this point of view, the US clearly distinguished itself from previous hegemonies (the Netherlands and the United Kingdom). Until the 1970, the newfound surplus was based on two main export trends: food aid (which represented an essential part of wheat and oil exports) and animal feed (corn and soybeans) mainly intended for Europe and Japan. But with the surplus in biomass trade came, almost year for year, a deficit in the trade of fossil energy (see Figure 6.1). A vertical frontier had clearly taken the place of the horizontal frontier. This is still the case today.

Post-hegemony ... and post-oil?

The American decline

Has the United States already lost its hegemonic position? This question raised a lot of debate in the seventies and beginning of the eighties in the context of the first oil shock, the (renewed) growth experienced in Europe and Japan, and American defeat in Vietnam. *After Hegemony*, published by Keohane in 1984, considered it a done deal. It was now a question of understanding the conditions for stability of international relations in a post-hegemonic world. However, the fall of the Berlin Wall and the collapse of the Soviet Block suggested that the diagnosis of decline had been premature. These events were interpreted as a victory for the United States heralding the advent of new unipolar world. The fact that developing countries had 'been brought into line' during the 1990s through the response to the debt crisis, that America reigned supreme in the field of information technology and communication, and that trade liberalisation as promoted by the World Trade Organization (WTO) was occurring, all seemed to confirm this victory.

But over the past few years, the tone of the discourse has changed again. There has been an accumulation of military setbacks by the United States, while China's share of global manufacturing has overtaken that of the United States. The decline of the American hegemony is once again forecast. For Arrighi, it is in this context that the current financialisation of the American economy should be interpreted, and that it is simply a repeat of what the United Provinces and the United Kingdom went through during their own phases of decline (Arrighi, 2005).

What developments can we read from the relationship to biomass? Can we already note the emergence of a new model of production and consumption of biomass on which a new hegemonic configuration may be founded?

It is certainly too early to consider that China could become the next hegemonic power and to look for a new relationship to biomass in that direction. For the time being, no biomass production or consumption model different to what has accompanied the US hegemony is discernible in China.

Conversely, it also makes little sense, particularly since India and China represent 36% of the world's population, to eliminate nation-states from the analysis and consider, following McMichael and several other authors, (Burch and Lawrence, 2009; Holt Giménez and Shattuck, 2011; Sage, 2013), that a corporate food regime – an agro-food regime of firms – has been established based solely on the intervention of multinationals.

For the moment, we can only make note of a number of transformations without claiming to describe the existence of new stabilised model.

Convergence of support to agriculture but non-reunification of the global market

The first observation that can be made involves agricultural policy. These policies were profoundly challenged both by WTO negotiations, and by the structural adjustment programmes that so-called 'developing' countries experienced. The last two decades have thus encouraged a certain convergence, towards the low end, in the levels of support accorded to agriculture by 'developed' and 'developing' countries (Figure 6.3). In Brazil, India and China, where at the end of the 1980s agriculture was still being taxed, the level of support has reached, or even surpassed, levels of support in Europe and the United States, where there has been a significant drop.

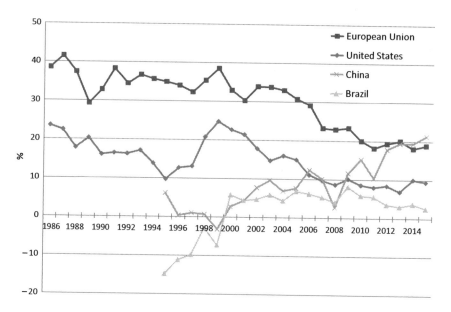

Figure 6.3 Producer support equivalent (PSE) as a share of gross farm receipt (GFR), 1986–2015.

However, the reunification of the global market is far from being achieved, at least as far as price formation is concerned. While in a country like France, as in the rest of the European Union, for the first time in generations prices of agricultural products now fluctuate with international prices, this is not the case in China and India. The sharp rise of international grain prices in 2007/8 was thus transmitted to European markets, but had little effect on domestic prices in China and India who maintained their goal of self-sufficiency in grains and continued to use international markets as a way to stabilise domestic prices by exporting surpluses or importing their deficit depending on their harvests.

This relative continuity in self-sufficiency strategies of India and China did not prevent the continued shift, or even the acceleration since the 2008 financial crisis, of global demand in imports of agricultural products towards Asia. In the past three decades, Europe has lost the central place it occupied for several centuries in the demand for imported biomass. Europe today accounts for less than 20% of global imports, while Asia accounts for more than 40%. In effect, China and India have maintained their goal of self-sufficiency for grain, but they have renounced such an objective for oilseeds and protein crops, and in the case of China for almost all non-food biomass (cotton, rubber, wood, etc.). In a little over a decade, China has become the biggest importer of soybean (64% of global imports in 2016), cotton (42%), wool (45%) and rubber (28%). The growth of Chinese raw materials imports has been so strong that it has generated 're-primarisation' of the economies in countries like Brazil or Argentina. Furthermore, sustainability issues concerning domestic agro-food chains, such as pressure on water resources or soil contamination, are becoming urgent and are imposing themselves on agricultural policy makers' agenda. These concerns could push China to rely even more on global value chains of biomass, externalising environmental and social costs of production (Cheng and Zhang, 2014; Zhang and Cheng, 2017). But as underscored earlier, the increasing support accorded to agriculture and the persistent self-sufficiency strategy for grains does not permit one to consider that China, regardless of its current status as the workshop of the world, is engaged on a trajectory similar to that of the United Kingdom in the 19th century with regards to its supply in biomass.

Radicalisation and spread of chemical agriculture

The conditions of biomass production and consumption remain relatively unchanged. From the point of view of uses of biomass, significant transformations are not yet a reality despite the promises of our governments to renounce use of fossil energies. In the field of agriculture, agro-food and agro-chemistry, these promises today take the shape of numerous research and investment projects carried out under the label of 'bioeconomy' (Chapter 9). For the moment, however, the sector of fuel for automobiles is the only sector where these promises have been translated into action, in Brazil, the United States and the EU. In the last two regions, this has involved policies guaranteeing markets, which have partially replaced supports that were dismantled following the WTO agreements.

Although a marginal activity, the substitution of petroleum-based fuels with ethanol or diester managed to generate very strong demand for agricultural raw materials (corn and oilseeds in particular), which contributed to a spike in international prices of food products in 2007/2008 (HLPE, 2011). This gives an idea of the consequences for global food security that vigorous policies to replace fossil resources with biomass may have. The chemical industry, which played such a crucial role in the emergence of the 20th century agricultural model, views biomass as a new reservoir of raw materials similar to what coal and oil were in the past, with the risk of transposing the same depletive logic.

On the side of biomass supply, we must first note that in the United States (see Figure 6.2) and in Europe (see Figure 6.4) growth in agricultural production was interrupted in the 1980s following reductions in support to the sector. But intensification continued, as is evidenced, for example, by the trend in corn yields in the United States, which have continuously grown at the same rate since 1970 (over 1.5% per year). On a global level, there is not, strictly speaking, a crisis of chemical agriculture. While there are certainly crises in specific sectors linked to price instability, as in Europe in the dairy sector since the end of dairy quotas, these are competitiveness crises resulting from the radicalisation of the 'American model' (ever bigger farms, ranches and machines) or from its spread to new areas. The issue is more fierce competition between champions of the model rather than a real calling into question of the model.

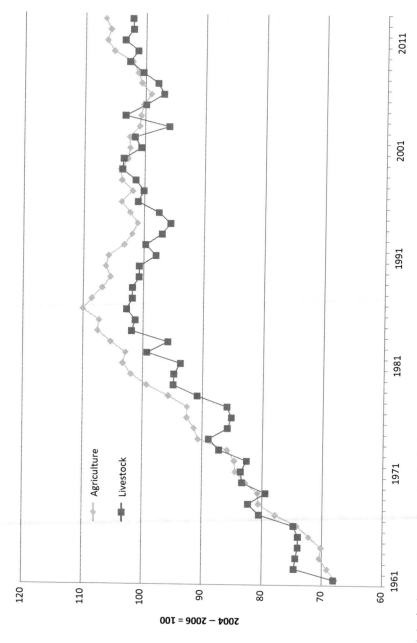

Figure 6.4 Agriculture and livestock production in Western Europe, 1960–2013.

This situation sometimes takes the shape of pioneer fronts, in most cases not involving the cultivation of territory previously occupied by hunter-gatherer or herder populations (though this still occurs in Indonesia), but rather the reallocation of already colonised lands to higher value uses. We are thus faced with the second round of a colonisation movement that started one or two centuries ago, even if the stock of 'virgin lands' has been drastically reduced. In fact, as Turner already described more than a century ago, pioneer fronts are more often than not characterised by a sequence of colonisation in which the trapper is replaced by the herder who is in turn replaced by the farmer (Turner, 1893). This is the kind of sequence that we are facing today in country like Argentina, where pastures are displaced by soybean fields, and on a global scale.

In parallel, the vertical exploitation or colonisation of living matter is continuing with GMOs, not those that enable the use of Monsanto pesticides (which appear at the end of their line) but those that produce molecules used as medicines, and with genomics, which in just a few years has radically transformed animal selection (Chapter 7) and medical biology.

This radicalisation and spread of chemical agriculture is of course increasingly contested for its consequences for the environment. This contestation has given rise to a multitude of initiatives and projects seeking to 'green' the model ('responsible agriculture') or to propose more or less radical alternatives (agroecology, organic agriculture, permaculture, etc.). This has generated, among other things, a flurry of new standards and labels aiming to promote these alternatives to consumers (organic farming) or simply to neutralise criticism of the model (sustainability standards such as RSPO). These initiatives, however, fail to really change the logic of biomass production that has been in play since the mid-20th century, either because they remain marginal on a global scale, or because they are actually accommodating the past logic.

Conclusion

The history of humanity is part of the continuum of the history of the cosmos and the earth, and as such, cannot escape the laws of thermodynamics. Dissipative structures, human societies have succeeded one another by mobilising ever-increasing quantities of energy and matter.

For the great part of human history, energy and matter was essentially provided by direct solar radiation through biomass, and supplemented by wind and water bodies. As shown by the hegemony of the United Provinces in the 17th century, wealth and power were based on the 'the happy combination' of these resources – in other words the capacity to use wind and water bodies to mobilise to one's advantage the maximum amount of nearby and distant biomass. But, with the United Provinces, this mobilisation was also based on the creation of numerous markets. Such are the factors on which their hegemony was founded.

The Industrial Revolution, by exploiting gigantic reservoirs of coal, oil and natural gas accumulated underground, prodigiously increased the amount of energy mobilised by societies and with this came the complexity of the latter. It gave birth to two hegemonic configurations – the United Kingdom and the United States – which had very different relationships with biomass. The hegemony of the United Kingdom saw an international division of labour in which the entire globe was mobilised to supply global biomass markets centred on Europe. Growth in biomass production was horizontal, based on multiple pioneer fronts, carried forward by railroad development and mass migration.

The hegemony of the United States involved a radical transformation of the place of agriculture in the socio-ecological metabolism. Specialised in food supply, agriculture became, after the Green Revolutions, a consumer of energy and no longer a supplier as it had always been throughout human history. At the same time, the international division of labour from the 19th century was redefined by the spread of self-sufficient strategies enabled by the use of fossil energy in agricultural production.

In the last two decades, the conditions of biomass production and consumption have been doubly contested. First, by the redefinition, under the WTO, of policies protecting agricultural markets. More recently and with increasing force, these conditions are contested for their environmental consequences, without this contestation however, 'revolutionising' the metabolism of industrial societies.

Bataille analyses this process and accuses excess energy, the accursed share, of being at the origin of human drama: war, industrial accumulation, and bureaucracy. But as we have pointed out, Bataille neglected the concept of entropy and the fact that any maximisation of energy dissipation must inevitably involve a maximisation of the production of entropy. In our world, waste, in all its material forms (nuclear, CO_2, active nitrates, pesticide residues, etc.) represents, along with heat, the very concrete form of this entropy. The production of waste, in excess of the absorption capacity of various sinks (oceans, atmosphere, water tables, etc.) and living cycles, is today the accursed share that our societies must deal with.

Notes

1 This idea was taken up by Ilya Prigogine in a different way with the concept of dissipative structure. For a recent discussion on the transposition of the perspective of Lotka and Prigogine on human affairs, see Adams (1975) and for a relatively accessible French version, Roddier (2012).

2 As shown by the archaeological works based on skeletal and dental analysis, this transition often involved deterioration in the health status of communities. See for example Larsen (2006), Mummert et al. (2011). This deterioration was linked to a significant increase in pressure from parasites induced by sedentarisation and the increased density of populations, as well as the reduction in dietary diversity which caused various nutrient deficiencies. Inequalities thus developed on the basis of access to food resources.

3 The concept of metabolism is widened from the biological organism to society. We do not find the term directly in Bataille's work, but we do find it in Marx and in the theoreticians of ecological economics.

4 By using the terms 'solar' and 'mineral' we are slightly betraying the works of Wrigley who talks of 'organic' and 'mineral' societies. Yet, coal and oil can be considered to be part of the organic world as they originate from the transformation of biomass. The use of the term 'mineral' to characterize economies using coal and oil on a massive scale, as well as numerous other minerals resolves the issue. Moreover, the use of the term 'solar' instead of 'organic' allows one to capture the importance of biomass, produced directly or indirectly by solar radiation, as a source of matter and of energy, and of the use of wind and water bodies, indirect products of solar radiation, as a source of energy.

5 The European 'colonization' drive can be considered even earlier if the Spanish Reconquista and the crusades led by Teutonic knights or even the large-scale clearing of lands undertaken between the eleventh and thirteenth centuries are included.

6 The very practical expression of 'Neo-Europe' was coined by Alfred Crosby to designate countries that benefited from European migration in the nineteenth century (mainly temperate America, Australia, New Zealand and South Africa) (Crosby, 1973).

7 The word Capitalocene has been also proposed by several authors on the other side of the Atlantic. See Moore (2016).

8 De Vries assumed a per capita consumption of 200 kg per year, which gives an estimated 600,000 people fed by imports.

9 In England imports per capita and per year grew from 1 kg to 10 kg between 1700 and 1800; that is for the latter date, a 4% contribution to nutritional needs. In France just 1 kg per capita per year was imported at the end of the eighteenth century (Sieferle, 2001: 98).

References

Adams R.N. 1975. *Energy and Structure: A Theory of Social Power.* Austin, University of Texas Press.

Arrighi G. 1994. *The Long Twentieth Century: Money, Power and the Origins of Our Times.* Londres, Verso.

Arrighi G. 2005. Global Governance and Hegemony in the Modern World System. In Ba A. D. and Hoffmann M.J. (dir.), *Contending Perspective on Global Governance: Coherence, Contestation and World Order*, Londres, Routledge, pp. 57–71.

Barbier E. 2011. *Scarcity and Frontiers: How Economies Have Developed through Natural Resource Exploitation.* Cambridge/New York, Cambridge University Press.

Bataille G. 1967 [1949]. *La part maudite, précédé de la notion de dépense.* Paris, Éditions de Minuit.

Block F.L. 1977. *The Origins of International Economic Disorder: A Study of United States International Monetary Policy from World War II to the Present.* Berkeley, University of California Press.

Board of Trade 1917. *The Food Supply of the United Kingdom.* Londres, His Majesty's Stationery Office, p. 35.

Bonneuil C., Fressoz J.B. 2016. *The Shock of the Anthropocene: The Earth, History and Us,* New York, Verso Books.

Bonneuil C., Fressoz J.B. 2017. Capitalocène: une histoire conjointe du système Terre et des systèmes Monde. In Allaire D. (Eds.), *Transformations agricole et agroalimentaires, Entre écologie et capitalisme*, Versailles QUAE, pp. 41–58.

Braudel F. 1979. *Civilisation matérielle, économie et capitalisme, XVe–XVIIIe siècle. Tome 3: Le temps du monde.* Paris, Armand Colin.

Bunker S.G. 1985. *Underdeveloping the Amazon: Extraction, Unequal Exchange, and the Failure of the Modern State.* Chicago, University of Chicago Press.

Burch D., Lawrence G. 2009. Towards a Third Food Regime: Behind the Transformation. *Agriculture and Human Values*, 26(4), 267–279.

Catton W.R. 1982. *Overshoot: The Ecological Basis of Revolutionary Change*. Champaign, University of Illinois Press.

Cheng G., Zhang H., 2014. *Chinal's Global Agricultural Strategy: An Open System to Safeguard the Country's Food Security*. Singapore, RSIS RSIS Working Paper Singapore, p. 19.

Clark G. 2007. *A Farewell to Alms: A Brief Economic History of the World*. Princeton/ Oxford, Princeton University Press.

Cohn T.H. 1993. The Changing Role of the United States in the Global Agricultural Trade Regime. In Avery W.P. (dir.), *World Agriculture and the GATT*, Londres, Lynne Rienner Publishers, pp. 17–39.

Crosby A.W. 1973. *The Columbian Exchange: Biological and Cultural Consequences of 1492*. Westport, Greenwood Press.

Crosby A.W. 1986. *Ecological Imprialism: The Biological Expansion of Europe*, Cambridge, Cambridge University Press.

Crosby A.W. 2006. *Children of the Sun: A History of Humanity's Unappeasable Appetite for Energys*. New York, W. W. Norton & Company.

Daly H.E., Cobb J.B., Cobb C.W. 1994. *For the Common Good: Redirecting the Economy toward Community, the Environment, and a Sustainable Future*. Boston, Beacon Press.

Daviron B. 2008. The Historical Integration of Africa in International Food Trade: A Food Regime Perspective. In Fold N. and Larson M. (dir.), *Globalization and Restructuring of African Commodity Flows*, Uppsala, Nordika Afrikain Institutet, pp. 44–79.

Daviron B., Voituriez T. 2006. Régimes internationaux et commerce agricole. In Berthaud P. and Kebadjian G. (dir.), *La Question Politique En Économie Internationale*, Paris, La Découverte.

Davis R. 1979. *The Industrial Revolution and British Overseas Trade*. Leicester, Leicester University Press.

De Decker K., 2015. Medieval Smokestacks: Fossil Fuels in Pre-industrial Times. *Low-Tech Magazine*, consultable en ligne: www.lowtechmagazine.com/2011/09/peat-and-coal-fossil-fuels-in-pre-industrial-times.html (consulté le 16 novembre 2016).

de Vries J. 1974. *Dutch Rural Economy in the Golden Age, 1500–1700.*, New Haven, Yale University Press.

de Vries J., Van Der Woude A. 1997. *The First Modern Economy: Success, Failure, and Perseverance of the Dutch Economy, 1500–1815*. Cambridge, Cambridge University Press.

Fischer-Kowalski M., Haberl H. 1998. Sustainable Development: Socio-economic Metabolism and Colonization of Nature. *International Social Science Journal*, 50(158), 573–587.

Fischer-Kowalski M., Haberl H. 2007a. Conceptualizing, Observing and Comparing Socio-ecological Transitions. In *Socioecological Transitions and Global Change: Trajectories of Social Metabolism and Land Use*, Cheltenham, Edward Elgar Publishing pp. 1–30.

Fischer-Kowalski M., Haberl H. 2007b. *Socioecological Transitions and Global Change: Trajectories of Social Metabolism and Land Use*, Cheltenham, Edward Elgar Publishing.

Friedmann H. 2005. From Colonialism to Green Capitalism: Social Movements and the Emergence of Food Regimes. In Buttel F.H. and McMichael P.D. (dir.), *New Directions in the Sociology of International Development. Research in Rural Sociology and Development*, Vol. 11, Amsterdam, Elsevier, pp. 227–264.

Friedmann H., McMichael P. 1989. Agriculture and the State System: The Rise and Decline of National Agricultures, 1870 to the Present. *Sociologia Ruralis*, 39(2), 93–117.

Guillaume M. 1988. Les limites de l'utilitarisme. *Revue européenne des sciences sociales*, 26(82), 99–108.

HLPE. 2011. *Price Volatility and Food Security: A Report by the High Level Panel of Experts on Food Security and Nutrition of the Commitee on World Food Security.* Rome, FAO.

Hohenberg P.M. 1967. *Chemicals in Western Europe, 1850–1914.* Chicago, Rand McNally & Company.

Holt Giménez E., Shattuck A. 2011. Food Crises, Food Regimes and Food Movements: Rumbling of Reform or Tides of Transformation. *The Journal of Peasant Studies*, 38(1), 109–144.

Hopkins R.F., Puchala D.J. 1980. *Global Food Interdependance: Challenge to American Foreign Policy.* New York, Columbia University Press.

Hornborg A. 2003. The Unequal Exchange of Time and Space: Toward a Non-normative Ecological Theory of Exploitation. *Journal of Ecological Anthropology*, 7(1), 4–10.

Hornborg A. 2012. Accumulation: Land as a Medium of Domination. In Hornborg A., Clark B. and Hermele K. (dir.), *Ecology and Power: Struggles over Land and Material Resources in the Past, Present, and Future*, Londres, Routledge, pp. 13–22.

Kander A., Malanima P., Warde P. 2014. *Power to the People: Energy in Europe over the Last Five Centuries: Energy in Europe over the Last Five Centuries.* Princeton, Princeton University Press.

Krausmann F., Fischer-Kowalski M. 2013. Global Socio-Metabolic Transitions. In Singh S., Haberl H., Chertow M., Mirtl M., Schmid M. eds. *Long Term Socio-Ecological Research*, New York, Springer, pp. 339–365.

Krausmann F., Fischer-Kowalski M. 2017. Transitions socio-métaboliques globales. In Allaire G. and Daviron B. (Eds.), *Transformations agricoles et agroalimentaires: Entre écologie et capitalisme*, Versailles, Quae, pp. 23–40.

Krausmann F., Fischer-Kowalski M., Schandl H., Eisenmenger N. 2008a. The Global Sociometabolic Transition: Past and Present Metabolic Profiles and their Future Trajectories. *Journal of Industrial Ecology*, 12, 637–656.

Krausmann F., Schandl H., Sieferle R.P. 2008b. Socio-Ecological Regime Transitions in Austria and the United Kingdom. *Ecological Economics*, 65(1), 187–201.

Larsen C.S. 2006. The Agricultural Revolution as Environmental Catastrophe: Implications for Health and Lifestyle in the Holocene. *Quaternary International*, 150(1), 12–20.

Lewontin R.C., Berlan J.-P. 1990. The Political Economy of Agricultural Research: The Case of Hybrid Corn. In Caroll C.R., Vandermeyer J.H. and Rosset P.M.P. (dir.), *Agroecology*, Nova York, McGraw-Hill, pp. 613–626.

Lotka A.J. 1922. Contribution to the Energetics of Evolution. Natural Selection as a Physical Principle. *Proceedings of the National Academy of Sciences of the United States of America*, 8(6), 147–151.

Martínez-Alier J. 1997. *Ecological Economics: Energy, Environment, and Society.* Oxford, Basil Blackwell.

Martínez-Alier J. 2014. *L'écologisme des pauvres: une étude des conflits environnementaux dans le monde.* Paris, Les petits matins.

Martínez-Alier J., Muradian R. 2015. Taking Stocks: The Keystones of Ecological Economics. In Martínez-Alier J. and Muradian R. (dir.), *Handbook of Ecological Economics*, Londres, Edward Elgar Publishing, pp. 1–25.

McMichael P. 2009. A Food Regime Genealogy. *Journal of Peasant Studies*, 36(1), 139–169.

McMichael P. 2013. *Food Regimes and Agrarian Questions*. Halifax, Fernwood Publishing.

McNeill W.H. 1992. *The Global Condition: Conquerors, Catastrophes and Community*. Princeton, Princeton University Press.

Moore J.W. 2003. The Modern World-System as Environmental History? Ecology and the Rise of Capitalism. *Theory and Society*, 32(3), 307–377.

Moore J.W. 2010. 'Amsterdam is Standing on Norway' Part I: The Alchemy of Capital, Empire and Nature in the Diaspora of Silver, 1545–1648. *Journal of Agrarian Change*, 10 (1), 33–68.

Moore J.W. 2016. *Anthropocene or Capitalocene?: Nature, History, and the Crisis of Capitalism*. Oakland, CA, Pm Press.

Mumford L. 1934. *Technics and Civilization*. New York, Harcourt.

Mummert A., Esche E., Robinson J., Armelagos G.J. 2011. Stature and Robusticity during the Agricultural Transition: Evidence from the Bioarchaeological Record. *Economics & Human Biology*, 9(3), 284–301.

Prigogine I. 1969. Structure, Dissipation and Life. *Theoretical Physics and Biology*, Vol. 23, p. 52.

Roddier F. 2012. *Thermodynamique de l'évolution: un essai de thermo-bio-sociologie*. Artignosc-sur-Verdon, Éditions Parole.

Ruggie J.G. 1982. International Trade, Transactions, and Change: Embedded Liberalism in the Postwar Economic Order. *International Organization*, 36(2), 379–415.

Sage C. 2013. The Interconnected Challenges for Food Security from a Food Regimes Perspective: Energy, Climate and Malconsumption. *Journal of Rural Studies*, 29, 71–80.

Sahlins M.D. 1976. *Âge de pierre, âge d'abondance: l'économie des sociétés primitives*. Paris, Gallimard.

Sieferle R.P. 2001. *Subterranean Forest: Energy System and the Industrial Revolution*. Cambridge, White Horse Press.

Slicher Van Bath N.H. 1963. *The Agrarian History of Western Europe 500–1850*. Londres, Edward Arnold Ltd.

Smil V. 2001. *Enriching the Earth: Fritz Haber, Carl Bosch, and the Transformation of World Food Production*. Cambridge (MA)/Londres, MIT Press.

Spash C.L. 1999. The Development of Environmental Thinking in Economics. *Environmental Values*, 8(4), 413–435.

Steffen W., Crutzen P.J., McNeill J.R. 2007. The Anthropocene: Are Humans Now Overwhelming the Great Forces of Nature, AMBIO. *A Journal of the Human Environment*, 36, 614–621.

Turner F.J. 1893. The Significance of the Frontier in American History. In Ridge M. (dir.), *Frederick Jackson Turner: Wisconsin's Historian of the Frontier*, Madison, State Historical Society of Wisconsin.

van Tielhof M. 2002. *The 'Mother of All Trades': The Baltic Grain Trade in Amsterdam from the Late 16th to the Early 19th Century*. Boston, Brill.

Vissac B. 2002. *Les vaches de la République: saisons et raisons d'un chercheur citoyen*. Versailles, Éditions Quæ.

Wallerstein I. 1974. *Capitalist Agriculture and the Origins of the European World-Economy in the Sixteenth Century*. New York, Academic Press.

Wallerstein I. 2006. *Comprendre Le Monde: Introduction À L'analyse Des Systèmes-Monde*. Paris, La Découverte.

Warde P. 2007. *Energy Consumption in England & Wales, 1560–2000.* Napoli, Consiglio nazionale delle ricerche.

Webb W.P. 1964. *The Great Frontier.* Austin, University of Texas Press.

Wilkinson R.G. 1973. *Poverty and Progress: An Ecological Model of Economic Development.* London, Methuen.

Wrigley E.A. 2010. *Energy and the English Industrial Revolution.* Cambridge, Cambridge University Press.

Zhang H., Cheng G. 2017. China's Food Security Strategy Reform: An Emerging Global Agricultural Policy. In Wu F. and Zhang H. (Eds.), *China's Global Quest for Resources: Energy, Food and Water*, London, Routledge, pp. 23–40.

7   Food Regime Analysis

A reassessment

John Wilkinson and David Goodman

Introduction

For some 30 years now, especially in Anglophone agri-food studies, Food Regime Analysis (FRA) has provided a macro-historical perspective to frame and consider the broad issues of the modern world agri-food system. Food Regime Analysis has not been without critics of its grand narratives and historiography (Goodman and Watts, 1994; Bernstein, 2015), and its influence has varied considerably in this period. Thus, at the turn of the millennium, it experienced a temporary eclipse, partly due to difficulties in convincingly mapping the dynamics of regime transition and identifying the institutional 'pivots' of a contemporary third food regime. This eclipse was accentuated by the shift in academic attention towards the micro-worlds of alternative food networks and new 'quality' markets (Whatmore and Thorne, 1997; Wilkinson, 1997; Murdoch, 1998; Goodman, 2003; Morgan et al., 2006)

However, Food Regime Analysis has enjoyed greater prominence since its 'resurrection moment' in 2005 (Le Heron and Lewis, 2009) and the resurgent interest in global issues – climate change, sustainability, food-fuel competition – and the severe and continuing food security crisis exposed by the inflationary upsurge and volatility of international food prices in 2006–08. Both macro interpretations of these changes and formulations of alternative agendas have claimed lineage to the historical account provided by the food regimes narrative. The narrative itself, from the outset a blend of different traditions of analysis, including world-systems theory, international regime analysis, and French Regulation Theory, has similarly evolved in the light of these contemporary discussions.[1] Its basic schema, however, continues to serve as a standard frame of reference for situating contemporary macro trends.

In this chapter, we argue that the food regimes account of historical developments in agriculture and food systems is flawed and discuss how its theoretical framework compromises its analysis of the changing international agrarian political economy in the early 21st century. Rather than undertake a critical textual exegesis,[2] we focus on the 'stylised' version of Food Regime Analysis that has become generalised in the literature of agri-food studies.[3] We suggest that the limitations of this version stem from its theoretical positioning, which views the world from the hegemonic centre and so places excessive weight on processes of hegemonic regime formation, crisis and succession. This has led to the corresponding neglect of multi-polarities in the evolving capitalist world system, as well as the historical continuities in the agri-food accumulation strategies pursued by other ascendant economies. For these reasons, we suggest that it is more accurate historically and more revealing analytically to discuss a plurality of regional food orders that increasingly share common scientific and technological platforms and institutional structures rather than retain the concepts of hegemony and international food regime. In the following discussion, we argue that a fundamental reassessment is needed of the central concepts, causal mechanisms, periodisations, and framing devices of food regimes theory.

Recent contributions to the burgeoning field of world history, as yet poorly acknowledged by Food Regime Analysis, have revealed serious shortcomings in its foundational historical account. The homogenising framework of 'regimes' and hegemonic strategy fails to recognise *alternative* agrarian developmental trajectories and highly *differentiated* institutional landscapes that were consolidated during the second half of the 19th century. Food Regime Analysis also struggles to explain the emergence of a territorial 'quality' food agenda in 'the North' since the 1980s, most prominently in the European Union. This neglect of regional and national difference and the multi-polarities it produces is also at the root of a questionable prognosis of political economic futures in the Global South. In short, subjecting the food regimes narrative to critical scrutiny has now become a precondition for an adequate characterisation of the global agri-food system and future trajectories.

Although our alternative account remains skeletal, we hope that it offers convincing evidence of the insights to be gained from a multi-polar, pluralistic approach to the emergence of global agri-food systems that draws out historico-spatial continuities and their analytical salience in understanding contemporary developments.

The following section critiques the historical account of the first food regime in terms of British hegemony. We argue that this interpretation obscures alternative policies articulated by the ascendant economies of continental Europe and how their agrarian trajectories structured European agriculture and food systems in the years of the second post-WWII food regime and into the current period. Food Regime Analysis does not capture these regional differences and accordingly fails to recognise their importance for understanding contemporary patterns of change. Analogously, with its over-emphasis on British hegemony and the role of the European settler economies, the food regimes narrative presents a truncated, partial treatment of colonialism, and under-estimates its devastating structural impacts on rural economies and societies and the corrosive effects on their capacity for autonomous peasant-based development. It similarly ignores the pre-emptive, resource-based colonialism undertaken by Japan, which so transformed the Asian landscape and underlies Japan's key role in shaping the contemporary global agri-food system.

The third section contrasts the different models of capital accumulation under-lying the first and second food regimes and interrogates the analysis of US food aid, the Green Revolution, and the internationalisation of US agribusiness. Our discussion criticises the food regimes perspective for its questionable general-isations and its corresponding neglect of historical continuities and geographical differences.

The fourth section examines recent debates on the driving forces in contemporary agriculture and food systems. We critique the premise of northern hegemony and argue that the rise of the BRICS (Brazil, Russia, India, China and South Africa) and other emerging economies is creating a polycentric world order, problematising the analytical and political economic salience of the North–South perspectives advanced by food regimes narratives. While recognising that agriculture and farmland have been increasingly incorporated into the processes of financialisation, notably since the 2008–09 global economic crisis, we argue that the new demands for food, feed and natural resources from the 'emerging' economies are the main drivers. The chapter concludes on a programmatic note by suggesting that agri-food systems be integrated analytically into a broad natural resources perspective with issues concerning resource access and security as the central dynamic.

The 'first food regime'

As mentioned earlier, the notion of a single world food regime structured around the interests of a hegemon obscures the divergent paths taken by competing ascendant economies as they responded to hegemonic ambition. This neglect of alternative trajectories leads to the problematic characterisation and periodisation of the first food regime, conventionally defined by the years 1870–1914.

In the Food Regime Analysis narrative, this period is represented as one of rapidly expanding global free trade under British hegemony based on the export from colonial and settler countries of wage-foods for the growing European industrial proletariat in exchange for manufactured goods. However, with the 'second' Industrial Revolution (Landes, 1969; Smil and Kobayashi, 2012) underway, Germany and the United States were mounting their ultimately successful challenge to Britain's industrial leadership and both resorted to protectionism to limit its hegemonic power. Japan, for its part, after expanding agricultural production into its northern island of Hokkaido, advanced into Manchuria and Formosa (Cwiertka, 2006). Moreover, food was subsumed under broader resource strategies, combining protectionist policies and colonialism, to ensure secure access to raw materials, especially cotton, timber and coal.

With its focus on hegemony, the food regime account fails to recognise the *unilateral* nature of Britain's adoption of free trade and its willingness, in Polanyi, 1957 [1944], 138) words, to 'sacrifice her agriculture'. The British model, consolidated by the abolition of the Corn Laws in 1846 (Schonhardt-Bailey, 2006) and the 1849 repeal of the Navigation Acts, was only briefly imitated by the leading European nations in the 1850s and following the Anglo-French Treaty of 1860. This interlude proved short-lived as the sustained, century-long decline in ocean freight rates after 1815 steadily yet inexorably enhanced the competitiveness of US wheat exports and threatened domestic agricultures throughout Europe (North, 1958).

Rather than replicating the British model, continental Europe embarked on a radically different course. Facing competition from the emerging international market for wheat,[4] aggravated by the onset of the agrarian crisis and Great Depression of 1873–96 (Hobsbawm, 1979), the major continental European powers re-instituted agricultural protection to nurture family-based mixed dairy farming (Tracy, 1982), while granting selective tariff exemptions for essential industrial raw materials as their colonial supply bases developed (Gerschenkron, 1989 [1943]). The relationship between industrialism, the 'scramble for raw material supplies' and the spread of imperialism play a central role in the analysis by Polanyi, 1957 [1944]) of the protectionist counter-movement against the 'self-regulating' world market in the industrialising nations of Western Europe (Silver and Arrighi, 2003: 332). As Polanyi, 1957 [1944]: 217) states, 'Imperialism and the half-conscious preparation for autarchy were the bent of Powers that found themselves more and more dependent upon an increasingly unreliable system of world economy'.

These protectionist agriculture and food policies were consolidated in the years before 1914, and reinforced in the inter-war period, and it is not too fanciful to see continuities in the post-WWII architecture of the Common Agriculture Policy (CAP). More recently, these continuities re-emerge in the territorial, 'quality' agri-food agenda and policymakers' appeals to the notion of Europe's 'rural exceptionalism' in international agricultural trade negotiations. In short, a world-historical narrative centred on a hegemonic interpretation of the British model is poorly equipped to trace the different historical paths of European agricultural and rural development, as well as their subsequent wider impacts on contemporary agriculture and food systems. It is similarly unable to capture the role of agriculture and food in the development of US capitalism in the 19th century (Aglietta, 1979; Brenner and Glick, 1991), or the very different strategies articulated by Japan.

Two additional limitations of the historiography of the first international food regime (IFR) arise from its treatment of colonialism, and the centrality it gives to Britain's trading relations with the white 'settler-colonial' societies of the post-bellum United States, Canada, Argentina and the Antipodes.

Colonialism enters the Food Regime Analysis narrative in the truncated form of the development of the European settler economies and their integration as world food 'baskets' and natural resource frontiers under British hegemony. This highly selective account ignores the impact of imperial expansion on India, the Far East, and the countries caught up in the 'scramble for Africa': that is, 'the "three continents" where the "peasant question" was manifested most sharply in colonial conditions and thereafter' (Bernstein, 2014: 1037). As Bernstein continues, 'peasantries are largely missing from the 1st century of … accounts of IFRs, other than as affected by the patterns of trade they established' since these settler states 'in effect … lacked peasantries' (ibid., 1037–1038). In turn, this partial focus underlies a key proposition of Food Regime Analysis, namely that the disarticulation of peasant economies in the Global South is regarded as a phenomenon *not* of the first but the second post-WWII food regime, with US food aid, the Green Revolution and the internationalisation of the US agri-food model as primary causes.

The account of the first food regime does recognise the significant contribution of slavery and New World plantation export economies to industrial development in Europe,[5] but the lack of a global analysis supports the tacit assumption that elsewhere domestic food systems and rural economies emerged basically intact from European imperialism. Recent contributions to world history emphasise that this assumption is simply untenable. These analyses expose the exploitative logics of colonialism and mercantile capitalism and their devastating impacts on the social and economic fabric of rural societies in the 'three continents', including the destruction of proto-industrial rural manufacture and measures to restrict the development of industrial factory production (Parthasarathi, 2011; see also Pomeranz, 2000; Beckert, 2014). In turn, rural deindustrialisation (Bagchi, 2010) or 'reprimarization' (Reinert, 2007) increased the pressure on land as alternative occupations and sources of household livelihood were foreclosed. Bagchi (op. cit.) argues the case in detail for India but identifies the same processes at work in China, Burma, the Philippines, Thailand, Egypt and the Ottoman Empire. The combination of colonialism and gunboat free trade capitalism destroyed the diversified rural economy, stripping it down to subsistence peasant production. Japan, for its part, pursued similar policies of rural disarticulation in this period in Manchuria, Formosa and later Korea.

A further limitation of the first Food Regime Analysis concerns the instrumental role attributed to the new settler countries, particularly the US, Canada, Argentina, Australia and New Zealand, in the formation of an international grains market and lowering world prices.[6] In fact, in the years 1870–1900, only the United States performed this role with the incorporation of its new western agricultural frontier. Russia was its only significant rival in this growing international trade, while India's contribution far exceeded that of Australia in this period (Tracy, 1982). Furthermore, although integrated into world markets, agricultural systems in the European settler economies were harnessed to national industrial development strategies. This was dramatically demonstrated by the 1828 United States Tariff Act, which protected its northern textile industry against British imports at the expense of Southern cotton exports, Britain's main source of raw material. Australia and New Zealand similarly promoted industrial development behind protective barriers to create alternative sources of employment, thereby preventing the continuous expansion of the agricultural frontier and the diminishing returns of what Reinert (2007) calls the 'primarisation' trap.

The periodisation of the first world food regime (1870–1914) is an unhappy choice since these years witnessed a generalised return to protectionism in continental Europe and the deepening of colonial domination. Indeed, the earlier period from the 1840s to the early 1870s marked the true heyday of the institutionalisation of the British food regime. The settler countries, for their part, subordinated agriculture and food to broader domestic industrial development goals. Rather than one hegemonic world food regime, we can identify the consolidation of a variety of different 'national-colonial' systems in negotiation with emerging world markets.

The 'second food regime'

In the framework of Food Regime Analysis, the movement from the first to the second world food regime is characterised simultaneously by rupture and hegemonic change – the shift from Britain to the United States – and by continuity in the shared notion of a world food regime. In practice, however, we are dealing with qualitatively very different processes. From 1850 until 1914, Britain promoted the international organisation of food production and trade in agricultural raw materials as an integral component of its accumulation model. In this specific case, unlike the rising economies of continental Europe, domestic agriculture and the vaunted system of 'high farming' were deliberately sacrificed to the needs of industry, and the British diet and food-wage relation therefore were constituted through international trade.[7]

In the second food regime under US hegemony, this relationship between international trade and wage goods is no longer central to the accumulation model. Unlike its hegemonic predecessor, this 'self-centred, largely self-sufficient, continent-sized economy' did not rely heavily on the world economy for supplies of foodstuffs and raw materials (Silver and Arrighi, 2003: 339). Thus, in his exposition of the regulation approach, Aglietta (1979) identifies a classic inter-sectoral relation between agriculture and industry within the domestic US economy as the characteristic feature of the 'Fordist' model of accumulation, whereby 'cheap' food and plentiful supplies of animal protein were combined with the mass consumption of manufactured goods. This interdependence is captured vividly in President Roosevelt's New Deal pledge to 'put a car in every garage and a chicken in every pot'. With Asian exports of vegetable oils and fats excluded by tariff protection to promote the American soybean complex (Berlan et al., 1977; Lockeretz, 1988), colonial or quasi-colonial supplies were basically limited to sugar and fruit.

In the US case, therefore, the international or global dimension of the food regimes narrative emerges from Cold War geo-politics. These were manifested initially in the distribution of US wheat surpluses to war-ravaged Western Europe under the 1948 Marshall Plan, and later in the strategy of food aid exports formalised under Public Law 480 of 1954, which, at least for a time, became a key component of the wage relation in some urbanising, industrial import-substituting developing countries. This global dimension is reinforced, it is argued, by the internationalisation of the US agro-industrial complex, popularised as the 'Green Revolution'. The dynamics of the second food regime are thus significantly different from those of the first regime. We look at each of these pivotal elements of Food Regime Analysis in the following paragraphs, beginning with the Marshall Plan in Europe and concessionary PL 480 US grain exports to Third World countries.

As agricultural production recovered rapidly in Western Europe, regaining pre-war levels in the late 1940s, the focus of the Marshall Plan shifted from food aid to a broader geo-political strategy to support re-industrialisation and economic growth in order to establish a buffer zone against the Soviet Union and the spread of communism.[8] However, rather than simply replicating the US agri-food model *tout court*, as the Food Regime Analysis account maintains, Western European countries continued on their historical agrarian trajectories, revitalising their specialised capabilities and developing domestic resources (Reinert, 2007). The related argument that the CAP launched in the early 1960s reproduces the institutional structures of the 1930s New Deal regulatory model is similarly flawed.

The emergence of the CAP needs to be seen against the background of the agricultural crises of the 1920s and 1930s and the protectionist responses adopted by national governments, including tariffs, import quotas and direct market intervention. Similar measures were applied in wartime and in the period of post-war reconstruction. It is these policy responses, overlaid with the harrowing wartime experiences of hunger, food rationing and import dependence, that indelibly marked the post-war settlement for agriculture in the major Western European countries (Tracy, 1982). These continuities and experiences are likewise imprinted on the CAP and its programmes, including the long-sustained commitment to small-scale farming,[9] which now finds contemporary expression in support for 'social agriculture' (Buller, 2004)

Only by giving full weight to the continuities and influences that have forged these alternative paths of agricultural development can we assess the significance of specific European policies and trajectories – territoriality, multifunctionality, geographical indications – and recognise their independent contribution to contemporary debates on the shape and dynamics of world food systems.

With food aid identified as the institutional 'pivot' of the second food regime (Friedmann, 2009), the key issue hinges on the long-term repercussions of concessionary wheat imports as the catalyst of the internationalisation of the US food system. As a geo-political strategy, food aid under Public Law 480 was a highly selective operation, and larger Third World countries – including those with a traditional wheat-based diet – were the main recipients of subsidised grain shipments. In the case of smaller countries, such as Colombia for example, these shipments undoubtedly disrupted the production of domestic staples and accelerated import dependence (Friedmann, 1982; Gaviria, 2011). However, many of the larger developing countries targeted by food aid have remained major grains producers and exporters and have successfully pursued their own strategic agri-food interests. It is difficult to reconcile this success with the food regime narrative in which food aid and Green Revolution technologies have combined to debilitate Third World agriculture.[10] This is equated with 'the peasantry', an undifferentiated and normative category, bereft of social class relations, a characterisation that has attracted strong criticism recently (Bernstein, 2014, 2015; Jansen, 2014). This account also begs the question of how a range of Asian countries were able to harness participation in food aid programmes to the implementation of effective agricultural development strategies.[11]

Similarly, national agricultural development programmes and policy institutions established when food aid was at its highest by industrialising Latin American countries later helped to position the Southern Cone as a global competitor of the United States following the world food crisis of the early 1970s. For example, Brazil was a major recipient of concessionary wheat aid in the 1950s, yet in the following decade it implemented an ambitious wheat import substitution policy of agricultural modernisation, including credit subsidies, protective tariffs, imports of tractors, and the promotion of co-operativism. It was this state-supported sector of highly capitalised, family-labour farms producing wheat-soy in rotation and organised in powerful cooperatives that allowed Brazil to respond so rapidly to the international market opening created by the US embargo on soy exports in 1973. Without these national policies, Brazil would not be challenging the US as the world's leading agricultural exporter in the 21st century.

The food regimes analysis of food aid is based on a highly generalised characterisation of the then Third World that is blind to the institutional capacity of the larger developing countries to pursue their own strategic agri-food priorities. Such an undifferentiated perspective makes it difficult to comprehend the rise of the BRICS economies and the challenge they now pose to the US and Europe.

The narrow focus on the internationalisation of the US agri-food model also ignores the case of Japan, which not only defended its traditional rice-based food system by controlling access to its domestic market but also developed powerful agricultural commodity traders and actively exploited 'new resource frontiers' abroad (Ciccantell, 2009) to meet domestic food supply needs. These new frontiers included the *cerrados* region in Brazil, which served to consolidate Brazil's position in the global grains and oilseed markets after 1970. Brautigam (2009) and Ciccantell (2009) both suggest that the resource access strategy pursued by Japan in its ascendancy may now be serving as a model for China's current overseas investments in oil, minerals and agriculture. According to Brautigam (2009: 253), in 1996, over a decade before current debates on 'land grabbing', 'the Japanese owned 12 million hectares of farmland overseas, triple the size of arable land at home' (Yamauchi, 2002; Reuters, 2007). Japan is again emerging as a key global player in agri-food development through the aggressive expansion of its traders and the promotion of new frontier regions, notably the Nacala 'corridor' in Mozambique.

In parallel with PL 480 food aid, Food Regime Analysis identifies the Green Revolution as a second dimension of the internationalisation of the US agribusiness model, with similarly disruptive effects on Third World peasantries. But these innovations should not be seen exclusively through the lens of a hegemonic US food regime. The main crops affected – wheat and rice – were developed and disseminated by the public network of international agricultural research centres or the Consultative Group for International Agricultural Research (CGIAR). Moreover, China, despite being cut off from world trade and investment in the 1960s, implemented its own Green Revolution on the same technology platform of irrigated, high-yielding, fertiliser-responsive plant varieties (Naughton, 2007).

It is more appropriate, therefore, to see these innovations as a generalised outcome of the existing globally available dominant technology paradigm.

This discussion has suggested that the characterisation of the first and second food regimes is based on questionable historiography and periodisation. These problems originate, we argue, in the theoretical positioning of Food Regime Analysis at the hegemonic centre, which produces broad generalisations lacking in nuance and complexity. No analytical space is found for the 'awkward' historical counter examples and plural paths of change that explicate the continuities and diversity so evident in the contours of the global agri-food system. These analytical difficulties are magnified in the food regime account of contemporary developments, as we see in the next section.

FRA and current global agri-food transformations

Resurrection' and a third food regime

In Food Regime Analysis, the multiple crises of the early 1970s – the collapse of the Bretton Woods system, the US-Soviet détente and massive grain deals, the US dollar devaluation, an extraordinary upsurge in international commodity prices, and the World Food Conference of 1974 – mark the demise of the second US-dominated food regime. However, subsequent efforts to identify an institutionally stable successor regime failed to gain wide acceptance. As Burch and Lawrence (2009: 267) observe, numerous attempts were made 'to explain the transition from a second to a third regime and to delineate the form and content of that putative regime' (see also Pritchard, 2009). This long *impasse* was addressed in the 'resurrection moment' of 2005, when Friedmann (2005) detected the outlines of an incipient 'corporate-environmental food regime', and McMichael (2005) argued that a third or 'corporate food regime' had been consolidated since the 1980s under the aegis of the global neoliberal political order.

Further attempts to establish the contours of a new world food regime followed in a 2009 symposium (Campbell and Dixon, 2009). The foundations of a third food regime are detected, for example, in the processes of financialisation and the subordination of food and agriculture to the dictates of finance capital (Burch and Lawrence, 2009), displacing the earlier emphasis by these authors on the globalisation of corporate food retailers and their supply chains (Burch and Lawrence, 2005). Campbell (2009) suggests that two contending but mutually constitutive regimes coexist within these transnationalised chains, one based on 'quality' foods for privileged consumers and the second on processed industrial food for low-income groups. Two similar competing trajectories in global food systems also frame Jane Dixon's (2009) reinterpretation of the nutrition transition using the tools of Food Regime Analysis.

However, these varying efforts to distinguish a new, structurally stable regime all share the same fundamental weakness: a virtually exclusive focus on northern interests, whether nation-states, US–EU rivalries, or transnational corporate actors, that is embedded in an outmoded global geo-political perspective. The central point of reference is an international order premised on North Atlantic hegemony, predominantly of the United States and transnational agribusiness, which turns on an asymmetric North–South axis and invites core-periphery framings of inter-state power. In short, the 'South' is generalised as the pawn of northern power brokers. Yet the case of Japan, the 'Asian geese', the nascent agribusiness complexes of the Southern Cone, and especially the rise of China, all belie the blanket characterisation of the South as the simple passive object of northern interests.

In the food regime narrative, the origins of the third or corporate food regime are found in the structural adjustment policies imposed by the IMF and the World Bank in response to the Third World debt crisis of the early 1980s. These policies of liberalised trade and foreign direct investment, privatisation and deregulation – the Washington Consensus – involved the 'rolling back' of rural development programmes, public subsidies and the regulatory powers of state agencies in the Global South. As this neo-liberalised world order became entrenched, the 'corporate food regime' emerged, 'organised around a politically constructed division of agricultural labour between northern staple grains traded for Southern high-value products (meats, fruit and vegetables)' under 'the free trade rhetoric associated with the global rule (through states) of the World Trade Organization (WTO)' (McMichael, 2009a: 148).[12]

A multi-polar global agrarian political economy

In our view, the 'core-centric' nature and focus on 'the Northern agenda' of the food regime framework robs it of the flexibility needed to accommodate the speed and new directions of change in the global political economy of agriculture and food. This rigidity arguably explains the failure to recognise the continuing prominence of the national developmental state, the diversity and interlocking nature of state–private-sector interests found in the Global South, and the 'unravelling' of the global neoliberal order (Arrighi, 2007; see also Cooper and Flemes, 2013; Storz et al., 2013). This diversity is forcefully demonstrated by recent analyses of the BRICS, which 'have reclaimed the role of the state in development far beyond the limits of the Washington Consensus framework' (Ban and Blyth, 2013: 242). Furthermore, 'common understandings about world trade as a largely North–North and North–South phenomenon have been definitively challenged by the BRICS boom of outward investment in both developed and developing countries' (ibid., 243).

These analyses emphasise the salience of diversity and difference in the global economy. Each of the BRICS economies adjusted in individual ways to the Washington Consensus while retaining a central role for the state, whether in the form of public investment and credit policies to support 'national champion industries', as in the case of China, or Brazil's hybrid response of 'liberal neo-developmentalism' (Ban, 2013), with state ownership in strategic sectors to achieve global competitiveness. As Wilkinson (2014: 3) observes, the BRICS 'all have strong developmentalist states, with public firms and national development banks as key strategic actors. They neither act nor look like prototypes of a neo-liberal regime'.

A vast related literature in international geo-politics and international political economy has marshalled abundant evidence of inter-state rivalry and the rise of a multi-polar world order, including analyses of the challenge to the WTO multilateral trade system posed by bilateralism and the 'new dynamic regionalism' (Schott, 2009),[13] the changing architecture of global governance (Golub, 2013; Mielniczuk, 2013), and the growing economic power of East Asia and China (Breslin, 2005, 2011; Arrighi, 2007; Strange, 2011). This scholarship buttresses the view that the emerging economies of the Global South are changing the hierarchy of the capitalist world system, in Golub's (2013) words, 'by restructuring it from within'. 'The vertical late-modern world system centred on the "West" is giving way to a polycentric international structure in which new regional and transnational South–South linkages are being formed' (ibid., 1,000).

These arguments are strongly complemented by a recent study of the world's leading grain traders, the so-called ABCD companies,[14] which suggests that 'The world of agriculture and food is in a new moment' (Murphy et al., 2012: 56; see also Baines, 2014). Noting the key role that Brazil, China and India now play in setting the WTO multilateral trade agenda, these authors indicate that

> A profound change is underway, changing the balance of power irrevocably: developing countries today account for 47% of all imports and more than half of all exports. The ABCDs have positioned themselves already for this change and are important players in Brazil, China, India and other emerging economies. But the change is bringing new competitors and new political tensions.
>
> (ibid., 56)

This competitive challenge is given concrete expression by recent IATP research on the changing role of China in the global animal protein complex – feed, pork, dairy and poultry – and how the US model of intensive livestock production has been adapted to local conditions by public and private domestic firms, although including joint ventures with global agribusiness corporations in some cases.[15] While there are commonalities in the diffusion of this model, the key finding of this research is to identify 'the rise of *agribusiness with Chinese characteristics*' (Schneider and Sharma, 2014: 10, original emphasis). This new reality was dramatically illustrated in 2013 by the takeover of the largest US pork producer, Smithfield Foods Inc., by the Chinese firm, Shuanghui, now the WH Group. As the authors observe, their research 'is an attempt to show how China's story ... is a global one, with global links and global impacts' (ibid., 8). These studies reinforce the point that the North–South axis at the heart of Food Regime Analysis misreads the transformations and rivalries generated by the nascent multi-polar global food economy.

The 'new moment' in the global political economy of agriculture and food and in the transnationalised corporate world of commodity trading is powerfully illustrated by the outward expansion of the Chinese state-owned agribusiness and food conglomerate, COFCO.[16] With the dietary transition raising increasingly acute food security concerns, state and private agribusiness firms have been encouraged since 2000 to 'go out' (*zou chuqu*) and invest overseas in order to insulate their supply chains from market uncertainty. In this respect, COFCO's expansion into global grain trading mirrors wider fears of resource insecurity more generally and utilises the 'more-than-market' instruments characteristic of China's resource policy – corporate acquisitions, joint ventures, long-term commodity supply contracts, and lease agreements – with a view to gaining autonomy from foreign transnationals[17] and challenging their control of global commodity markets.[18]

The 'global links and global impacts' of COFCO's growth as it responds to state food security imperatives are exemplified by the massive scale of its sudden entry into the strategic soybean sector in Brazil and the Southern Cone, the global centre of export production and the traditional sphere of influence of the ABCD corporations. After operating mainly in the Chinese domestic market for 50 years,[19] COFCO in 2009 diversified its activities overseas with two major acquisitions – majority holdings in Noble Group's agribusiness, with global assets in grain sourcing, processing and distribution, and in the Dutch firm, Nadira, a global diversified agribusiness with strong interests in Latin American grain and oilseeds. At a stroke, COFCO has gained 'direct entry into the origination and marketing of soy in the whole of the Southern Cone. With these acquisitions, COFCO's turnover reaches US\$ 63.3 billion, in the same league … as the Big Four' (Wilkinson et al., 2015: 18).

Although the ABCDs retain their dominance in soya marketing in Brazil and the Southern Cone, the meteoric rise of COFCO emphasises the importance of analytical frameworks that are sufficiently flexible to capture the fluidity of the forces shaping global food systems. A brief vignette of the spatial migration of soybean production from the Centre-South of Brazil provides further illustration of this point. Under a Brazil–Japan cooperation agreement, PRODECER, major Japanese traders played an instrumental role in opening up the Centre-West *Cerrado* region in the late 1970s and 1980s. However, the movement of the grains frontier to the North and North-East over the past decade has exposed the critical lack of transport, storage, crushing capacity and logistics infrastructure and created investment opportunities for new actors, including some players whose presence epitomises the recent financialisation of agri-food systems.

These opportunities have encouraged new forms of investment by complex alliances partnering established commodity traders – the ABCDs, Marubeni, Mitsubishi, Mitsui and other Japanese trading companies (*sogo shosha*) – and a host of newcomers (Wilkinson and Pereira, 2015). These include land leasing and farm management companies, Brazilian agribusiness firms, such as the Amaggi Group, Brazilian investment funds, foreign private equity groups, such as Blackstone, commodity traders diversifying into agriculture, as in the case of Glencore, and leading global Brazilian companies from other sectors, including Companhia do Vale do Rio Doce (CVRD) and construction companies, such as Odebrecht (ibid.) As Wilkinson and Pereira (2015: 25) conclude, 'Whatever the outcome, the Big

Four, who still dominate the soy trade in Brazil, will face ... the wide range of new actors muscling in on the downstream phases of this strategic commodity chain'.

As the case studies of the global grains-livestock complex and the ABCD companies reveal so plainly, China since the turn of the 21st century has become the fulcrum of global agri-food restructuring by virtue of its capacity to redefine the rules of the game in global commodity markets. The gradual shift towards greater reliance on these markets, already discernible in the 1990s, accelerated rapidly following China's accession to the WTO in 2001. Initially, this dependence was limited to imports of non-food primary products – feed grains (soy), pulp, cellulose, timber, cotton and tobacco – as China pursued its long-established geo-political objective of self-sufficiency in staple food grains. However, with rising incomes, rapid urbanisation and the dietary transition towards 'western' patterns of food consumption, notably of animal protein, China reached a strategic crossroads in 2012 with the decision to relax import quotas on maize, rice, wheat and other grains, stimulating its growing industrial livestock complex (Naughton, 2007; Fan, 2012; Aglietta and Bai, 2013).

The scale of China's food and natural resource demand and projected dependence on external supplies has huge implications for the world food system. The concept of food security, previously associated with the Millennium Development Goals, now defines the strategic national policy of a country whose levels of demand, even with modest import dependence (5–10%), will place an enormous strain on world trade. Observing that 'China has outgrown its resource base' (Farooki and Kaplinsky, 2012: 103) and that a 'remarkable shift in food consumption from cereals to meat products in general, and pork products in particular' (71) has occurred, these authors expect the pressure on global supplies of food and feed crops to intensify.

Schneider and Sharma (2014: 15) convey some idea of scale by citing USDA estimates that China's imports of pork represent about 1% of domestic consumption but 'close to 12% of the world's exports'. In the case of soy, China only became a net importer in 1996 but was importing half of the world's soybeans by 2005. As these floodgates open, China's impact on its trading partners and the directions of world trade will be profound. For example, 'In 2011–12, nearly 82% of Brazilian soy exports went to China' (Sharma, 2014: 24), and China also is becoming a key market for beef from Brazil, the world's leading exporter. The global dairy market is also beginning to feel the tremors of China's growing demand, (Sharma and Rou, 2014). To paraphrase Schneider and Sharma (2014), 'Chinese characteristics' are now woven into the fabric of the global 'corn-soy-livestock complex', the defining feature of modern agri-food systems.

These stylised facts indicate how patterns of global agricultural trade and investments are being reconfigured as China negotiates what are still the early stages of transition towards becoming a major net food importer. China's pursuit of food security is triggering a fundamental re-alignment of the world food economy as its trade and investment strategies incorporate resource frontiers, old and new. These dynamics will be decisive in the reorganisation of the world food system and carry far greater significance than the various 'drivers' that rise only to be discarded in the debates on the form and content of a putative third global food regime (Burch and Lawrence, 2005, 2009). The influence of China and its trading partners in re-aligning the geo-political axes of global agricultural trade and investment flows needs to be at the forefront of analytical frameworks that seek to elucidate the relationships between capitalist development and the structuring of world food systems.

We realise that this emphasis on China and other emerging actors risks being taken as implicit support for a hegemonic world-systems perspective. We are clearly opposed to such an interpretation as it would perpetuate the limitations of Food Regime Analysis. Rather we are making the simpler, far more prosaic point that these actors are representative of the multi-polarity of the contemporary global food system and their role in its reorganisation should be acknowledged.

Conclusion

In summary, this review has revealed a number of shortcomings of Food Regime Analysis relating to its historiography and periodisation, its partial account of colonialism, the neglect of developmental state strategies and the multi-polarity of the global agri-food system. These limitations are rooted, we have suggested, in the framing concepts of regime formation, hegemony and hegemonic succession. In turn, this positioning is expressed in terms of a simplified characterisation that pits the 'North' against the 'South', loosely identified with US agribusiness and undifferentiated peasant-based food systems. As we have argued at some length in this paper, this binary opposition misreads the dynamics of change in the global political economy of agriculture and food and should be abandoned.

Rather than elaborate these points at greater length, we prefer to conclude on a programmatic note by taking up some implications of the global food crisis of 2006–08. The multi-polarity of the global food system was revealed in truly transparent and unmistakable terms by this crisis and its aftermath. While scholars dispute the relative causal importance of biofuel policy mandates and financial speculation (Evans, 2008; HLPE, 2011, 2013; UNCTAD, 2012), this crisis has certainly heightened perceptions of global biophysical constraints and intensified the contested access to food and natural resources more generally (Harvey and Pilgrim, 2011). The dimensions of food and resource security issues have been further complicated by the technological convergence of the food, feed and other land-based sectors as general sources of biomass, which potentially can provide feedstocks for the production of food, feed, fibre, renewable energy, and a variety of industrial products, including plastics and chemicals (Langeveld et al., 2010; Richardson, 2010; White House, 2012; McKay et al., 2015). This convergence prompts (Baines, 2014: 106) to ask 'whether the concept of the world food system as a distinct political economy arena still has analytical currency'.

Macro-level analyses of change in global agri-food systems will need to take account of the increasing technological substitutability of renewable biomass raw materials and explore the theoretical and material implications of the revaluation of land and the conceptual meaning of 'agriculture' in this brave new bioeconomic world of polyvalent biomass resources. In future, the wider perspective of *resource access and security* appears to offer a fruitful point of entry for the analysis of the political economic restructuring of contemporary agriculture and food systems.[20,21] In the spirit of Food Regime Analysis, it is vital that agri-food studies continue to engage with these global themes and track their complex unfoldings to the full.

Notes

1 For example, in its newly resurrected form, attention has shifted from the historical institutionalisation and subsequent collapse of successive food regimes to a meta-theoretical analysis of the food-wage relation in the Braudelian *longue durée* of capitalist world development in which food regimes become merely conjectural 'moments' (McMichael, 2009b). On geo-historical structuralist epistemology, see Brenner et al. (2010).

2 Close readings of the genealogy of Food Regime Analysis that trace the contributions of its main protagonists can be found in Araghi (2003), Araghi (2009), Bernstein (2014, 2015) and Jansen (2014).

3 For a recent example of this version, which reproduces the binary formulations of Food Regime Analysis, see Levidow (2015). Thus his analysis of the contemporary agri-food transition in Europe pitches 'a nascent "corporate-environmental regime"' and its 'neo-productivist narrative' against 'agroecological alternatives'.

4 The 1870s saw the emergence of a single international market for wheat but, from an institutional perspective, the British regime was established in the 1840s and 1850s.

5 On the fundamental role of land-intensive exports of food and fibre from the circum-Caribbean plantation zone in the 'dramatic easing' of ecological constraints to British and European industrialisation, see Pomeranz (2000), especially Chapter 6. For complementary analyses, though with different emphases, see Parthasarathi (2011) and Beckert (2014).

6 Bernstein (2015) notes that Friedmann's (2005) reworking of the first food regime places greater emphasis on the waves of European migration to these settler societies and the formation of a 'new class' of family-labour commercial farms dependent for their reproduction on international export markets. This first regime is now designated as the 'colonial-diasporic food regime'.

7 Grains and later beef imports are given pride of place in the Food Regime Analysis account, but sugar, tea, milk, butter, lard, potatoes, fish and beer were also key elements in the formation of the British urban diet (Oddy, 2003). Thus Pomeranz (2000: 274) cites estimates by Mintz (1985) 'that sugar made up roughly 2% of Britain's caloric intake by 1800, and a stunning 14% by 1900'.

8 In the case of Germany, this policy reversed the so-called Morgenthau Plan proposed in 1944 by the then US Secretary of the Treasury to de-industrialise the country in order to eliminate its future military-industrial capability.

9 This support is epitomised by the limited success of the 1968 Mansholt Plan, an early CAP reform initiative, which proposed to encourage some five million small farmers to leave agriculture and consolidate their holdings into 'more economically sized' farms.

10 In this respect, the ravages caused by the post-WWII colonial independence struggles and the subsequent influence of the Soviet bloc and China on rural development in the nations supporting the Non-Aligned Movement are simply not addressed in the food regime narrative. In many countries, particularly in Africa, collectivisation policies arguably undermined traditional practices and customary land rights in the countryside to a much greater extent than food aid.

11 South Korea was the seventh largest recipient of food aid in 1960, rising to third and fourth place in 1970 and 1980, respectively. It is now no longer among the fifteen most important recipients.

12 Friedmann (2009) is not persuaded that the WTO and its Agreement on Agriculture is 'the founding institution of a corporate food regime' and supports Pritchard's (2009) view that it is a vestige of the second food regime. Pritchard (2009) suggests that the 2008 collapse of the Doha Round of trade negotiations is symptomatic of the inadequacy of the WTO's institutional framework 'to accommodate the complex political machinations associated with a multi-polar world' (our emphasis). This collapse 'should put an end to speculation of a WTO-led transformation of global food politics towards unfettered market rule, the supposed basis for a neo-liberalised "third food regime"' (ibid., 297).

13 The most recent expressions of this shift away from multilateralism are the Trans-Pacific Partnership (TPP) and the EU-Canada Comprehensive Economic and Trade Agreement (CETA). Despite President Trump's 2017 announcement of US withdrawal from the TPP, this agreement may go ahead without US participation, possibly under the leadership of China. CETA was approved by the European Parliament in February, 2017 and now requires approval by national parliaments in the EU.

14 That is, Archer Daniels Midland, Bunge, Cargill, and Louis Dreyfus.

15 See Schneider and Sharma (2014), Sharma (2014), Sharma and Rou (2014), and Pi et al. (2014).

16 COFCO: the China National Cereals, Oils and Feedstuffs Corporation, according to Schneider and Sharma (2014). This paragraph draws closely on their account.

17 These concerns are revealed by the conditions imposed by the Chinese competition authorities in granting their approval of the acquisition in July, 2013 of the US-based grain trader, Gavilon, formerly ConAgra, by Marubeni, Japan's leading trading house and the largest supplier of soya to China in 2012. As reported by the *Financial Times*, 24 April, 2013, Marubeni and Gavilon will be required to 'continue selling soya to China as separate companies, with two different teams and with firewalls between them blocking the exchange of market intelligence'.

18 On China's resource strategy and diplomacy in the energy and minerals sectors, see Taylor (2012), Burgos and Ear (2012), Ciccantell (2009), Beeson et al. (2011), Ferchent (2011), and Gallagher and Porzecanski (2010).

19 For details of COFCO's diversification into sugar production and processing and biofuels production via recent corporate takeovers in Australia in competition with Bunge, among others, see Murphy et al. (2012: 41–42).

20 The adoption of a resource security perspective in agri-food studies is suggested by analyses of contemporary Chinese resource policy and its parallels with the earlier Japanese model of resource access. These analyses focus mainly on minerals and energy (Bunker and Ciccantell, 1995; Ciccantell, 2009; Beeson et al., 2011) but see Hofman and Ho (2012). For a longer historical analysis of ascendant economies in terms of resource access and the incorporation of natural resource peripheries, both external and internal, see Ciccantell and Bunker (2004).

21 A broader resources perspective might be loosely read into Food Regime Analysis approaches to land grabbing and biofuels (McMichael, 2010, 2012) and discussion of the food sovereignty movement in terms of agro-ecological bioregionalism (McMichael, 2013).

References

Aglietta M., 1979. *A Theory of Capitalist Regulation*. London, New Left Books.

Aglietta M., Bai G., 2013. *China's Development: Capitalism and Empire*. London, Routledge.

Araghi F., 2003. Food Regimes and the Production of Value: Some Methodological Issues. *Journal of Peasant Studies*, 30(2), 41–70.

Araghi F., 2009. The Invisible Hand and the Visible Foot: Peasants, Dispossession and Globalisation. In Akram-Lodi A.H. and Kay C.H. (Eds.), *Peasants and Globalisation: Political Economy, Rural Transformation and the Agrarian Question*, London, Routledge, pp. 111–147.

Arrighi G., 2007. *Adam Smith in Beijing: Lineages of the Twenty-First Century.* Cambridge, Verso.

Bagchi A.K., 2010. *Colonialism and the Indian Economy.* Oxford, Oxford University Press.

Baines J., 2014. Food Price Inflation as Redistribution: Towards a New Analysis of Corporate Power in the World Food System. *New Political Economy*, 19(1), 79–122.

Ban C., 2013. Brazil's Liberal Neo-developmentalism: New Paradigm or Edited Orthodoxy? *Review of International Political Economy*, 20(2), 298–331.

Ban C., Blyth M., 2013. The BRICS and the Washington Consensus: An Introduction. *Review of International Political Economy*, 20(2), 241–255.

Beckert S., 2014. *Empire of Cotton.* London, Penguin/Allen Lane.

Beeson M., Soko M., Yong W., 2011. The New Resource Politics: Can Australia and South Africa Accommodate China? *International Affairs*, 87(60), 1365–1384.

Berlan J.-P., Bertrand J.-P., Lebas L., 1977. The Growth of the American Soybean Complex. *European Review of Agricultural Economics*, 4(4), 395–416.

Bernstein H., 2014. Food Sovereignty via the "Peasant Way": A Skeptical View. *The Journal of Peasant Studies*, 41(6), 1031–1063.

Bernstein H., 2015. *Food Regimes and Food Regime Analysis: A Selective Survey.* BICAS Working Paper 2 Brazil, Netherland, South Africa, China, BRICS Initiative for Critical Agrarian Studies (BICAS).

Brautigam D., 2009. *The Dragon's Gift: The Real Story of China in Africa.* Oxford, Oxford University Press.

Brenner N., Peck J., Theodore N., 2010. Variegated Neoliberalization: Geographies, Modalities, Pathways. *Global Networks*, 10(2), 182–222.

Brenner R., Glick M., 1991. The Regulation Approach: Theory and History. *New Left Review*, 188, 45–119.

Breslin S., 2005. Power and Production: Rethinking China's Global Economic Role. *Review of International Studies*, 31(4), 735–753.

Breslin S., 2011. The "China Model" and the Global Crisis: From Friedrich List to a Chinese Mode of Governance? *International Affairs*, 87(6), 1323–1343.

Buller H., 2004. The "espace productif", the "théâtre de la nature" and the "territoires de développement local": The Opposing Rationales of Contemporary French Rural Development Policy. *International Planning Studies*, 9(2–3), 101–119.

Bunker S., Ciccantell P., 1995. Restructuring Markets, Reorganising Nature: An Examination of Japanese Strategies for Access to Raw Materials. *Journal of World-Systems Research*, 1(3) 201–242.

Burch D., Lawrence G., 2005. Supermarket Own-Brands, Supply Chains and the Transformation of the Agri-Food System. *International Journal of Agriculture and Food*, 13(1), 1–16.

Burch D., Lawrence G., 2009. Towards a Third Food Regime: Behind the Transformation. *Agriculture and Human Values*, 26(4), 267–279.

Burgos S., Ear S., 2012. China's Oil Hunger in Angola: History and Perspective. *Journal of Contemporary China*, 21, 35–367.

Campbell H., 2009. Breaking New Ground in Food Regime Theory: Corporate Environmentalism, Ecological Feedbacks and the 'food from somewhere' Regime. *Agriculture and Human Values*, 26, 309–319.

Campbell H., Dixon J. 2009. Introduction to the Special Symposium: Reflecting on Twenty Years of the Food Regimes Approach in Agri-Food Studies. *Agriculture and Human Values*, 26, 261–265.

Ciccantell P., 2009. China's Economic Ascent and Japan's Raw Material Peripheries. In Hung H.-F. (dir.), *China and the Transformation of Global Capitalism*, Baltimore, Johns Hopkins University Press, pp. 109–129.

Ciccantell P., Bunker S., 2004. The Economic Ascent of China and the Potential for Restructuring the Capitalist World-Economy. *Journal of World-Systems Research*, 10(3), 565–589.

Cooper A.F., Flemes D., 2013. Foreign Policy Strategies of Emerging Powers in a Multipolar World: An Introductory Review. *Third World Quarterly Review*, 34(6), 943–962.

Cwiertka K.J., 2006. *Modern Japanese Cuisine: Food, Power and National Identity*. London, Reaktion Books.

Dixon J., 2009. From the Imperial to the Empty Calorie: How Nutrition Relations Underpin Food Regime Transitions. *Agriculture and Human Values*, 26, 321–333.

Evans D., 2008. *Rising Food Prices: Drivers and Implications for Development*. Briefing Paper. London, Chatham House.

Fan S., 2012. Walk the Walk, Talk the Talk: Major Food Policy Developments in 2012. In *Global Food Policy, 2012*, Washington, DC, IPFRI.

Farooki M., Kaplinsky R., 2012. *The Impact of China on Global Commodity Prices: The Global Reshaping of the Resource Sector*. London, Routledge.

Ferchent M., 2011. China-Latin America relations: Long-Term Boon or Short-Term Boom? *The Chinese Journal of Politics*, 4, 55–86.

Friedmann H., 1982. The Political Economy of Food: The Rise and Fall of the Post-War International Food Order. *American Journal of Sociology*, 88(Supplement), 246–286.

Friedmann H., 2005. From Colonialism to Green Capitalism: Social Movements and the Emergence of Food Regimes. In Buttel F.H. and Mcmichael P. (Eds.), *New Directions in the Sociology of Global Development: Research in Rural Sociology and Development*, Vol. 11, Amsterdam, Elsevier, pp. 227–264.

Friedmann H., 2009. Discussion: Moving Food Regimes Forward: Reflections on Symposium Essays. *Agriculture and Human Values*, 26, 335–344.

Gallagher K., Porzecanski R., 2010. *The Dragon in the Room: China and the Future of Latin American Industrialization*. Stanford, CA, Stanford University Press.

Gaviria C., 2011. The Post-War International Food Order: The Case of Agriculture in Colombia. *Lecturas De Economia*, 74, 119–150.

Gerschenkron A., 1989 [1943]. *Bread and Democracy in Germany*. Ithaca, NY, Cornell University Press.

Golub P., 2013. From the New International Economic Order to the G20: How the "Global South" is Restructuring Capitalism from Within. *Third World Quarterly*, 34(6), 1000–1015.

Goodman D., 2003. The Quality 'turn' and Alternative Food Practices: Reflections and Agenda. *Journal of Rural Studies*, 19, 1–7.

Goodman D., Watts M., 1994. Reconfiguring the Rural or Fording the Divide? Capitalist Restructuring and the Global Agro-Food System. *The Journal of Peasant Studies*, 22(1), 1–49.

Harvey M., Pilgrim S., 2011. The New Competition for Land: Food, Energy and Climate Change. *Food Policy*, 36, 540–551.

HLPE., 2011. *Price Volatility and Food Security*. A report by the High Level Panel of Experts on Food Security and Nutrition of the Committee on World Food Security Rome, FAO/CWFS.

HLPE., 2013. *Biofuels and Food Security*. A report by the High Level Panel of Experts on Food Security and Nutrition of the Committee on World Food Security Rome, FAO/CWFS.

Hobsbawm E., 1979. *The Age of Empire, 1848–1914*. Harmondsworth, Penguin.

Hofman I., Ho P., 2012. China's 'developmental outsourcing': A Critical Examination of Chines Global 'land grabs' Discourse. *The Journal of Peasant Studies*, 39(1), 1–48.

Jansen K., 2014. The Debate on Food Sovereignty: Agrarian Capitalism, Dispossession and Agroecology. *The Journal of Peasant Studies*, 42(1), 213–232.

Landes D., 1969. *The Unbound Prometheus: Technological Change and Industrial Development in Western Europe from 1750 to the Present*. Cambridge, Cambridge University Press.

Langeveld H., Sanders J., Meeusen M., 2010. *The Biobased Economy: Biofuels, Materials and Chemicals in the Post-Oil Era*. London, Earthscan.

Le Heron R., Lewis N., 2009. Discussion: Theorising Food Regimes: Intervention as Politics. *Agriculture and Human Values*, 26(4), 345–349.

Levidow L., 2015. European Transitions Towards a Corporate-environmental Food Regime: Agroecological Incorporation or Contestation? *Journal of Rural Studies*, 40, 76-89.

Lockeretz W., 1988. Agricultural Diversification by Crop Introduction: The US Experience with the Soybean. *Food Policy*, 13(2), 154–166.

McKay B., Sauer S., Richardson B., Herre R., 2015. The Political Economy of Sugarcane Flexing: Initial Insights from Brazil, Southern Africa and Cambodia. *The Journal of Peasant Studies*, 43(1), 195–223.

McMichael P., 2005. Corporate Development and the Corporate Food Regime. In Buttel F.H. and McMichael P. (Eds.), *New Directions in the Sociology of Global Development. Research in Rural Sociology and Development*, Vol. 11, Amsterdam, Elsevier, pp. 265–299.

McMichael P., 2009a. A Food Regime Genealogy. *Journal of Peasant Studies*, 36(1), 139–169.

McMichael P., 2009b. A Food Regime Analysis of the 'world food crisis'. *Agriculture and Human Values*, 26, 281–295.

McMichael P., 2010. Agrofuels in the Food Regime. *The Journal of Peasant Studies*, 37(4), 609–629.

McMichael P., 2012. The Land Grab and Corporate Food Regime Restructuring. *The Journal of Peasant Studies*, 39(3–4), 681–701.

McMichael P., 2013. *Food Regimes and Agrarian Questions*. Halifax and Winnipeg, Fernwood Publishing.

Mielniczuk F., 2013. BRICS in the Contemporary World: Changing Identities, Converging Interests. *Third World Quarterly*, 34(6), 1075–1090.

Mintz S. 1985. *Sweetness and Power: The Place of Sugar in Modern History*. New York, Penguin.

Morgan K., Marsden T., Murdoch J., 2006. *Worlds of Food: Place, Power and Provenance in the Food Chain*. Oxford, Oxford University Press.

Murdoch J., 1998. The Spaces of Actor-Network Theory. *Geoforum*, 29(4), 357–374.

Murphy S., Burch D., Clapp J., 2012. *Cereal Secrets: The World's Largest Grain Traders and Global Agriculture*. Oxford, Oxfam Research Reports.

Naughton B., 2007. *The Chinese Economy: Transitions and Growth*. Cambridge, Mass, The MIT Press.

North D., 1958. Ocean Freight Rates and Economic Development. *Journal of Economic History*, 18(4), 537–555.

Oddy D.J., 2003. *From Plain Fair to Fusion Food.* Woodbridge, Boydell Press.

Parthasarathi P., 2011. *Why Europe Grew Rich and Asia Did Not: Global Economic Divergence, 1600–1850.* Cambridge, Cambridge University Press.

Pi C., Rou Z., Horowitz S., 2014. *Fair or Fowl: The Industrialization of Poultry Production in China.* Minneapolis, MN, IATP. Available online www.iatp.org.

Polanyi K., 1957 [1944]. *The Great Transformation.* Boston, Beacon Press.

Pomeranz K., 2000. *The Great Divergence: China, Europe and the Making of the Modern World Economy.* Princeton, Princeton University Press.

Pritchard B., 2009. The Long Hangover from the Second Food Regime: A World-Historical Interpretation of the Collapse of the WTO-Doha Round. *Agriculture and Human Values*, 26, 297–307.

Reinert E., 2007. *How Rich Countries Got Rich. Why Poor Countries Stay Poor.* London, Constable.

Reuters, 13 November, 2007. "Japan Mitsui Affiliate Buys Brazilian Farmland".

Richardson B., 2010. *From a fossil-fuel to a bio-based economy? Reframing the third wave of biotechnology.* IPEG Papers in Global Political Economy No. 45, International Political Economy Group, British International Studies Association.

Schneider M., Sharma S., 2014. *China's Pork Miracle? Agribusiness and Development in China's Pork Industry.* La Hague, Institute of Agriculture and Trade Policy. consultable en ligne www.iatp.org/files/2014_03_26_PorkReport_f_web.pdf (consulté le 11 octobre 2016).

Schonhardt-Bailey C., 2006. *From the Corn Laws to Free Trade.* Cambridge, Mass, The MIT Press.

Schott J., 2009. America, Europe and the New Trade Order. *Business and Politics*, 11(3), 1–22.

Sharma S., 2014. *The Need for Feed. China's Demand for Industrial Meat and Its Impacts.* Global Meat Complex: The China Series La Hague, Institute of Agriculture and Trade Policy. consultable en ligne www.iatp.org/files/2014_03_26_FeedReport_f_web.pdf (consulté le 11 octobre 2016).

Sharma S., Rou Z., 2014. *China's Dairy Dilemma: The Evolution and Future Trends of China's Dairy Industry.* La Hague, Institute of Agriculture and Trade Policy. consultable en ligne www.iatp.org/files/2014_02_25_DairyReport_f_web.pdf (consulté le 11 octobre 2016).

Silver B. J., Arrighi G.; 2003. Polanyi's "double movement": The belle époques of British and US hegemony compared. *Politics & Society*, 31(2), 325-355.

Smil V., Kobayashi K., 2012. *Japan's Dietary Transition and Its Impacts.* Cambridge, Mass, The MIT Press.

Storz C., Amable B., Caspar S., Lechevalier S. (dir.), 2013. Bringing Asia into the Comparative Capitalist Perspective. *Socio-Economic Review*, 11(2), 217–232.

Strange S., 2011. China's Post-Listian Rise: Beyond Radical Globalisation Theory and the Political Economy of Neoliberal Hegemony. *New Political Economy*, 16(5), 539–559.

Taylor I., 2012. China's Oil Diplomacy in Africa. *International Affairs*, 82(5), 937–959.

Tracy M., 1982. *Agriculture in Western Europe: Challenge and Response, 1880–1980.* London, Granada.

UNCTAD, 2012. *Don't blame the physical markets: Financialization is the root cause of oil and commodity price volatility*, Briefing Paper No. 25, United Nations Conference on Trade and Development, Geneva.

Whatmore S., Thorne L., 1997. Nourishing Networks: Alternative Geographies of Food. In Goodman D. and Watts M. (dir.), *Globalising Food: Agrarian Questions and Global Restructuring*, London, Routledge, pp. 287–304.

White House., 2012. *National Bioeconomy Blueprint*. Washington, DC, The White House.

Wilkinson J., 1997. A New Paradigm for Economic Analysis? Recent Convergences in French Social Science and an Exploration of the Convention Theory Approach, with a Consideration of its Application to the Analysis of the Agro-Food System. *Economy and Society*, 26(3), 305–339.

Wilkinson J., 2014. *Challenges to the Dominant Food System*. Unpublished paper prepared for QUNU Genève, QUNU.

Wilkinson J., Pereira P., 2015. Brazilian Soy: New Patterns of Investment, Finance and Regulation. Unpublished paper.

Wilkinson J., Wesz Jr., V.J., Lopane A., Munz L., 2015 "Brazil, the Southern Cone and China: The agribusiness connection", Unpublished paper.

Yamauchi A., 2002, "Towards Sustainable Agricultural Production Systems: Major Issues and Needs in Research", Proceedings of the Forum on Sustainable Agricultural Systems in Asia, Nagoya, Japan, June.

8 The Holstein cow as an institution of the agricultural modernisation project

Commodity or common good?

Julie Labatut and Germain Tesnière

Introduction

'Holstein' is the common name for the most familiar breed of black-and-white dairy cows that produce the milk found on our tables in the morning. It is the most widespread cattle breed on farms in the world and in our collective imagination, and is associated with the successes and the crises experienced by dairy farming and the milk industry Europe. The Holstein is one of the breeds that has undergone the most selection, particularly with the development of animal genetics industry in the 20th century, and still remains a pioneer breed for the dairy industry. Likewise, with the recent development of genomics in animal selection, Holstein was the first dairy breed to benefit from this modern biotechnology that can 'read' an individual's DNA and instantaneously identify sires or brood dams with high-performance potential.

This cow, the ultimate 'machine cow' (Ruet, 2004), has become one of the symbols of the industrialisation of agriculture and the commodification of living organisms. The market for Holstein bull sperm, embryos and cattle for reproduction purposes is estimated at €335 million in financial transactions worldwide (2015, from a professional trade source).

There are many specialised companies in this sector and the Holstein breed has gone global. This breed is just as suitable for French medium-sized pasture-based dairy farms as the giant dairy farms in California that house several thousand cows. Despite climates considered to be unfavourable for this ultramodern breed, emerging countries are also turning to the Holstein breed through the process of 'holsteinisation', whereby Holstein genetics are spread around the world (Theunissen, 2012) and to other breeds, improving their milk production. Holstein cattle are associated with the dairy mass-market development.

However, although Holsteins are emblematic of the commodification of living organisms, the breed itself nevertheless is a common-pool resource, belonging as much to dairy farmers as to breeders, and nowadays to other stakeholders. There are currently no (exclusive) intellectual property rights on animal breeds and, as such, no limitations on the access to these animals, nor, for the time being at least, to the genetic products derived from them. In 2017, an animal breed is the common property of all the farmers that use it. Drawing on Hess and Ostrom

(2003: 121), who make the distinction between flow and resources in common-pool systems whereby even if the resource units produced (flow) by the resource (here, the animal breed) are commodities (e.g. animals, embryos, semen), the actual 'breed' is a common-pool resource, possibly threatened if its management is left entirely up to market mechanisms. For example, there is a risk of inbreeding or spread of genetic anomalies when the best individuals of an animal population are overused for reproduction purposes (leading to deterioration of the resource).[1] One study has shown that the number of ancestors that have contributed half the genes found in the Holstein population, dropped by a factor of 3.5 between 1988 and 2003 (Mattalia et al., 2006). In contrast, breeds with small population sizes are threatened by the non-use of the breed (extinction) or breeders that no longer contribute to the breeding programme (Labatut et al., 2012).

The Holstein breed is a particularly relevant case study for analysing the paradox of management of a common-pool resource faced with growing commodification of the resource units produced from the resource. This issue is becoming increasingly important in the context of the recent changes in the animal genetics market: growing globalisation, deregulation, withdrawal of government funding from genetic selection programmes (particularly in France where government funding was previously high), and emergence of new technologies that dramatically accelerate genetic progress (genomics, semen sexing, etc.).

Here, we will examine the evolution of the Holstein breed to explore the different forms of industrialisation in agriculture. First, we will cover some of the steps in the genetic and marketing process that led to the biological and institutional creation of the Holstein breed and to the holsteinisation of French dairy cattle. Then, we will use the concept of 'breeding regime' (Labatut et al., 2011, 2013) to shed light on the evolution of the Holstein breed. A breeding regime is an institutional regime made up of political, scientific, technical, informational and organisational measures that determine the dynamics of an animal population and its genetic progress. We carry out this analysis in the particular case of France. We will show how holsteinisation in France is part of a dual cooperative-public breeding regime based on Fordist-type industrialisation, and we describe the measures implemented to manage the tension between common-pool resources and commodification dynamics.

From the (Holstein-)Friesian to the modern Holstein, an account of the construction of the breed

Although zootechnical studies have explored the genetic evolution of the Holstein and socio-economic studies have analysed the changes in the dairy market and industry, there are no studies that trace the genetic history of the breed in light of the history of the dairy industry. Although we do not claim to exhaustively bridge this gap, we attempt to identify the parallelism of both historical trajectories.

In contrast to most French breeds, the Holstein, whose official name in France is 'Prim'Holstein', is found in all the dairy regions of the country and is not specific to any one of them. As underlined by Pellegrini (1999), the Holstein breed 'no longer evokes the cow from the German region of the same name, but now refers to a high-performing dairy breed, selected in North America and now found worldwide through commercial distribution of semen'. The breed has a long history of selection and crosses with various branches of black-pied cattle.[2]

The history of the Prim'Holstein cannot be traced back to a single country. This breed was introduced in France from the Netherlands and was then crossed with various other types to provide 'new blood', first from the Netherlands and then from the United States in 1965–70. The change in the name of this breed in France reflects the various influences that have contributed to its construction: *'Hollandaise'* (Dutch), *'Française Frisonne Pie Noire'* (French Black-Pied Friesian), *'Française Frisonne'* (French Friesian) and finally Prim'Holstein. But let's start from the beginning.

The so-called 'black-pied' cattle found around the world seem to all come from the same region along the coast of the North Sea, an area covering the present-day Dutch provinces of Friesland and North Holland (Netherlands), the Jutland peninsula (Denmark) and the German state of Schleswig-Holstein.[3] In the mid-19th century, exports rose and European buyers (England, Belgium, Prussia, etc.) 'attached great importance to the characteristics and the purity of the breed, leading to the need for pedigree records' (Denis, 2010). Two herdbooks[4] were then created: one (NRS) for Dutch cattle in general, primarily dairy, and the other (FRS) specifically for Friesian cattle, relatively dual-purpose.[5] It was only in 1905 that the Hollander-Friesian black-pied cow breed was officially defined as such in the Netherlands. Overseas, the breed was renowned for its dairy production and qualified as 'Hollander' or 'Friesian'. The first exports to North America, Canada and the United States date back to 1852, giving rise to the North American branch of the breed that developed under the name of 'Holstein-Friesian'. Since that time, selection in North American has been carried out with almost no 'new blood' brought in from other countries. The ensuing phase of unrestricted trade and genetic exchanges between countries was followed in 1905 by a phase during which North America closed its borders for animal health safety reasons. From the beginning, this breed, with its own herdbook and a dedicated association created in 1885 (Holstein-Friesian Association of America), was selected based exclusively on dairy criteria (drinking milk, low in butterfat and protein) and for long body length.

In the mid-19th century, animals from the dairy branch of the Hollander-Friesian black-pied cattle from the Netherlands were not common in France. Imports only became significant in 1830–40 (in Normandy). This Dutch breed was tall, low in muscle mass, had pronounced hook bones and, above all, boasted high milk production (Spindler, 2002). The breed gradually spread throughout France, first under the name of 'Hollander'. It became established in 'sustenance' farms, dairy farms located around large cities at the time, particularly Paris (Denis, 2010). There were some farms with Hollander cows in highly populated industrial areas such as northern France (Nord-Pas-De-Calais, Picardie), the Parisian region (Ile-de-France) and the Bordeaux area (Gironde). However, other attempts to introduce the breed in rural areas were not very successful, particularly due to the poor adaptation of Hollander cattle to the most common farming conditions at that time.

In the Netherlands, the control of dairy performance and the creation of bull stud farms occurred in the early 20th century. According to Flamant (2011),

> the Dutch and the Danish played a pioneering role in this field in the early 20th century by systematically recording individual cow production and by controlling milk quality in herds to provide clear accurate data that could be used by all dairy farmers, for example, for comparison in livestock competition.
>
> (Flamant, 2011)

France only followed this example after WWI (*Contrôle Laitier Beurrier*, dairy unions in eastern France). In parallel, in the late 19th century and early 20th century, a veritable dairy industry began to develop owing to the progress in transportation means and storage techniques. Thus, with the development of the railroad and the generalisation of pasteurisation, the Paris milk distribution network expanded to nearly 300 km (Vatin, 1996).

The importance of the Dutch population and the desire to improve the breed led breeders in northern France to create, in Lille in 1922, the genealogical pedigree for the Hollander breed created under the name of '*Herd-Book français de la race Hollandaise*'. At that time the role of the herdbook was to record those animals that met a 'breed standard' and establish their pedigree (recording of births and publication of directories). Between WWI and WWII, the population increased significantly, from 200,000 cows in 1918 to 600,000 cows in 1938 (Denis, 2010). The population reached 840,000 in 1943, representing only 5.2% of the entire French cattle population.[6] However, after WWI, the French population of the breed had been devastated, and new imports were brought from the Netherlands to rebuild the herds. The breed spread mainly in northern France, north-eastern France, the Parisian region, and in the south-west. The services of the breed association (who maintained the herdbook), now based in Cambrai (northern France), rapidly grew in demand. Milk records became mandatory for members of the breed association as of 1948, marking the beginning of the relationship between objective, scientific measurements of performance and the pedigree records, the two pillars of genetic selection.

Until the end of WWII, the animals of the two main branches (American and European) had the same production and morphological characteristics. American breeders then began to intensify the selection of Holstein-Friesians according to dairy-production and udder-quality criteria, while in Europe selection concentrated on the butterfat content[7] and conformation.[8] In the Netherlands, the Hollander-Friesian black-pied breed, originally renowned for its milk production, was steered from 1945 towards 'a balanced dual-purpose model, primarily at the behest of FRS that resulted in a smaller body frame and better muscle development' (Denis, 2010). In France, selection turned to a dual-purpose breed and the Hollander breed then changed its name to 'French Black-Pied Friesian' (*Française Frisonne Pie Noire*, FFPN) in 1952, following the FRS Friesian model, but the breed was still primarily known as a dairy cow. The population increased to 1,500,000. This trend was not copied in other countries; for example, the United Kingdom, Italy and the North Holland province in the Netherlands chose to maintain the predominantly dairy breed.

After WWII, the progression of the FFPN breed continued in France, partly encouraged by the growing consumption of dairy products, and specialisation in production. It is effectively during this period that dairy companies (cooperatives) began to invest heavily and companies such as Danone and Chambourcy launched the production of yogurt (Vatin, 1996). For Vatin, by the early 1960s, Paris had an 'authentic dairy industry'. However, the analysis of this evolution requires a look back at French agriculture after the end of WWII.

During the reconstruction period, agriculture in France was criticised for its low efficiency compared with agricultural systems in other countries, such as Denmark, the Netherlands or the United States (see article by Pierre Fromont in the *Le Monde* newspaper, 28 May 1946). For Fromont, who had published a rural economy treatise,

> the agricultural technical revolution is not just about replacing the horse or oxen with a tractor; that is only one aspect—undoubtedly the most spectacular, but not the most important. The most important instrument in agriculture production is the living organism, plant or animal … [they] are the real agricultural machine-tools.
> (Pierre Fromont, 28 May 1946, *'La révolution technique en agriculture et la politique' Le Monde*, cited in Cranney, 1996)

This excerpt illustrates the appeal of industrial processes to improve the efficacy and the yield of the agricultural tools and techniques, and one in which living organisms are considered as machines. The American Holstein breed was to become the emblem of this logic, whose development is today criticised because the 'price to pay' is the fragility of the breed, imposing changes in the tasks of the breeder, who must implement a multitude of animal health measures (Ruet, 2004: 66).

For Pierre Fromont in 1945,

> as in the industrial sector, we have witnessed the ageing and planned for the renewal of our equipment and machine-tools, like in agriculture, it is important to consider the efficacy of our biological tools. However, it must be admitted that, overall, these tools have not benefitted from the same improvement efforts as exerted in many other countries. One must continuously address the issue of efficacy of the living machine-tool … The work towards constructing a breed has been carried out on a smaller scale in France than in Denmark, Great Britain and the United States … Thus, to take but just one example, the average annual production of a dairy cow is evaluated at 1800 litres in France and exceeds 3000 litres in Denmark.
> (ibid.)

However, for Fromont, and where fate contradicts him, the solution does not lie in importation:

> we must get to work immediately. Other than the fact that improvement of living organisms is a necessarily slow process, because the cycle must be completed and cannot be rushed, it is almost impossible to have recourse to the method that is used for our industrial machine-tools: imports from other countries.
>
> (ibid.)

Thus, as Flamant (2011) explains, INRA researchers in the 1960s adjusted the Friesian cow breeding programme to meet the 'actual situation of farms':

> In the French context of small farms, the economic outcome obviously relies on the sale of milk, which provides for a monthly source of revenue, but also on the added value that meat provides ... they propose a bull breeding programme targeting the French ideal of the Dutch-origin Friesian breed, the *Française Frisonne Pie Noire* with a goal of increasing herd milk production to gain in competitiveness at a level similar to that of other European Community countries and nonetheless maintain the meat production qualities.
>
> (Flamant, 2011: 2)

However, the importation of Holstein genetics will play a major role in the development of the French cattle population.

In parallel, during the 1960–70s, a second dairy revolution (Vatin, 1996) was based on the development of intensive dairy farming, with the 'creation of efficient stables, the increase in dairy cow milk yield, the intensification of the farm-factory through the introduction of refrigeration on farm premises and the payment for production according to milk quality' (Vatin, 1996). The economic context was favourable to dairy production, providing incentive for European farmers (and abroad, —Harris and Kolver, 2001) to massively import Holstein bulls from the United States and Canada to improve the milk yield of their cows and thus augment their productivity. The holsteinisation process began in several European countries and in France under the influence of crosses carried out using North American Holstein strains, the dual-purpose FFPN became specialised in milk production and grew in size and in udder quality.

The first introductions of Holstein-Friesian cattle from North America occurred in 1965 and 1966 in the Isere department. In 1972, the FFPN was the leading French breed in terms of population with 6 million head of cattle, exceeding that of the Normande breed (5.7 million, at its peak). On the European scale, in the 1970s, eight national Friesian black-pied populations made up the large majority of the 23 million dairy cows in the EEC (Vissac, 2002: 174) and were to constitute the 'target for absorption by the American Holstein' (ibid.). In 1979, the breed became 'French Friesian' (Française Frisonne) encompassing at that time all Hollander-type animals, Holsteins born in France and the Friesian-Holstein hybrids.

The holsteinisation process was both biological and institutional. It occurred on different time scales and with different dynamics in each European country. For example, Vissac (2002) indicates that British farmers only became interested in American Holsteins much later, and attributes their initial disinterest in the selection for individual milk production to their large herd sizes (the United Kingdom had chosen to breed for milk yield when imports of Dutch dairy cows began). Likewise, in the Netherlands, the massive infusion of the local population with American Holsteins occurred long after France began importing (Theunissen, 2012). Vissac (2002) suggests that the Netherlands were in a defensive position, being the cradle of the Holstein breed. The percentage of the North American Holstein strain in the French cattle population increased from 40% in 1970 to 78% in the 1990s (Boichard et al., 1993, 1996). Boichard et al. (1993) also indicate that the percentage of Holstein genes in the black-pied bulls used for artificial insemination (AI[9]) was low prior to 1970, greatly increasing thereafter. From the 1980s, the proportion of Holstein blood reached nearly 100% for bulls used for AI. Thus, the percentage of Holstein genes in females rose from 5% in 1970 to 83% in 1990, which was clearly greater than the upper estimation that had been predicted by statistical geneticists at INRA (Colleau and Tanguy, 1984) and attests to the sharp and unexpected rise in holsteinisation. By the time milk quotas were introduced, the absorption of the local population by the American branch was practically irreversible. Nevertheless, the increase in productivity was accompanied by other changes, such as the increase in stature, a change in conformation, improvement in udder morphology and a decrease in female fertility. Moreover, the arrival of the Holstein in France did not take place as peacefully as it may seem. In rural areas there were conflicts, sometimes vehement, between supporters and detractors. Duroselle (1980) notes that the Holstein was the 'be-all and end-all of modern selection in terms of milk for its supporters' and 'depicted as a calamity for its detractors'.

Several years after the French Livestock Act (1966), the Union for the Selection and Promotion of the Holstein Breed (*Union pour la sélection et la Promotion de la Race*, UPRA) (French Friesian at the time) was created in 1975 (as an association under the French Act of 1901). A collective organisation that determined the scope of the breed as a common-pool resource (Labatut, 2013a), this breed association aimed to define the targets of a breeding programme and also provide services for breeders, maintain the herdbook for registered animals and ensure the promotion of the breed. Since 1989, the UPRA headquarters have been located in St. Sylvain d'Anjou (Maine et Loire), the epicentre of farms in western France in terms of density. In 1990, with the goal to 'better make known the efforts with regard to breed genetics and the breed's considerable population size in France', the UPRA decided to 'abandon the terms "French" and "Friesian" and chose a new name: Prim'Holstein'.

The structure of the UPRA breed association was created as a hybrid structure, both a parliament for the breed and an organisation that provides advice and services to its members (farmers). This ambiguous mixture between sovereign functions and extension service functions 'caused all the same a certain number of concerns' (Bieri, Director of PHF, 2014 interview). Following the 2006 Agricultural Guidance Law (*Loi d'Orientiation Agricole*, LOA), an amendment of the 1966 Livestock Act (2006), Prim'Holstein France (PHF) became on 1 July 2008 the French Prim'Holstein Breeders Association (*Association des Eleveurs de la race bovine Prim'Holstein*), whose main vocation was to offer 'independent services and counsel for dairy farmers on breed genetics and the management of their herd' (PHF). The regulatory missions (maintenance of the herdbook, breed policy) were entrusted to a new organisation, a selection organisation 'OS Prim'Holstein'. This OS is the 'breed parliament' with members from PHF representing the member farmers/breeders, selection companies and AI cooperatives representing the stakeholders in the creation and dissemination of genetic progress, and finally other partners (milk records operators, EDE[10] CNIEL[11]). In 2009, based on a decision by Ministry of Agriculture, a representative of the Red-pied breed (formerly called *Pie Rouge des Plaines*) was integrated as the fourth member of this OS because the breed harbours a large proportion (90–95%) of Holstein ('red') genes (Bieri, 2014 interview). At the departmental and regional levels, breeder associations were created to carry out local promotional activities for the breed by organising, for example, livestock competitions. Although they are independent from PHF, PHF funds and provides technical support for the organisation of these activities. In 2014, PHF assembled roughly 6,700 member farmers/breeders and various unions or departmental breed associations. In the past few years, the emphasis has been placed on functional criteria, i.e. reproduction and health, without neglecting the currently high-performance level. The various components of the global merit index (called *index de synthèse global* in France, or ISU) that assigns a value to individual sires are the translation of the breeding strategy chosen jointly by the various members of the OS. Thus, during its last revision in 2012, the ISU weights were the following: milk production (35%), morphology (15%), fertility (22%), udder health (18%), lifespan (5%) and milking rate (5%).

In 2010, the breed represented more than 60% of all dairy cows in France, attesting to the strong development of this breed in the country and its hegemony over other dairy breeds. On 1 January 2013, the French Prim'Holstein population included 2,422,000 cows or 31% of all cows (dairy and suckling) in France (Base de Données Nationale de l'Identification, BNDI, data, Idele data processing). Each year, more than 2,500,000 cows are inseminated with a pure Holstein breed (source: Prim'Holstein France). The average milk yield of Holstein cows in France is 9,329 kg of milk in 355 days of production (2014 milk records data, raw values) with an average butterfat content of 39.1% and an average protein content of 31.9%.

This genetic and socio-economic history of the holsteinisation of the French cattle population occurred in a Fordist system of animal selection that we detail below.

Holsteinisation in the dual cooperative-public selection regime: Between common-pool resource and tradable goods

With the 1966 Livestock Act, the French government set up a national centralised selection programme for animal breeds, particularly for cattle, sheep and goat breeds based on an alliance between breeders, researchers (INRA geneticists) and government agencies. For Flamant (2011), the selection of animal breeds 'whose genetic heritage is considered to be of public interest' justifies the 'public investments all along the selection chain' (Flamant, 2011). The period from the 1960s until the early 2000s make up a 'cooperative and public selection regime' (Labatut et al., 2013). The sharing of resources and the 'collegial' management of the breed were at the heart of this regime, whose national genetic policy was heavily funded by the government, in an effort to prevent inbreeding and to ensure that genetic progress was accessible to all breeders across the nation. In the interest of the national economy and food security, genetic progress can be considered a common-pool resource (Allaire et al., 2018). To meet this goal, the law defined the roles of the various partners involved in the selection process. The National Commission for Genetic Improvement (*Commission Nationale d'Amélioration Génétique*, CNAG), whose members are agents from the Ministry of Agriculture, scientists and genetic selection stakeholders, supervised the activities related to the genetic selection policy (validation of UPRA certification, definition of selection targets, and regulation of the sale of semen, etc.).

The government designated INRA and the Institut de l'Elevage to manage the national genetic databases, which are shared among all the breed stake-holders, and constitute a platform of public information. The assessment of sires and brood dams, the calculation of the genetic value of animals (indexes), was delegated to INRA (in addition to its research mission) and to the Institut de l'Elevage. In contrast to the plant selection system, where the targets and performance criteria for varieties are defined by the private companies that produce them, animal selection is carried out in a dual cooperative-public regime, where the selection targets for each breed, translated into genetic indexes that evaluate the genetic merit of each animal based on these targets, are defined collectively within the UPRAs. In this regime, as we have demonstrated elsewhere, 'the private market, cooperative associations and public agencies are not opponents but are collaborative partners' (Labatut et al., 2013). The recognition of the public stakes on these common-pool resources was not actually established until after the implementation of market schemes that ensured the distribution of the benefits from genetic progress and the sustainability of the resource (Labatut et al., 2013). Thus the 1966 Livestock Act regulated the semen market and Artificial Insemination (IA) market by limiting competition, particularly by establishing monopolies according to geographical zone (territories) for AI cooperatives. This territorial organisation of the semen market sought to ensure access to AI services as well as to genetic progress at fair prices for all livestock farmers. AI cooperatives were thus vested with a public service mission for the transfer of genetic progress.

Recognisably, this breeding regime has 'allowed [livestock farmers and cooperatives] to invest in long-term breeding programmes at no risk. Today, it is acknowledged that the implemented strategy, set up to foster cooperation, has paid off' (CSAGAD seminar, 18 October 2006). This breeding regime accompanied the development of the Holstein breed in France, a country characterised by small herds raised in many different types of production systems and local terroirs, features that differ considerably to the countries that have historically exported their Holstein genetic products. Although the French Holstein strain was little exported during the years that followed the 1966 Livestock Act, the French breed had sufficiently progressed to compete with the North American genetics in France. Thus, this dual cooperative-public regime was the background for the development of Fordist-type industrialisation of cattle genetics. On the one hand, the selection industry relied on the interconnection between cooperatives and the 'mass consumption' of genetic progress (common selection targets and thus a genetically uniform product offer, with dissemination of semen by IA services nationwide). On the other hand, collegial socio-economic associations were constituted to represent the stakeholders (UPRAs). As in the Fordist-monopolistic type of industrialisation identified by the theory of regulation (Boyer, 2002), the international trade and outlook of French genetics was poor and the State controlled the market (through the CNAG). In this breeding regime, the tasks of design and execution of breeding programmes were clearly divided between the public institutes that devised the breeding programmes and the genetic evaluation tools (indexes) and the farmer-owned cooperatives that implemented the programmes and marketed the semen evaluated by INRA and the Institut de l'Elevage.

In France, this form of industrialisation contributed to the development of a market for Holstein genetics as in other countries, while participating nonetheless in a certain degree of standardisation and therefore some degree of reduction in domestic biodiversity (i.e. less-productive breeds were abandoned (Audiot, 1995)). Historically, France has enjoyed a large diversity of animal breeds that varies with the locality and traditional regional products. One of the stakes at hand was to avoid an unreserved influx of Holsteins into all the other French dairy breeds used in various economic sectors. Vissac called attention to the decrease in the number of breeds in the early 1970s: 'the number of cattle breeds with more than 100,000 breeding females has dropped from 21 in 1945 to 7 in 1971' (Vissac, 2009: 136). Nevertheless, compared with other countries, the French dual breeding regime has more recently been recognised as acting in favour of diversity:

the French system is unanimously considered to be highly efficient because it allowed, although its initiators did not realise it, the preservation of our livestock animal and breed diversity. This is the strong point in French breeding, due to its history and geography. France is undoubtedly the country with the most breed diversity in the world. The cooperation between stakeholders in selection has fostered work in many species and breeds simultaneously.

(Giroud, 2009)

As we have noted elsewhere (Labatut et al., 2013), this collegial system is linked to the role of the State and professionals from the agriculture sector in the implementation of selection policy. Thus, although the Holstein largely dominates the cattle population, other breeds remain nonetheless firmly rooted in economically important industries (Normande, Montbeliarde, French Brown, Simmental, Tarentaise, Abondance, Vosgienne, etc.). Some of these are even expanding in historically Holstein areas, such as the Montbeliarde due to its more robust features (Courdier et al., 2012).

The genomics selection regime: A trend towards segmented industrialisation?

Since 2006, profound political and technological changes have radically disrupted the animal selection landscape and have led to the emergence of a new breeding regime (Labatut, 2013b; Labatut et al., 2013; Allaire et al., 2018). In 2006, the 1996 Livestock Breeding Act was amended as part of the Agricultural Guidance Law (*Loi d'Orientation Agricole*, LOA) that reorganised the genetic selection infrastructure and the role of its stakeholders.

Several years prior, the territorial monopoly of selection cooperatives had been criticised by the French Competition Council (*Conseil de la concurrence*), which fined[12] the genetics sector for obstruction to fair competition following complaints filed by veterinarians and foreign private operators that wanted to set up business in France. The French government and those involved in implementing the selection policy thus made a move to reorganise the genetics sector through the 2006 LOA, which abolished the territorial monopolies of AI cooperatives and thus encouraged the deregulation of the genetics market, but preserved the democratic access to genetic progress by setting up a Universal Artificial Insemination Service (*Service Universel d'Insémination Artificielle*, SUIA). Each livestock farmer is now free to choose his/her own semen collection centre or semen store. The SUIA also issues calls for tender to ensure coverage in areas with low cattle farm density and the distribution of semen for breeds with smaller population sizes.

The government has decreased its funding of animal genetics activities partly because it has reached the goal set in 1966 (and due to a decrease in agriculture public budget), and French genetics now enjoys the same reputation as its competitors. It now delegates the authority and the responsibility for managing the national selection system to a trade association made up of selection stakeholders (specialised organisations and livestock farmers), 'Livestock Genetics France' (*France Génétique Elevage*, FGE). The responsibilities of the CNAG have been largely reduced and the management of the genetic selection industry is now mainly in the hands of trade professionals. In the 2006 LOA, selection cooperatives have become animal selection businesses (*Entreprises de selection*, ES) and UPRAs have become breeding organisations (*Organismes de selection*, *OS*), which theoretically are more inclusive with regard to livestock farmers that use the breeds selected and managed by the OS. In certain cases (as for the French Brown breed), stakeholders have joined forces to create combined selection organisation-businesses (OESs) that design and carry out the breeding programme. As noted previously (Labatut et al., 2013), a central aspect of this new organisation is that the State nevertheless maintains a monopoly on the production of 'official' indexes mandated to public research and development (R&D) institutes (INRA, Institut de l'Elevage), and the management of the genetic databases remains in the public domain. However, in conjunction with new technologies (to a similar degree as those that occurred when AI was developed), other changes have rapidly taken place and will overwhelm this approach. Getting rid of the territorial monopoly helped accelerate the merging of operators and fusions between cooperatives. In 2015, there were three main animal selection firms with a Holstein genetic selection programme in France: Evolution, Gènes Diffusion and Origenplus.

As of 2009, a radical technology change restructured the organisation of the genetics sector: genomics. This innovation was first developed for the three main dairy breeds (Holstein, Montbeliarde, Normande) based on a new form of public-private partnership (set up in the early 2000s) compared with the previous modes of cooperation for innovation. It involves a consortium between public research and some partner in genetic selection businesses. With DNA chips, genomics can assess almost instantaneously to the genetic potential of an animal, without the multiple steps of progeny testing, for which the value of a bull was only determined when a sufficient number of his daughters had been tested, i.e. after four or five years. We have described how genomic evaluation technology works in more detail in a previous publication (Labatut et al., 2014). The issue here is to identify the various changes that shed light on this new breeding regime and the concomitant change in the genetics industrialisation trajectory.

The complex system of common-pool resources on which selection activities rely in agriculture has two components (Labatut et al., 2013):

- Genetic resources: the genome of the breed population, taken in its entirety (thus difficult to separate and privatise).
- Information resources: the database used to design the selection programme, recording animal performances and genetic indexes. It can be managed by a breeder association or declared a public good (as is the case in France since the 1966 Livestock Act), or developed by private companies (most recent cases).

The changes in the new regime operate on both of these components along with a transformation of knowledge systems due to new technologies (genotyping, sexing, OPU-IVF,[13] etc.). The production, treatment and transfer of data (phenotypes and genotypes) has become a major strategic issue, differentiating among operators in a competitive environment. Genetic data that rely on monitoring a high (but limited) number of animals on farms to determine the genetic value of a sire based on its descendants are produced in conditions that are totally different to those based on genomic data obtained directly at the embryo stage and require state-of-the-art techniques that are generally patented. Therefore, the organisations that produce this data have structured themselves. Consequently, the Labogena laboratory carried out genotyping for almost all the selection stakeholders after it invested in 2008 in an Illumina sequencing platform (Illumina also manufactures DNA chips) for research and genomic selection. This Economic Interest Group (GIE) created in 1994 included, until 2013, INRA and professional members from the selection sector (including the National Union of Animal Breeding and Insemination Cooperatives (UNCEIA), the Federation of Chambers of Agriculture (APCA), the *Races de France* federation of animal breed associations and the Institut de l'Elevage). In 2013, following various financial problems and disagreements on the governance of the laboratory, Labogena was sold and bought by Evolution, a group borne from the merging of several French selection companies, today one of the main cattle selection companies in the world.

In parallel, although INRA still has the regulatory monopoly on the production of official genetic indexes according to the 2006 LOA, part of the indexing service has become a commodity. Thus, companies or regional structures develop and offer breeders their own genomic evaluation tools. *Gène Diffusion*, a selection company in northern France has joined forces with the *Institut Pasteur* in Lille and Wageningen University (NL) to develop its own evaluation system, GD Scan, based on its own criteria (foot health) for the Holstein breed. *Ingenomix*, a biotechnology company created by the French Limousin association has specialised in 'genome-wide association studies between phenotypes and genotypes and in engineering of DNA tests to offer genomic technology',[14] and has developed Evalim®, a private (branded) genetic evaluation tool. Thus, genomic technologies replace the labour-intensive collective and public progeny testing with a private service – that of genotyping animals using DNA chips that provide genomic information on individual animals at a low cost (Labatut et al., 2013). INRA and the Institut de l'Elevage are no longer the only R&D partners for selection stakeholders, although these two institutes still continue to carry out indexing activities on common and historical selection criteria.

Some private companies are increasingly turning to an integrative system that covers the various steps in data production and processing, such as buying out Labogena or projects to incorporate performance testing organisations in selection companies. Thus the National Genetic Database (*Système National d'Information Génétique*, SNIG), previously entirely public, is now in a transition phase where, even if data collection continues to be shared among all members, some parts of regional databases lend themselves to privatisation for use in R&D and genetic evaluation. Thus, some private partners can invite public research institutes to work with them on certain data or selection criteria without sharing and transferring the results to all the stakeholders in selection. Moreover, this integration of data collection with a breeding programme goes far beyond the historical partners. Thus, since the development of genomics, technology for sexing semen has boomed, leading to optimal and profitable use of genomics promoting the production of females in crosses between high-genetic-value animals. An American company, Sexing Technologies, holds the monopoly on this technology for which it has bought all of the patents and is now equipping selection companies around the world. Owing to the profits brought by this technological success, this company is now investing in the genetic selection of sires. As the owner of bulls, it is in a position to sell semen and invests in large experimental farms to produce the amount of data required for selection based on specific differentiation criteria.

Although this technological innovation disrupts, as we have just illustrated, the property rights regime of genetic information, it is also accompanied by important changes in terms of the second component of the common-pool resource system: that of genetic resources and the market for indexed semen. In the former regime, the genetics marketed for each breed was not differentiated (beyond individual variations from one bull to another); the selection targets, translated into ISU (a global merit index) were collegially determined within the breed organisation and common to all selection companies. In the new competitive context of the 2006 LOA and due to the possibilities offered by genomics (possibility to create new 'private' selection criteria independently of the previous, labour-intensive progeny testing method), the selection companies now seek to distinguish themselves from their competitors through a segmented and diversified offer. These companies hire marketing consultants and invest in the creation of a brand image. They engage in studies that identify farmer typologies, and glean 'behavioural segmentations' or user profiles from them to target their genetics products. Thus, in an introductory speech, one of the CEOs of these businesses, used as his catch phrase a citation from Christophe Lafougère (CEO of Gira Food, a market consultancy firm): 'the future lies in segmentation', adding that 'investment in the brand, the image, is truly an investment for the future' (*France Agricole*, 21 May 2014).

These companies are not selling Holstein bulls, but a segmented supply of semen from their 'brand' of Holstein with an image of their own construction. They promote 'cumulative' and not 'corrective' genetic crosses to produce bulls that correspond to specific segments: the 'production' segment, the 'quality' segment, the 'health' segment, the 'endurance' segment. Some businesses have begun to sell 'packs' of bull semen that correspond to these segments. The genetics products are thus no longer centred on the breed or the individual bull but rather on a breeder profile (particularly because in genomics, the bulls are replaced much faster in the product catalogues and are more numerous and thus less well known by farmers).

All of these elements lead us to identify the switch from a mass-market structure in which a low number of 'star' bulls was put on sale for all livestock farmers to a segmented market (in which the 'star' bulls are still featured), the switch from a Fordist-type industrialisation to a flexible industrialisation with the creation of Holstein diversity answering to various segments. Thus this new regime is not fostering a change in animal selection towards a goal of enhanced sustainability but towards segmentation in which some of the segments will target the creation of more productive Holstein cows while others will target more 'sustainable' or 'robust' Holstein cows (e.g. in terms of disease resistance). This outlook, which gave rise to a consensus among several types of stakeholders, supposes that the various models are compatible with the corresponding production.

These observed changes will likely be enhanced by a new profound change in policy: the European Animal Breeding Regulation that is currently in the validation process and planned for implementation in 2018 This text, drafted to 'simplify and align the conditions for sharing data and genetic material between different European countries' (Dantin, speech at the 2015 Paris International Agricultural Show), aims to replace the former selection policy organisations in each Member State in terms of maintaining the herdbooks, implementation of selection programmes, performance testing and genetic evaluation for cattle, sheep, goat, pig and equine species. Reinforcing the deregulation trend for the genetics market that has already begun, this regulation consists in 'moving from a system that is still rather strongly administered through government agencies to a contract-based system' (FGE press kit, 16 January 2015) and covered by its 'own liability regime' (Dantin, 2015 PIAS speech).

The regulation is structured around 'breed societies' that combine the missions of maintaining the herdbook and the implementation of a selection programme, and the related activities of performance testing and genetic evaluation (until now, these activities called for different operators that worked together in synergy in a collegial system). The certified breed societies can thus choose their service providers, for performance testing as well as for genetic evaluation, through calls for tender. This regulation also makes it possible to certify several breed societies for a single breed (until now only one breed organisation was authorised to define the selection targets for the whole breed). Thus, we hypothesise that each breed society, by integrating all the steps involved in selection and breeding and choosing their own selection targets and evaluation index, will participate in the momentum that has initiated differentiation among stakeholders in France since the 2006 LOA and the advent of genomics, and will encourage the development of a polycentric breed authority and increased segmentation of genetics products.

At the time of this writing, the regulations are being debated within the French trade organisations and scientific institutes, with very divergent views. For the commercial trade stakeholders, this regulation will be the opportunity 'to place the breeder at the centre of the system and create the conditions to restructure and streamline the organisations and companies that gravitate around the genetic sphere, to augment the competitiveness of European livestock producers' (FGE press kit, 16 January 2015). An audit carried out on the behalf of the National Livestock Commission in 2016 sees this regulation as the opportunity to change the previous system that was considered by some to be complex and not very dynamic. However, the audit suggests keeping some of the collegial aspects, with a strong inter-professional governance system. For some European countries, this regulation provides the opportunity to set up 'a deregulated landscape, autonomous but under government control, [giving] a freer range to creativity' (Michel Dantin, interview PIAS 2015). The scientists and public institutes who have until now been responsible for the regulatory aspects of evaluation worry that R&D activities will become be uncoordinated and that scientific innovation in service of the breeder will decrease in efficacy over the long term. Each group of stakeholders is working on defining various scenarios of application of the regulation. The coming years will be critical for observing the trends in stakeholder positions and in the practical application of these profound changes that affect the organisation and implementation of selection policies.

Conclusion

This short historical recount of holsteinisation and breeding regimes that accompanied it lifts the veil on the various issues at stake in the management of common-pool resources and new forms of agricultural industrialisation. That which is a 'common-pool resource' in the Holstein as a breed is an intangible good (in the sense that the production of genetic resources is the result of how the flow of produced resources is used) and it can be considered 'uncontrollable'. Given its systemic dimension, this institution (the breed) cannot be controlled by the government, particularly as the current breeding regime, centred on genomics, seems to favour a polycentric system of governance. In the near future, will there be several Holstein breeds within the one and same country? Are we headed to several breed 'brands' and an upheaval of the 'breed' concept? What are the stakes behind the maintenance of breed diversity? Would it be even more threatened by the development and spread of multiple Holstein 'brands' ('long-lived' Holstein, 'rustic' Holstein, etc.) that will perhaps be able to better compete on the same markets as the more rustic breeds with small population sizes? The cultural dimension of traditional local breeds will likely continue to help protect some degree of biodiversity. Genomics is often touted as an opportunity to select for more sustainable animals (Institut de l'Elevage and INRA, 2011). The first observations tend to show that genomics is above all used to accelerate genetic progress and augment the market shares for the Holstein breed for the companies that segment their genetics (thus, the addition of a 'health' criterion may well be accompanied by an increase in the weight given to the 'milk yield' trait in a private composite index).

Although the previous dual cooperative-public model was widely criticised by some 'alternative' breeders involved in movements to promote traditional varieties for the selection of plant resources, the French national scheme being considered too complex (Bessin, 2012), we should reflect on the way these stakeholders will react to the current deregulation trends with regard to selection and flexible industrialisation. What aspects of animal selection will remain in the public domain in the future? Are we experiencing the emergence of initiatives to rehabilitate common-pool resources?

Notes

1 The strong decrease in fertility of the Holstein cattle, as well as the recent emergence of new hereditary diseases, is a sign that inbreeding is becoming a serious threat in the short term.
2 'Pied' is the term used to describe the coat of animals with large patches of two or more colours, one of which is usually white; it is also used to designate the animal itself. Usually used in conjunction with the other dominant, non-white, colour (e.g. black-pied, red-pied, etc.).
3 Source: Prim'Holstein France, http://primholstein.com/.
4 Breed registries of the male and female parents of an animal and their pedigrees. This term can also designate the organisation that is mandated to maintain this registry.
5 Dual-purpose breeds provide good yields of both meat and milk.
6 Source: Prim'Holstein France.
7 Fat content in milk, expressed in grams per kg of milk.
8 Physical appearance of a livestock animal, scored according to production type (dairy, beef or dual-purpose).
9 Formerly called artificial insemination, now called animal insemination.
10 Departmental Livestock Identification Agency.
11 Centre National Interprofessionnel de l'Economie Laitière.
12 *Decision no. 04 D-49 of 28 October 2004 on the antitrust practices in the cattle artificial insemination sector.*
13 Method of harvesting oocytes and in vitro fertilisation used for embryo transfers, consisting in collection of oocytes in a live animal with the aid of an ultrasound probe.
14 www.ingenomix.fr/english.html, consulted on 7 May 2015.

References

Allaire G., Labatut J., Tesnière G., 2018. Complexité des communs et régimes de droits propriété: Le cas des ressources génétiques animales. *Revue d'Economie Politique*, 128(1), 109–135.

Audiot A., 1995. *Races d'hier pour l'élevage de demain*. Versailles, Inra éditions, Collection Espaces ruraux.

Bessin J., 2012. *Maintien de la biodiversité animale domestique: pratiques paysannes et points de vue d'éleveurs sur les obstacles et leviers d'action dans les dispositifs de gestion des races animales*. Mémoire de fin d'étude d'ingénieur. Montpellier, SupAgro.

Boichard D., Bonaiti B., Barbat A., 1993. Effet du croisement Holstein sur les caractères laitiers en population Pie Noir. *Inra Productions animales*, 6(1), 25–30.

Boichard D., Maignel L., Verrier E., 1996. Analyse généalogique des races bovines laitières françaises. *Inra Productions animales*, 9(5), 323–335.

Boyer R., 2002. Aux origines de la théorie de la régulation. In Boyer R. and Saillard Y. (dir.), in: *Théorie de la régulation: l'état des savoirs*, Paris, La Découverte, pp. 21–30.

Colleau J., Tanguy D., 1984. Modélisation de la diffusion des gènes Holstein à l'intérieur de la population bovine Pie Noir Française. *Génétique sélection évolution*, 16(3), 335–354.

Courdier M., Moureaux S., Mugnier S., Gerard A., Gaillard C., Verrier E., 2012. *extension des races bovines montbéliarde et simmental dans l'ouest de la France: motifs et enjeux pour les éleveurs*. Paris, J3R.

Cranney J., 1996. *inra. 50 ans d'un organisme de recherche*. Versailles, Inra éditions.

Denis B., 2010. *Races bovines: histoire, aptitudes, situation actuelle*. Riaucourt, Castor & Pollux.

Duroselle M., 1980. *La Holstein, miracle ou mirage?* Puylaurens, M. Duroselle.

Flamant J.-C., 2011. *La sélection génomique: entre promesses et interrogations.*Castanet-Tolosan, Mission Agrobiosciences.

Giroud J., 2009. *Semences et recherche: des voies du progrès.* rapport au Conseil économique, social et environnemental Paris, Conseil économique, social et environnemental.

Harris B. L., & Kolver E. S. (2001). Review of Holsteinization on intensive pastoral dairy farming in New Zealand. *Journal of dairy science*, 84, E56-E61.

Hess C., Ostrom E., 2003. Ideas, Artifacts, and Facilities: Information as a Common-Pool Resource. *Law and Contemporary Problems*, 66(1–2), 111–146.

Institut de l'élevage Inra., 2011. *La révolution génomique animale.* Paris, France agricole.

Labatut J., 2013a. *Construire la biodiversité: Processus de conception de «biens communs».* Paris, Presses des Mines.

Labatut J., 2013b. Emerging Markets, Emerging Strategies under the Genomic Era. In: *Annual Meeting of the European Association for Animal Production*, Nantes, EAAP.

Labatut J., Aggeri F., Bibé B., Girard N., 2011. Construire l'animal sélectionnable. *Revue d'anthropologie des connaissances*, 5(2), 302–336.

Labatut J., Allaire G., Aggeri F., 2013. Étudier les biens communs par les changements institutionnels: régimes de propriété autour des races animales face à l'innovation génomique. *Revue de la régulation*, 14(2), revue en ligne https://regulation.revues.org/ 10529 (accessed October 23rd 2016).

Labatut J., Astruc J.-M., Barillet F., Boichard D., Ducrocq V., Griffon L., Lagriffoul G., 2014. Implications organisationnelles de la sélection génomique chez les bovins et ovins laitiers en France: analyses et accompagnement. *Inra Productions animales*, 27(4), 303–316.

Labatut J., Bibé B., Aggeri F., Girard N., 2012. Coopérer pour gérer des races locales: conception, rôles et usages des instruments scientifiques de sélection. *Natures sciences sociétés*, 20, 143–156.

Mattalia S., Barbat A., Danchin-Burge C., Brochard M., Le Mezec P., Minery S., Jansen G., van Doormaal B., Verrier E., 2006. La variabilité génétique des huit principales races bovines laitières françaises: quelles évolutions, quelles comparaisons internationales? *In: 13e Rencontres recherches ruminants*, 6–7 décembre, Paris, 239–246.

Pellegrini P., 1999. De l'idée de race animale et de son évolution dans le milieu de l'élevage. *Ruralia*, 5, revue en ligne http://ruralia.revues.org/112 (accessed May 10th 2014).

Ruet F., 2004. De la vache machine en élevage laitier. *Quaderni*, 56(1), 59–69.

Spindler F., 2002. Les races bovines en France au xixe siècle, spécialement d'après l'enquête agricole de 1862. *Ethnozootechnie*, 3, 17–57.

Theunissen B., 2012. Breeding for Nobility or for Production? Cultures of Dairy Cattle Breeding in the Netherlands, 1945–1995. *Isis*, 103(2), 278–309.

Vatin F., 1996. *Le lait et la raison marchande.* Rennes, Presses universitaires de Rennes.

Vissac B., 2009. Une seconde révolution de l'élevage. In Inra, (dir.), *In: Dans les pas de bertrand vissac, un bâtisseur: de la génétique animale aux systèmes agraires*, Paris, Inra, pp. 136.

9 Transitions towards a European bioeconomy

Life Sciences versus agroecology trajectories

Les Levidow, Martino Nieddu, Franck-Dominique Vivien and Nicolas Béfort

Introduction

Since at least 2005, inter-governmental organisations have promoted a 'bioeconomy' as crucial for societal progress. They imply that a bioeconomy already exists, linking natural processes with new biotechnologies, which thereby warrant extra support to achieve further advances. A bioeconomy denotes 'the aggregate set of economic operations in a society that use the latent value incumbent in biological products and processes to capture new growth and welfare benefits for citizens and nations' (OECD, 2006: 1).

Priorities for a bioeconomy vary across countries (Levidow, 2015a). Although the US policy framework mentions diverse future products (White House, 2012), its R&D funds have prioritised efforts to turn lignocellulosic biomass into liquid fuel, known as second-generation biofuels. With a broader scope, the European Commission relaunched the Life Sciences as essential tools for a Knowledge-Based Bio-Economy (DG Research, 2005). Its KBBE vision combines environmental sustainability with economic advantage through more flexible, eco-efficient uses of biomass.

The EU's KBBE itself has been framed in various ways. In the dominant agenda, natural resources offer renewable biomass amenable to conversion into industrial products via a diversified biorefinery, thus horizontally integrating value chains (Becoteps, 2010). Environmental sustainability becomes dependent upon markets to stimulate technological innovation. In an alternative agenda, diverse knowledges inform agroecological methods as the basis for a truly knowledge-based bioeconomy (e.g. Niggli et al., 2008).

In this chapter, we analyse the following questions: What forces drive diverse agendas for a European bioeconomy? What are their inter-relationships, e.g. symbiotic, competitive or even antagonistic? How does each one link food and non-food products? How does it link food, energy and other sectors?

To explore those questions, the chapter examines debates around a transition towards a bioeconomy. As above, this concept has at least two trajectories: the dominant one drawing on Life Sciences and a marginal one drawing on agroecology. This chapter explores how the different bioeconomy trajectories relate to knowledge production and product quality as dual aspects which together can shape different futures. The first section introduces the theoretical concept of 'food regime', its potential 'greening' and its precursors in early concepts of a bioeconomy. The second section introduces the European Commission's bioeconomy agenda encompassing divergent trajectories, further analysed in subsequent sections – Life Sciences (third section) and agroecology (fourth section). In conclusion, the fifth section summarises prospects for a change in the dominant regime through forms of bioeconomy.

Regime transitions: What innovation trajectories?

Across the history of agricultural change, a food regime in general has been defined as a 'rule-governed structure of production and consumption of food on a world scale' (Friedman, 1993: 30–31). This global perspective identifies a hegemonic role played by an historical series of three global regimes (McMichael, 2009), despite significant national variations within each regime (Wilkinson and Goodman, chapter 7). Starting from the post-1990s regime, this section discusses current variants and alternatives, innovation paradigms potentially realising such alternatives and early bioeconomy conceptions as precursors of today's trajectories.

Greening the neoliberal agro-industrial regime?

Corresponding to wider neoliberal policy frameworks, since the 1990s, the dominant food regime has promoted 'liberalization of trade and empowerment of transnational corporations' (Friedmann, 1993: 55). The current regime replaced the post-War regime of national markets, which was increasingly in tension with export surpluses and differentiation of food markets. In the 1980s and 1990s, the debate on a successor regime arose from rule changes, especially in the Uruguay Round of the General Agreement on Tariffs and Trade (GATT). 'The implicit rules evolved through practical experiences and negotiations among states, ministries, corporations, farm lobbies, consumer lobbies and others, in response to immediate problems of production, distribution and trade' (Friedmann, 1993: 31). Through pressures for liberalisation of agricultural policies and markets, the GATT framework was superseded by the World Trade Organization in 1994.

In the North the agro-industrial regime has subsidised intensive production methods for surpluses that can be globally exported and undermine less-intensive methods elsewhere. This pressure has pushed farms everywhere to adopt intensive methods, expelling rural populations to cities, especially shantytowns of the Global South (Friedmann, 1993, Friedmann et al. 2016).

This neoliberal regime has served to globalise the agro-industrial methods that had already become prevalent in the US, Europe and parts of Brazil by the 1990s. Such methods had already involved technological change, e.g. substitution of animal energy by fossil energy, substitution of animal waste by chemical fertilisers, in turn facilitating national specialisations such as links between soya and animal feed for intensive livestock. Agribusiness developed transnational circuits among such national specialisations within such value chains.

What drivers and prospects for the neoliberal agro-industrial regime to undergo change? The potential comes from tensions within the regime, as prophetically described just prior to the WTO:

> These prefigure alternative rules and relations. One is the project of corporate freedom contained in the new GATT rules. The other is less formed: a potential project or projects emerging from the politics of environment, diet, livelihood, and democratic control over economic life.
>
> (Friedmann, 1993: 51)

In this perspective, a regime's trajectory is an endogenous result of collective action. Such a transition 'can be defined as a period of unresolved experimentation and contestation', and that such periods 'are full of multiple possibilities' (Friedmann, 2009: 335–336). The new dominant regime poses threats that have provoked ecological, social, ethical, culinary and cultural contestations from diverse movements such as SlowFood, La Via Campesina, indigenous peoples, etc.

Some analyses have focused on strategies for 'greening' agro-food systems and supply chains as the potential basis of a new agro-food regime, with diverse potential trajectories. For example, 'green' industrial farms could replace agrochemical inputs with permissible bio-inputs for 'organic' certification, as happened in California (Guthman, 2004). By contrast, social movements promote new forms of production and consumption, linked through demands for food sovereignty, thus reducing market transactions and undermining capital accumulation. In some early visions of biofuels, non-food uses of biomass could support strategies for local energy autonomy and ecological re-industrialisation. As the dominant pattern, however, biomass energy extends global value chains of Malaysian palm oil and Brazilian sugar cane for biorefineries in Europe or North America (Nieddu et al., 2014).

The above corporate strategies can potentially overcome social or environmental tensions through an agro-food green capitalism, theorised as a 'corporate-environmental food regime' (Friedmann, 2005). Through capital-intensive processes, food ingredients are decomposed and then recombined for food or non-food products, drawing on knowledge of such recombinations. Second, markets are differentiated, e.g. between high-quality fresh products and chemically recombined ingredients. (Friedmann, 2005: 258).

This incipient regime shifts agro-industrial production methods in ways that reduce harmful environmental effects and accommodate consumer demand for 'green' products. A regime shift depends on higher quality standards, some of which were associated with alternatives to the dominant regime, e.g. organic food and functional foods. To be commercially successful, the change must devise new norms for product identity and quality. Some businesses maintain such a capacity from their power over global supply chains, as a basis to impose environmental and quality standards that can gain consumer support.

Through market differentiation, there have been efforts to market 'food from somewhere', by contrast with the dominant regime of 'food from nowhere', i.e. capital-intensive agro-industrial production (Campbell, 2009). As a different strategy, organic farming has become conventionalised in many places. Within food regime theory, some perspectives have emphasised prospects for the dominant regime to incorporate such alternatives, which thereby play symbiotic roles, for example as competitive options in the food marketplace.

But such a symbiotic perspective downplays tensions of many kinds. Despite its alternative products, the dominant food regime encounters socio-political resistance from agendas for agro-food localisation and its different quality basis. Non-commercial solidarity aims, alongside multifunctionality, have symbolised the *altermondial* agenda. The latter spatially reorganises rural production; this enhances ecological sustainability by protecting 'endangered food, biodiversity and local traditional knowledge' (Fonte, 2013). Such tensions between ideal-type trajectories lie within an incipient corporate-environmental food regime.

Innovation paradigms of knowledge and quality

Given the many strategies for diversifying food systems, how do they complement or contest the neoliberal agro-industrial regime? These dynamics can be explored through concepts of product quality and their knowledge-bases. Drawing on the generic concept of technoscientific paradigms (Dosi, 1982; Malerba, 2002), theoretical typologies distinguish between innovation paradigms of food systems, as follows (see Table 9.1).

As an overview, bioeconomy agendas can be theorised as two ideal-types of innovation. Along with strong techno-economic promises, one type characterises and develops compositional identities of semi-finished or intermediate products at various stages of value chains, e.g. through traceability or life-cycle analysis of bio-based chemical products from decomposable biomass. The other type constructs an integral identity of a product – including a territorial identity and labels, producer-consumer proximity, engagement with ecological processes, etc. Each pathway constitutes food through different forms of quality.

More specifically, Life Sciences emerged from 'new biotech' companies, in turn dependent on specific complementarities between institutionalised forms of science, Intellectual Property Rights and finance (Coriat et al., 2003). Life Sciences aim to modify plants to enhance productivity in adverse conditions, e.g. caused by pests, pathogens, drought, saline environments and unfertile soils, or to design plants for new objectives such as altered nutritional content or carbon chains for non-food uses (Vanloqueren and Baret, 2009: 972). These aims are often linked with a decomposability paradigm, identifying single traits or functional attributes (based on genetic characteristics) which can be extracted, decomposed and recomposed. Emphasising computable data, technoscientific knowledge seeks to characterise such components, for selectively recombining them into novel products (Allaire and Wolf, 2004). Through its various names, such as green or blue biotech, innovation agendas seek to link food and non-food uses within the same paradigm.

Table 9.1 Divergent agendas of a European bioeconomy (each agenda combines paradigms from the upper part of the table)

	Dominant agenda	Marginal agenda
Agri-innovation paradigm		
Technoscientific (Vanloqueren and Baret, 2009)	Genetic engineering and Life Sciences: modifying plants and animals for greater productivity or for new objectives, e.g. nutritional content.	Agroecological engineering: designing agricultural systems that minimise need for external inputs, instead relying on ecological interactions.
Quality (Allaire and Wolf, 2004)	Decomposability (via converging technologies) for re-composing qualities into novel combinations for extra market value.	Integral product identity via holistic methods and quality characteristics recognisable by consumers, as a basis for their support.
Knowledge (Allaire and Wolf, 2004)	Computable data for novel inputs and/or outputs that can gain market advantage, especially by matching compositional qualities with consumer preferences.	Knowledge systems for validating comprehensive product identities, e.g. organic, agroecological production methods, territorial characteristics, specialty products.
R&D agendas	**Life Sciences**	**Agroecology**
Problem-diagnosis: agro-economic threats to be overcome	Inefficient production methods disadvantaging European agro-industry, which falls behind in global market competition for technoscientific advance.	Agro-industrial monoculture systems – making farmers dependent on external inputs, undermining their knowledge, distancing consumers from agri-production knowledge, etc.
Solution in sustainable agriculture	More efficient plant-cell factories as biomass sources for diverse industrial products, thus substituting for fossil fuels and expanding available resources.	Agroecological methods for maintaining and linking on-farm resources (plant genetic diversity and bio-control agents), thus minimising usage of external resources.
Intensification of renewable resources	More efficient biomass conversion from lab knowledge and more decomposable qualities.	Eco-functional intensification via farmers' knowledge of agroecological methods.
Agri-energy linkages	Redesigning plants and processing methods for more efficiently converting biomass	Converting agricultural waste into bioenergy in on-farm small-scale units, thus substituting for external inputs.

(Continued)

Table 9.1 (Cont.)

	Dominant agenda	Marginal agenda
	into energy and other industrial products.	
Knowledge-Based Bio-Economy (KBBE)	Sustainable production and conversion of biomass (or renewable raw materials) for various food, health, fibre, energy and other industrial products.	Agroecological processes, in mixed and integrated farming, for optimising use of energy and nutrients, so that producers gain from the value that they add.
Scientific knowledge	Standard databases of lab knowledge (from converging technologies, esp. genomics) to integrate agriculture with other industries.	Scientific research to explain why some agroecological practices are effective, as a basis to intensify and apply them more widely.
Product quality	Verifiable compositional changes for better nutritional content, agronomic characteristics and/or extractable substances.	Sustainable cultivation methods and/or territorial identity recognised by consumers via food distribution systems.

Source: adapted from Levidow et al. (2013).

In a paradigm of decomposability, innovation identifies simple traits to convert biomass into intermediate products, such as bottled ethylene from biosourced bioplastic, by contrast with fossil-based biomass, which differs only in the age of the carbon utilised. Highlighting calculable facts, technoscientific knowledge characterises these semi-products (e.g. through life-cycle analysis), to be recombined selectively in global value chains. Although each entreprise may develop commercially confidential information, the construction of quality here depends on comprehensive standardised information about compositional characteristics and norms (e.g. 'bio-based content' in Europe or ASTM D6866 in the US). Decomposability must be supported by collective access to up-to-date databases (cf. Debref, 2014).

Diverse greening efforts have other trajectories. Some actors seek to establish novel products from local resources for non-food uses, sometimes by rediscovering the potential of neglected plants in a period of agriculture standardisation. 'Bioeconomy in Champagne-Ardenne', a regional project linking research and food industries, seeks to revive peasant agroecological knowledge of traditional plants such as high-protein vegetables. Other projects develop a 'doubly green chemistry' for utilising the complex structures provided by nature, rather than cracking agricultural feedstock as in an oil refinery (Nieddu et al., 2014).

The former strategy attempts to produce the same outputs, e.g. fuels and intermediate chemicals, which otherwise would come from fractionating oil.

This strategy comes from extending collective production heritages, either traditional or modern, such as fractionation techniques for decomposing feedstock. For example, the 19th century Fischer-Tropsch thermochemical process degraded coal into small molecules for biogas production; this heritage was later extended through thermochemical pathways for biodiesel or biochemical pathways for ethanol. More generally, collective production heritages are cognitive tools that create a basis for learning between producers and users.

Although today's innovations are portrayed as radically novel (e.g. catalysis, biotechnology), they necessarily rely on those earlier production heritages (Nieddu et al., 2014; Nieddu and Vivien, 2016). Building on these, innovators seek to make transition pathways economically and environmentally viable. For example, they seek a localised territorial production and/or a redesign minimising environmental burdens at the end of each product's life. As the overall context of research agendas, there has been a lock-in of genetic engineering and related 'new biotechs', alongside a lock-out of agroecology (Vanloqueren and Baret, 2009).

By contrast with the former, agroecology aims to redesign agricultural systems to minimise dependence on agrochemicals and energy inputs. Farms develop agroecosystems, whereby greater biodiversity performs various ecological services within and beyond food production. Such services include recycling nutrients, regulating microclimate and local hydrological processes, suppressing undesirable organisms and detoxifying noxious chemicals (Altieri, 1999; Nicholls et al., 2016).

Ecological interactions among biological components enable agricultural systems to boost their own soil fertility, productivity and crop protection (Vanloqueren and Baret, 2009: 972). In practice this is linked with an integral product identity paradigm, seeking to valorise distinctive comprehensive qualities that can be socially validated for/by consumers in various forms, e.g. organic certification, territorial characteristics, specialty labels or farmers' markets (Allaire and Wolf, 2004).

The agroecology agenda is more than an ensemble of techniques. It has three different aspects: agroecology of production systems in the strict sense, applying Odum's principles of systematic ecology; agroecology of alternative food systems; and agroecology as knowledge of relationships between agri-production and society (Van Dam et al., 2012: 27). This corresponds with another tripartite definition, namely, agroecology as a scientific discipline, an agricultural practice and a social movement. Linking those three forms is essential for transforming the dominant food system (Wezel et al., 2009: 28).

As a different ideal type, a paradigm of integral product identity seeks to produce social innovations that help to achieve a systemic coherence, through standards fulfilling claims for a moral economy (Busch, 2000). For example, if taken separately, organic certification or food localisation do not fulfil this paradigm. Actors want organic and agroecological agriculture to facilitate the construction of food sovereignty and locally sustainable trajectories, with products accessible to low-income social strata (Goodman and Goodman, 2009; McEntee, 2010). Such aims need a reflexive localism in the process of boundary-making and object design, where the process can be more important than the consequent standards or conventions (Fonte, 2013).

In such a reflexive perspective, the 'bio' is a site of new alliances between producers, distributors and social movements (see Chapter 10). The latter can hold products accountable for marketing claims, e.g. about 'preserving the family farm' or providing public goods. Identity-based differentiation depends upon collective resources for accommodating diverse private demands and public norms (Allaire and Wolf, 2004: 449, 454).

In that sense these socio-technical innovations constitute an integral product identity; this seeks to valorise distinctive comprehensive qualities which can be socially validated for/by consumers in various forms, e.g. organic certification, territorial characteristics, specialty labels or farmers' markets. Comprehensive-identity supply chains can valorise agroecological methods: '*Agroecologists favour alternative food systems operating at a regional scale or based on closer farmer-consumer relationships, or product networks that mobilise localized resources and have strong identities*' (Vanloqueren and Baret, 2009: 981).

Likewise the above tensions arise for non-food products: through life-cycle analysis or principles of green chemistry, efforts towards a territorial production seek conditions acceptable for ecosystem protection at the end of each product's life, but separate criteria fail to bring a systematic coherence. Tensions also arise around rival forms of knowledge from collective production heritages (Nieddu and Vivien, 2016, as above).

Section 2 will use those typologies to analyse the two bioeconomy agendas – Life Sciences and agroecology. Both have precursors in earlier concepts of a bioeconomy, as explained next.

Before the EC's bioeconomy, two other conceptions

From the history of 'bioeconomy' agendas, the earliest conception has contributed to ecological economics (Vivien, 1998) and somewhat to agroecology. Georgescu-Roegen became known for his work on peasant economy, contrasted with unsustainable energy in industrial agriculture (Vivien, 1999): 'the survival of humanity poses a totally different problem than any other species because it is not only biological nor only economic; it is bio-economic' (Georgescu-Roegen, 1975: 130; also 1971). Rene Passet's book, *L'économique et le vivant*, presents a renown schema of three spheres whereby the economic sphere lies in a of the social sphere, itself a sub-system of the biosphere; biological cycles become integrated at the heart of economic reason (Passet, 1979: 11, 2011).

The second conception is the biotech revolution. From the discovery of the DNA triple helix in 1953, further research elucidated the regulation of protein synthesis in 1961, enzymes' capacity to dissect the DNA molecule in a pre-dictible pathway in 1962, and isolation of the gene in 1969. This discovery was quickly understood – by Monsanto since 1972 – as not only a paradigmatic revolution in biology but also a great Schumpeterian rupture in pharmacy, medecine, agrosciences and chemistry. European Commission strategists warned that Europe was missing the global opportunity:

The Industrial Revolution took place in Europe, but the promises of Bio-technology and of its spin-offs were gradually moving away from our European horizons to the USA and some emerging economies. European leaders realised that Europe was facing a maybe unique chance to support its science base and to develop the potential represented.

(Aguilar et al., 2013: 10)

In the name of fulfilling the promise, public policy has been mobilised for research programmes and 'technology-driven initiatives' to overcome disciplinary frontiers and path dependencies.

Such support measures have made the techno-economic agenda more visible, thus provoking controversies. It illustrates a wider 'economics of technoscientific promises', which facilitate investment, mobilisation, circulation, and accumulation of resources (Joly, 2010). Such promises instrumentalise the Life Sciences in the service of explicitly seeking to industrialise biology (NRC, 2015). A key aim has been intellectual property rights (Birch et al., 2010), by commoditising knowledge through new markets for technology (Arora et al., 2001; Birch, 2017). Technological promises led policymakers to promote an institutional change granting property rights to 'biotechnological innovations', (EC, 1998), thus broadening the scope of discoveries or techniques that could be privatised. Such broader claims for intellectual property have led to controversy over its ethical and economic aspects: Should Life be patented? Who owns genes?

In those ways, the Life Sciences agenda promotes a linear model of innovation, whereby new products apply basic research for commercial uses. This agenda links technoscientific advance, their mass-media promotion and an economics of promises, including intellectual property rights. The latter is inscribed in various legitimation devices, e.g. validation by venture capital, alliances with incumbent firms or public support through technoscientific research programmes.

Bioeconomy: divergent agendas

The term 'bioeconomy' encompasses at least two different conceptions. The Life Sciences conception has gained a prominent role in the bioeconomy agenda of the European Commission since 2005; this exemplifies efforts towards a corporate-environmental food regime (Friedmann, 2005). Critical responses led to an agroecological conception of bioeconomy. After a brief survey below, each conception is elaborated in subsequent sections.

Life Sciences globally flexibilising agro-industries

Extending 'the new biotechs', the dominant trajectory of a bioeconomy reconceptualises agro-industrial production as decomposable and recomposable biomass. As an early symbol of a future bioeconomy, edible biomass was converted to biofuels; controversy provided a stimulus for greater ambition. The Life Sciences agenda redesigns and converts food, feed and non-food biomass into diverse industrial products, towards horizontally integrating various industrial sectors, while substituting for fossil fuels in the name of sustainably replacing fossil fuels. (OECD, 2006, 2017). It promotes flexibility of biomass feedstocks – their sources, types, conversion processes and end products – especially through novel biorefineries. R&D seeks more efficient techniques for converting biomass to cellulosic bioethanol and other industrial products, while also expanding opportunities for proprietary knowledge (Murphy et al., 2007).

According to a later report by the World Economic Forum, biorefinery strategies anticipate a competitive advantage for companies becoming 'backward-integrated' into multiple feedstocks and flexibly converting them into multiple products:

> The newly established value chain will have room for non-traditional partnerships: grain processors integrating forward, chemical companies integrating backwards, and technology companies with access to key technologies, such as enzymes and microbial cell factories joining them.
>
> (WEF, 2010: 20)

More flexible uses will give the Global South greater business opportunities to supply raw materials:

> a new international division of labour in agriculture is likely to emerge between countries with large tracts of arable land—and thus a likely exporter of biomass or densified derivatives—versus countries with smaller amounts of arable land.
>
> (ibid.: 21)

Complementing this globally flexible division of labour, a Life Sciences trajectory envisages a future 'value web', developing more flexible value chains through more interdependent, interchangeable products and uses, thus promoting horizontal integration of industrial sectors (Becoteps, 2010). Various examples provide flex-crops and flex-commodities, whereby raw material suppliers can be thrown into more intense competition for supplying upper parts of value chains (Borras et al., 2016). Such integration shifts power relations towards global markets and land uses serving such markets.

Tensions of the European bioeconomy agenda

Towards a European bioeconomy, the prevalent agenda claims to promote a 'holistic approach' – meaning especially Life Sciences, including genomics within converging technologies such as infotech and nanotech. This trajectory promises to enhance environmental sustainability, global economic competitiveness and thus European prosperity, according to the European Commission's narrative of technoscientific promise. This would flexibly accommodate rising global demand for food, feed and fuel, thus promising to alleviate constraints on natural resources (DG Research, 2005).

Yet tensions among bioeconomy perspectives have arisen during conferences, bringing together all stakeholders to formulate a technological roadmap. For alternative agendas, holistic means articulating between agroecological practices and industrial eco-design. In contrast, for Life Sciences, holistic means flexibly integrating material flows across industrial sectors.

Prevalent trajectories depend on a knowledge base of decomposability, conversion and recomposition through biochemical pathways: other trajectories develop thermochemical techniques (see Section 3). Both neglect food production. A different 'bioeconomy' trajectory seeks to preserve and improve extensive agriculture by linking different agro-innovation pathways; this links agroecology with an integral product identity for environmental sustainability and food quality (Levidow et al., 2013; see fourth section).

Those divergent trajectories have been accommodated within a multi-stakeholder compromise, the Knowledge-Based Bio-Economy (KBBE). This framework has extended the Lisbon agenda, which sought greater R&D investment in a knowledge-based economy to make Europe '*the globally most competitive knowledge-based economy by 2010*' (European Council, 2000). The European Commission defines the KBBE as '*the sustainable, eco-efficient transformation of renewable biological resources into health, food, energy and other industrial products*' (DG Research/ FAFB, 2006: 3). Food and non-food trajectories have been kept within the same agenda. On the one hand, EC policy documents emphasise food in order to portray 'the bioeconomy' as already enormous; on the other hand, R&D funds have gone mainly to Life Sciences, especially for non-food products (Schmid et al., 2012).

The next two sections examine the Life Sciences and agroecology agendas in more detail.

Life Sciences-based bioeconomy

In the Life Sciences trajectory of bioeconomy, agriculture becomes 'oil wells of the 21st century' (BiomatNet, 2006). This metaphor of biomass as 'biocrude', as feedstock for a biorefinery, naturalises the decomposability paradigm. 'The seed oils of plants are structurally similar to long chain hydrocarbons derived from crude oil' (EPOBIO, 2007: 10). This trajectory seeks to mimic an oil refinery cracking oil, as a basis for a like-for-like substitution of petrol products by building blocks from biomass (Nova-Institut, 2017). This follows the US roadmap of the Top Ten (Bozell and Petersen, 2010): '*New developments are ongoing for transforming the biomass into a liquid "biocrude", which can be further refined, used for energy production or sent to a gasifier*', according to an antecedent of the European Biofuels Technology Platform (Biofrac, 2006: 21).

Life Sciences emphasise promises that have been impeded by non-food 'biomass recalcitrance', to be overcome through biotech innovations, e.g. novel crops or microbial enzymes for easier biomass conversion. Some define bioeconomy as converting biomass into building blocks that can substitute like-for-like the products from oil for non-food uses such as energy, chemistry and biofuels (McCormick and Kautt, 2013). Other trajectories seek to substitute the functions brought by these products, rather than strictly identical renewable carbon substituting for chemical structures and fossil carbon (Colonna et al., 2015).

Alongside that decomposition-recomposition paradigm, the European context has alternative trajectories of a bioeconomy. Some favour extensive cultivation methods to provide biomass feedstock for 'biorefineries without biofuels' or alternative value chains (Gallezot, 2010). For example, Italy's Novamont company seeks feedstock for biorefineries producing chemicals, thus substituting for oil; novel plants are locally grown on poor soils, thus complying with principles of a circular economy (Nieddu et al., 2013; Bastioli, 2008).

In this Life Sciences agenda, knowledge production and economic activity function differently than in the 'biotech revolution'. Central to this is a Great Transition towards renewable resources for energy, chemicals and materials through a biorefinery, a transitional technological object. Although the biotech revolution has great narrative power, it does not unify this agenda because its actors seek to keep open their options, including alternatives to thermochemical or biochemical conversion trajectories.

Diversified biorefinery for horizontally integrating industries

To address many limitations of the early products, especially conventional biofuels, bioeconomy visions have promoted the concept of biorefinery producing second-generation biofuels (Banse et al., 2011; CEC, 2012; Huang et al., 2012). This would convert non-food components of plants, or non-food plants or from waste. Looking beyond biofuels, the European Biofuels Technology Platform develops strategies to optimise valuable products from novel inputs. It requests funds to '*develop new trees and other plant species chosen as energy and/or fiber sources, including plantations connected to biorefineries*' (EBTP, 2008: SRA-23).

More ambitiously, the 'integrated diversified biorefinery' has been envisaged to diversify inputs and outputs, especially through novel enzymes and processing methods, generating diverse by-products including biofuels:

> the integrated diversified biorefinery—an integrated cluster of industries, using a variety of different technologies to produce chemicals, materials, biofuels and power from biomass raw materials agriculture—will be a key element in the future. And although the current renewable feedstocks are typically wood, starch and sugar, in future more complex by-products such as straw and even agricultural residues and households waste could be converted into a wide range of end products, including biofuels
>
> (EuropaBio, 2007: 6)

This seeks horizontal integration of agriculture with the oil, chemical and transport industries, thus optimising the market value of resources and intellectual property. Inputs and outputs can be flexibly adjusted according to temporary market advantage, thus throwing suppliers into greater competition with each other and intensifying agri-production systems.

According to a lobby group for biofuel innovation, a successful diversified biorefinery depends on government subsidies for research and development and demonstration (R&D&D) plants. According to the European Biofuels Technology Platform, the necessary investment is too costly and commercially risky for the private sector, which therefore requests much more public funds to cover the risks. Testing commercial viability requires an expensive scale-up: 'With an estimated budget of €8 billion over ten years, 15–20 demonstration and/or reference plants could be funded' (EBTP, 2010: 26).

This vision justified allocating €4.7 billion to the bioeconomy in Horizon 2020, the EU's research framework for 2014–20, as well as potential diversion of other funds. '*Various funding sources, including private investments, EU rural development or cohesion funds could be utilised to foster the development of sustainable supply chains and facilities*' (CEC, 2012: 7). Towards future 'advanced integrated biorefineries' that could compete with fossil counterparts, the EU's Framework Programme 7 funded numerous projects totalling €50m and involving 68 European partners between 2010 and 2014. All these depend on an economics of technological promises: '*Biorefineries converting feedstock into chemicals and materials will become the backbone of the future production of sustainable products*' (Horizon 2020 call BBI.2017.F1, 2017).

To mobilise such investment, the model of innovative start-up depends on promises to become the 'Google' of the bioeconomy, in turn justifying broader intellectual property rights as a prerequisite. Yet any such ambition has institutional constraints. Given the large fixed-capital commitments and the complexity of knowledge to be integrated (Dubois, 2011, 2012), biorefinery development starts from agro-industrial sites (e.g. Pomacle-Bazancourt) or paper mills in Scandinavian countries. These depend on shared knowledge through open innovation platforms. These industrial complexes imply a complementary relation between food and non-food; in particular, as a co-product of biofuels, animal feed contributes to the economic viability of biorefineries. Yet such co-products have attracted criticism for disguising waste which potentially harms animal health, as well as for potentially supplementing the animal feed supply (cited in Levidow, 2015a).

Novel food recomposition trajectories

A decomposability innovation paradigm likewise informed the food industry's early research priorities and their incorporation into Framework Programme 7. The European food industry federation (CIAA) has led a European Technology Platform (ETP) Food for Life. Its research agenda has sought to link food innovation with future markets: 'consumer demands will drive the R&D and innovation needs' (FfL, 2005: 13; also FfL, 2007: 6). To avoid difficulties in marketing, it is '*essential to build effective systems of product tracing and identification that consumers can have trust in*' (FfL, 2007: 6).

To overcome consumer resistance to novel food products, the food industry has combined technoscientific innovation with appeals to 'natural' foods: '*Most of the novel food processing technologies carry the promise to deliver safe food without sacrificing naturalness and nutritional benefits*' (FfL, 2005: 25). Yet 'naturalness' claims became difficult to justify, so the concept is almost absent in subsequent documents.

Later such a claim was revived as '*less refined, more natural food ingredients to be used in minimal or gentle processing*'. As it turns out, this meant '*food ingredients by tailored fractionation of the raw material into classes, which are not pure isolates, but which consist of mixtures of structures and components with very good functionalities*' (FfL, 2016: 57). This trajectory must be somehow reconciled with consumer understanding of naturalness in order to gain commercial success.

Public health claims about novel products have been likewise a task for technoscientific innovation: '*These products, together with recommended changes in dietary regimes and lifestyles, will have a positive impact on public health and overall quality of life*' (FfL, 2008: 3). However, further research was needed to support such health claims: '*Therefore knowledge built up in the priority areas is aimed at reformulating a wide range of foods and designing new foods, and making them eligible for health claims*' (ibid.: 18). The industry also advocates life-cycle analysis of food production, e.g. to demonstrate eco-efficiency benefits in using natural resources, as a basis for informed choices by consumers (ibid.: 40).

To clarify public health benefits of novel foods, the agenda seeks to extend existing databases. These already had linked '*the composition and biological effects of nutrients and non-nutrients with putative health benefits*' (FfL, 2008: 51). By combining several information sources, a key aim has been '*harmonised national databases on food composition and consumption patterns, including ethnic and traditional foods*' (ibid.: 32).

Indeed, the food industry faces demands for locally familiar and speciality foods, based on an integral product identity. In response, the industry has sought to incorporate food traditions into technoscientific innovation: '*The integration of the rich traditions of European cuisine with the innovation-driven market place represents a great and constant challenge*', which can be addressed through '*innovation in and industrialisation of regional gastronomy*' (FfL, 2005: 9, 22), especially by '*using modern media and new digital technologies*' (FfL, 2016: 8). Thus its research agenda attempts to appropriate consumer desires by translating the product-identity paradigm into the decomposability paradigm (Levidow et al., 2013).

At the same time, the innovation agenda ruptures any food tradition. Through recomposition techniques, it optimises functional foods 'towards achieving the right metabolic effect' (FfL, 2017: 37). It also seeks to synthesise food components 'from non-food materials or the use of non-traditional resources such as insects or microalgae'. These sources provide means to overcome resource limits and thus make claims for environmental sustainability (FfL, 2017: 10).

In all these ways, technoscientific innovation remains central. Indeed, it is a solution searching for a problem: '*Many of the weaknesses identified could be solved technologically*' (FfL, 2008: 7). Pervasive tensions arise around claims for health or environmental benefits vis a vis unprocessed foods, which can retain an integral product identity.

Agroecology-based bioeconomy

Partly in response to the dominant agenda, organic agro-food organisations formed a stakeholder network to advocate organics and agro-ecosystems research for a 'knowledge-based bioeconomy' (Ifoam-Europe, 2006). They built broad stakeholder support including relevant commercial actors across the agro-food value chain as well as environmental NGOs.

Eventually they published a *Vision for an Organic Food and Farming Research Agenda to 2025* (Niggli et al., 2008), with the aim to set up Technology Platform Organics.

This was followed by a *Strategic Research Agenda*, which linked the term 'innovation' with public goods, efficiency, farmers' knowledge, learning and competitive advantage. It elaborated 'eco-functional intensification':

> The weakness of organic agriculture so far remains its insufficient productivity and the stability of yields. This could be solved by means of appropriate 'eco-functional intensification', i.e. more efficient use of natural resources, improved nutrient recycling techniques and agroecological methods for enhancing diversity and the health of soils, crops and livestock.
>
> (Niggli et al., 2008: 34; cf. Schmid et al., 2009: 59)

Horizontal integration between agriculture and energy production, partly from waste materials, provides means to shorten organic cycles as well as to substitute for external inputs:

> Diversified land use can open up new possibilities for combining food production with biomass production and on-farm production of renewable energy from livestock manure, small biotopes, perennial crops and semi-natural non-cultivated areas. Semi-natural grasslands may be conserved and integrated in stockless farm operations by harvesting biomass for agro/bioenergy and recapturing nutrients from residual effluent for use as supplementary organic fertiliser on cultivated land.
>
> (Schmid et al., 2009: 26)

This strategy develops new knowledge for a reflexive localism (cited above) around multi-stakeholder alliances broader than organic producers: '*Stakeholders along the whole food chain are able to participate in this development and civil society must be closely involved in technology development and innovation*' (Schmid et al., 2009: 16). The research strategy emphasises cooperation among all stakeholders in producing knowledge: '*The joint production of knowledge model transgresses the boundary between knowledge generators and users, so that* all *partners involved may be undertaking research*' (Padel et el., 2010: 58).

Indirect support came from changes in research policy. As a new opportunity for agroecological agendas, the EU's Food, Agriculture, Fisheries and Biotechnology (FAFB) research programme hosted expert foresight studies exploring wider knowledges for agricultural innovation. The exercises were commissioned by the EU's Standing Committee on Agricultural Research (SCAR), with support from some national agencies promoting farmers' knowledge. According to the first expert report, farmers often develop modest innovations but these are readily dismissed or ignored (SCAR FEG, 2007: 8). As a more fundamental diagnosis, research agendas have become more distant from producers' knowledge, instead favouring specialist laboratory knowledge for agricultural inputs and processing methods (SCAR FEG, 2007: 11).

As a way forward, the expert group advocated agroecological approaches, *in situ* genetic diversity, farmers' knowledge, etc. It also advocated new kinds of Agricultural Knowledge Systems (AKS) beyond the formal research system: '*The AKSs that have been developed outside the mainstream, to support organic, fair trade, and agroecological systems, are identified ... as meriting greatly increased public and private investment*' (SCAR FEG, 2008: 42). Agroecological approaches should be given priority:

> Approaches that promise building blocks towards low-input high-output systems, integrate historical knowledge and agroecological principles that use nature's capacity and models nature's system flows, should receive the highest priority for funding.
>
> (SCAR FEG, 2007: 8; also EU SCAR, 2012: 92)

The report linked agroecology with a sufficiency perspective, counterposed to the dominant productivist one.

In response to such expert reports and TP Organics' proposals, FP7 eventually gave greater prominence to agroecological themes, which reached a total budget of €20 million by 2010 and increased thereafter. Drawing on proposals from TP Organics, there were calls for the following production methods, generally as substitutes for external inputs: ecological services based on eco-functional intensification, enhancing soil management and recycling organic waste via mixed farming, replacing chemical or copper pesticides with bio-control agents, enhancing on-farm production of renewable energy, etc.

Some research topics have sought to facilitate public reference systems necessary for embedding agroecological methods within wider institutions, for example through community-supported agriculture, agricultural extension services, food retailers and territorial labels. Knowledge for/about closer producer-consumer relations was the focus of a new topic, '*Short chain delivery of food for urban-peri-urban areas*', *whereby food localisation brings producers closer to consumers* (circuits courts). *The topic emphasises 'sustainable solutions for water management and nutrient recycling' as a task for institutional interactions, for example in 'the relation between peri-urban pressures and the participation of farmers and other stake-holders in rural development measures*' (DG Research/FAFB, 2011: 31).

Despite those successes in influencing the KBBE programme, the Commission's senior officials still exclusively promoted the Life Sciences vision of a bioeconomy. This dominated documents for a public consultation that was meant to inform future research priorities for a European bioeconomy, especially in the successor to FP7 (DG Research, 2010). As a shift in strategy, TP Organics now highlighted divergent accounts of a European bioeconomy.

In responding to the public consultation, it criticised the Commission for favouring '*specific new technologies (such as genetic modification) and capital-intensive "innovation" at the expense of agriculture*' (TP Organics, 2011: 7). It counterposed agroecological methods and agro-food relocalisation for a different bioeconomy: government should value agricultural knowledge that has been already developed over many decades, especially in co-producing agriculture with public goods (ibid.: 10). Likewise it ambitiously advocated '*a network of agroecological innovation centres in farming communities across Europe*' for transdisciplinary and participatory approaches (TP Organics, 2011), potentially transforming relations between researchers and farming.

In all these ways, the strategy sought an explicit place for an agroecological vision in EU policy documents and long-term resources for stakeholder-knowledge networks. Supporters have intervened in EU agendas by promoting agroecological perspectives and expertise, especially through the European Innovation Partnership for Agricultural Productivity and Sustainability (EIP-Agri). This comes under the EC's Rural Development Regulation, which includes the aim of '*working towards agroecological production systems and working in harmony with the essential natural resources on which farming and forestry depend*'.

As its overall method, '*EIP-AGRI pursues the "interactive innovation model" which focuses on forming partnerships: using bottom-up approaches and linking farmers, advisers, researchers, businesses, and other actors in Operational Groups that engage in practical projects*' (EIP-AGRI, 2013). Operational Groups facilitate farmers' joint knowledge production with experts, including agroecological methods (Levidow, 2015b; TP Organics, 2017). Although initially marginal, this agenda has sought to transform European agriculture through agroecological practices.

Conclusion: divergent trajectories of a European bioeconomy

The liberalised agro-industrial system, globally dominant since the 1990s, has been destabilised by multiples crises and resistances, in turn creating new opportunities for transitions. This article has identified two European trajectories of a bioeconomy: one industrialising life through Life Sciences and the other promoting agroecological methods. The latter trajectory treats the economic system as a sub-system of a finite living world, whereby biological cycles are integrated at the heart of economic reason (Passet, 2011). That relation is inverted by the Life Sciences trajectory, which treats biological materials as a sub-system of a globally integrated market competition for commodity exports and intellectual property. These trajectories coexist within the EC's institutional compromise around the KBBE.

By comparing the two trajectories, it is possible to identify tensions within and across them. Those divergent trajectories have been explored here through the following questions. What forces drive different agendas for a European bioeconomy? Given these different agendas, what are their inter-relationships – symbiotic, competitive or even antagonistic? How does each one link food and non-food products? How does it link food, energy and other sectors?

'Life Sciences'

This trajectory exemplifies wider moves towards a corporate-environmental regime, as a variant of the dominant food regime. The latter links capital accumulation with 'green' innovations, some of which had been previously associated with alternatives, such as renewable resources (Friedmann, 2005). For the past decade, a KBBE has been jointly promoted by EU policymakers, capital-intensive industry and its public sector research base. A decomposability paradigm informs R&D for 'quality' novel foods, especially functional foods, as well as for simulating traditional specialty foods; this illustrates edible commodities from recombined ingredients (Friedmann, 2005: 258).

More importantly, the Life Sciences are extended far beyond food by linking economic and environmental sustainability through non-food biomass. The dominant bioeconomy trajectory aims less at decarbonising society and more at substituting renewable biomass for fossil carbon. The Life Sciences agenda seeks larger genomic databases to inform novel processes and non-food products, for example by cracking biomass into co-products or intermediate products, to provide greater global flexibility for input-output chains. Hence agriculture is promised to generate the black gold or El Dorado of the 21st century, especially through intellectual property.

Global competitive pressures of the neoliberal food regime are being extended for globally integrating several industrial sectors (agriculture, chemistry, energy). As new business opportunities increase resource burdens, especially through land use and biomass processing, the decomposition trajectory may not enhance environmental protection, nor livelihoods. Such benefits may potentially come from other trajectories relying on compositional knowledge, such as doubly green chemistry.

'Agroecology'

The agroecology trajectory seeks to go beyond the dominant food regime through multi-stakeholder networks and alternative research agendas. It links eco-functional intensification, agroecology and an integral product identity for remunerating producers through short food supply chains (also known as *circuits courts*). These innovations depend on a different knowledge base than the Life Sciences agenda.

In sum, invoking a European bioeconomy, rival stakeholder networks contend for influence over research priorities, innovation trajectories and wider policy agendas. Each promotes its own innovation niches and protections through institutional support measures, alongside its own narrative of economic and environmental promises. Each elaborates divergent meanings of the same terms (see Table 9.1). For example, 'building blocks' can mean either simulating oil-based materials through new compositional techniques (Nova-Institut, 2017) or else simulating nature's flows through agroecological methods (SCAR FEG, 2011). Such tensions warrant further analysis (Levidow et al., 2014; Levidow, 2015b), as a basis to identify multiple possibilities, their contestations and potential outcomes for food regimes (Friedmann, 2009: 335–336).

As shown here for bioeconomy agendas, it is necessary to study the complex relationships between Life Sciences and agroecology trajectories, including different relationships between food and non-food production within different paradigms. A sharper conceptualisation is necessary to analyse divergent enactments of agricultural innovation. Each involves distinct technoscientific trajectories, knowledge-bases, economic activities and their interlinkages. Such analysis can identify contestations and incorporations vis a vis the dominant regime, within and/ or beyond a corporate-environmental food regime.

References

Allaire G., Wolf S., 2004. Cognitive Representations and Institutional Hybridity in Agrofood Innovation. *Science, Technology & Human Values*, 29(4), 431–458.

Altieri M.A., 1999. The Ecological Role of Biodiversity in Agroecosystems. *Agriculture, Ecosystems and Environment*, 74(1), 19–31.

Arora A., Fosfuri A., Gambardella A., 2001. Markets for Technology and Their Implications for Corporate Strategy. *Industrial and Corporate Change*, 10(2), 419–451. doi: 10.1093/icc/10.2.419.

Banse M., Van Meijl H., Tabeau A., Woltjer G., Hellmann F., Verburg P.H., 2011. Impact of EU Biofuel Policies on World Agricultural Production and Land Use. *Biomass and Bioenergy*, 35(6), 2385–2390.

Bastioli C., 2008. Renewable Raw Materials and the Transition from a Product-based Economy to a System-based Economy. In: *Conferral of her Honorary Degree in Industrial Chemistry by University of Genoa*, 4 July, Genoa, Italy.

Becoteps 2010. *Bioeconomy 2030: Towards a European Bioeconomy that Delivers Sustainable Growth by Addressing the Grand Societal Challenges*. Brusssels, Bio-Economy Technology Platforms (Becoteps). www.epsoweb.org/file/560.

Biofrac 2006. *Biofuels in the European Union – A Vision for 2030 and Beyond*. Brussels, Commission of the European Communities.

BioMat Net, 2006. 1st International Biorefinery Workshop [website defunct].

Birch K., 2017. *Innovation, Regional Development and the Life Sciences: Beyond Clusters*. London, Routledge, Taylor & Francis Group.

Birch K., Levidow L., Papaioannou T., 2010. 'Sustainable Capital'? The Neoliberalization of Nature and Knowledge in the European 'Knowledge-Based Bio-economy'. *Sustainability*, 2(9), 2898–2918. www.mdpi.com/2071-1050/2/9/2898/pdf. doi: 10.3390/su2092898.

Borras Jr., S.M., Franco J.C., Isakson R., Levidow L., Vervest P., 2016. The Rise of Flex Crops and Commodities: Implications for Research. *Journal of Peasant Studies*, 43(1), 93–115. doi: 10.1080/03066150.2015.1036417.

Bozell J.J., Petersen G.R., 2010. Technology Development for the Production of Biobased Products from Biorefinery Carbohydrates—The US Department of Energy's "Top 10" Revisited. *Green Chemistry*, 12(4), 539–554.

Busch L., 2000. The Moral Economy of Grades and Standards. *Journal of Rural Studies*, 16, 273–283.

Campbell H. (2009). Breaking new ground in food regime theory: corporate environmentalism, ecological feedbacks and the 'food from somewhere'regime? *Agriculture and Human Values*, 26(4), 309-319.

CEC, 2012. *Innovating for Sustainable Growth: A Bioeconomy for Europe* {SWD(2012) 11 final} Brussels, Commission of the European Communities.

Colonna P., Tayeb J., Valceschini E., 2015. Nouveaux usages des biomasses. In *Le Déméter 2015*. Paris, Club Demeter. pp. 275–305.

Coriat B., Orsi F., Weinstein O., 2003. Does Biotech Reflect a New Science-Based Innovation Regime? *Industry and Innovation*, 10(3), 231–253.

Debref R., 2014. *Le processus d'innovation environnementale face à ses contradictions: le cas du secteur des revêtements de sol résilients.* thèse de doctorat en sciences économiques. Reims, Université de Reims-Champagne-Ardenne.

DG Research, 2005. *New Perspectives on the Knowledge-Based Bio-Economy: Conference Report*. Brussels, DG-Research, Commission of the European Communities.

DG Research/FAFB., 2006. *FP7 Theme 2: Food, Agriculture, Fisheries and Biotechnology*. 2007 Work Programme. Brussels, Commission of the European Communities.

DG Research, 2010. *European Strategy and Action Plan towards a Sustainable Bio-based Economy by 2020*. Brussels, DG-Research, Commission of the European Communities.

Dosi G., 1982. Technological Paradigms and Technological Trajectories: As Suggested Interpretation of the Determinants and Directions of Technical Change. *Research Policy*, 11(3), 147–162.

Dubois J.-L., 2011. Requirements for the Development of a Bioeconomy for Chemicals. *Current Opinion in Environmental Sustainability*, 3(1–2), 11–14. doi: 10.1016/j.cosust.2011.02.001.

Dubois J.-L., 2012. Refinery of the Future: Feedstock, processes, products. In Aresta M., Dibenedetto A. and Dumeignil F. (Eds.), *Biorefinery: From Biomass to Chemicals and Fuels*. Berlin, De Gruyer, pp. 20–47.

EBTP, 2008. *European Biofuels Technology Platform: Strategic Research Agenda & Strategy Deployment Document*. Newbury CPL Scientific Publishing. www.biofuelstp. eu/srasdd/080111_sra_sdd_web_res.pdf.

EBTP, 2010. *Strategic Research Agenda 2010 Update: Innovation Driving Sustainable Biofuels*. Newbury, CPL Scientific Publishing. www.biofuelstp.eu/sra.html.

EC, 1998. Directive 98/44/EC of the European Parliament and of the Council on Protection of Biotechnological Inventions. *Official Journal of the European Union*, 30(July), L 213 13.

EIP-Agri, 2013. *Strategic Implementation Plan: European Innovation Partnership, Agricultural Productivity and Sustainability*. Brussels, EIP-Agri. http://ec.europa.eu/agricul ture/eip/pdf/strategic-implementation-plan_en.pdf.

EU SCAR, 2012. *Agricultural Knowledge and Innovation Systems in Transition: A Reflection Paper*. Brussels, Standing Committee on Agricultural Research (SCAR) of the European Union. https://scar-europe.org/images/AKIS/Documents/AKIS_reflection_paper.pdf

EuropaBio, 2007. *Biofuels in Europe: EuropaBio Position and Specific Recommendations*. Brussels, European Association for Bioindustries.

European Council, 2000. *Presidency Conclusions, 23–24 Mars*. Lisbonne, Lisbon European Council.

FfL, 2005. *European Technology Platform Food for Life: Vision for 2020 and Beyond*. Brussels, CIAA.

FfL, 2007. *European Technology Platform Food for Life: Strategic Research Agenda.* Brussels, CIAA.

FfL, 2008. *European Technology Platform Food for Life: Implementation Action Plan.* Brussels, CIAA.

FfL, 2016. *Food for Tomorrow's Consumer: Strategic Research and Innovation Agenda.* Brussels, European Technology Platform Food for Life (FFL). Brussels: CIAA.

FfL, 2017. *European Technology Platform Food for Life: Implementation Action Plan.* Brussels, CIAA.

Fonte M., 2013. Reflexive Localism: Toward a Theoretical Foundation of an Integrative Food Politics. *International Journal of the Sociology of Agriculture and Food*, 10(3), 397–402.

Friedmann H., 1993. The Political Economy of Food: A Global Crisis. *New Left Review*, 1(197), 29–57.

Friedmann H., 2005. From Colonialism to Green Capitalism: Social Movements and Emergence of Food Regimes. *Research in Rural Sociology and Development*, 11, 227–264.

Friedmann H., 2009. Discussion: Moving Food Regimes Forward: Reflections on Symposium. *Agricultural and Human Values*, 26, 335–334.

Friedmann H., Daviron B., Allaire G., 2016. Political Economists have been Blinded by the Apparent Marginalization of Land and Food. *Revue De La Régulation*, [En ligne] 20, |2e semestre/Autumn 2016, mis en ligne le 20 décembre 2016. https://regulation.revues.org/ 12145.

Gallezot P., 2010. Alternative Value Chains for Biomass Conversion to Chemicals. *Topics in Catalysis*, 53(15), 1209–1213.

Georgescu-Roegen N., 1971. *The Entropy Law and the Economic Process.* Cambridge, (Mass.), Harvard University Press.

Georgescu-Roegen N., 1975. Energy and Economic Myths. *Southern Economic Journal*, 41(3), 347–381.

Goodman D., Goodman M.K., 2009. Food Networks, Alternative. In Kitchin R. and Thrift N. (Eds.), *International Encyclopedia of Human Geography.* Amsterdam, Elsevier Science, pp. 208–220.

Guthman J., 2004. *Agrarian Dreams: The Paradox of Organic Farming in California.* Berkeley, University of California Press.

Huang J., Yang J., Msangi S., Rozelle S., Weersink A., 2012. Biofuels and the Poor: Global Impact Pathways of Biofuels on Agricultural Markets. *Food Policy*, 37(4), 439–451.

Ifoam-Europe, 2006. *Principles of Organic Agriculture.* Bonn, Ifoam.

Joly P.-B., 2010. On the Economics of Techno-Scientific Promises. In Akrich M., Barthe Y. and Muniesa F. (Eds.), *Débordements. Mélanges Offerts À Michel Callon.* Paris, Presses des Mines, pp. 203–221.

Levidow L., 2015a. Eco-Efficient Biorefineries: Techno-Fix for Resource Constraints? *Économie Rurale*, [En ligne] 349–350, 31–55. |septembre–novembre 2015 http://econo mierurale.revues.org/4729 https://economierurale.revues.org/4718mis en ligne le 15 décembre 2017.

Levidow L., 2015b. European Transitions towards a Corporate-Environmental Food Regime: Agroecological Incorporation or Contestation? *Journal of Rural Studies*, 40, 76–89, doi: 10.1016/j.jrurstud.2015.06.001.

Levidow L., Birch K., Papaioannou T., 2013. Divergent Paradigms of European Agro-Food Innovation: The Knowledge-Based Bio-Economy (KBBE) as an R&D Agenda. *Science, Technology and Human Values*, 38(1), 94–125. doi: 10.1177/0162243912438143.

Levidow L., Pimbert M., Vanloqueren G., 2014. Agroecological Research: Conforming – Or Transforming the Dominant Agro-food Regime? *Agroecology and Sustainable Food Systems*, 38(10), 1127–1155.

Malerba F., 2002. Sectoral Systems of Innovation and Production. *Research Policy*, 31, 247–264.

McCormick K. and Kautt N., 2013. The Bioeconomy in Europe: An Overview, *Sustainability* 5(6),2589-2608, https://doi.org/10.3390/su5062589

McEntee J., 2010. Contemporary and Traditional Localism: A Conceptualisation of Rural Local Food. *Local Environment*, 15(9–10), 785–803.

McMichael P., 2009. A Food Regime Genealogy. *The Journal of Peasant Study*, 36(1), 139–169.

Murphy A.M., Van Moorsel D., Ching M., 2007. *Agricutural Biotechnology to 2030: Steady Progress on Agricultural Biotechnology.* Paris, OECD.

National Research Council 2015. *Industrialization of Biology: A Roadmap to Accelerate the Advanced Manufacturing of Chemicals.* Washington, DC, National Academy of Sciences. www.nap.edu/read/19001.

Nicholls C., Altieri M., Vazquez L., 2016. Agroecology: Principles for the Conversion and Redesign of Farming Systems. *Journal of Ecosystem Ecography*, S5, 010, doi: 10.4172/2157-7625.S5-010.

Nieddu M., Garnier E., Bliard C., 2014. Patrimoines Productifs Collectifs Versus Exploration Exploitation: Le cas de la bioraffinerie. *Revue économique*, 67(6), 957–987.

Nieddu M., Van Niel J., Youssef A., 2013. Novamont: Un modèle de bioraffinerie sans biocarburants. *Biofutur*, 32(344), 52–59.

Nieddu M., Vivien F.-D., 2016. La bioéconomie: entre enjeux économiques et projets de société. *Biofutur*, 35(378), 60–61.

Niggli U., Anamarija S., Schmid O., Halberg N., Schlüter M., 2008. *Vision for an Organic Food and Farming Research Agenda 2025 – Organic Knowledge for the Future.* Brussels, Ifoam Regional Group European Union.

Nova-Institut, 2017. *Biobased Building Blocks and Polymers: Global Capacities and Trends 2016–2021.* Hurth, Germany, Nova-Institute GmbH. www.bio-based.eu/reports.

OECD, 2006. *The Bioeconomy to 2030: Designing a Policy Agenda Scoping Paper.* Paris,

OECD, 2017. *Towards Bio-Production of Materials: Replacing the Oil Barrel.* (No. DSTI/STP/BNCT(2016)17/FINAL) Paris, OECD. www.innovationpolicyplatform.org/system/files/DSTI-STP-BNCT%282016%2917-FINAL.en__0.pdf.

Padel S., Niggli U., Pearce B., Schlüter M., Schimd O., Cuoco E., Willer E., Huber M., Halberg N., Micheloni C., 2010. *TP Organics, Implementation Action Plan for Organic Food and Farming Research.* Brussels, Ifoam-EU Group.

Passet R., 1979. *L'économique et le vivant.* Paris, Economica.

Passet R., 2011. L'avenir est à la bio-économie, *Libération*, 23 mai, www.liberation.fr/economie/2011/05/23/l-avenir-est-a-la-bio-economie_737500.

Research/FAFB, D.G., 2011. *FP7 Theme 2: Food, Agriculture, Fisheries and Biotechnology.* 2012 Work Programme. Brussels, Commission of the European Communities.

SCAR FEG, 2007. *Foresight Expert Group, FFRAF Report: Foresight Food, Rural and Agri-Futures.* Standing Committee on Agricultural Research. Brussels, Consultative Expert Group.

SCAR FEG, 2008. *2nd Foresight Exercise: New Challenges for Agricultural Research: Climate Change, Food Security, Rural Development, Agricultural Knowledge Systems.* Brussels: Standing Committee on Agricultural Research (SCAR), Foresight Expert Group (FEG).

SCAR FEG, 2011. *Sustainable Food Consumption and Production in a Resource-Constrained World*. Brussels: Standing Committee on Agricultural Research, Foresight Expert Group.

Schmid O., Padel S., Halberg N., Huber M., Darnhofer I., Micheloni C., Koopmans C., Bügel S., Stopes C., Willer H., Schlüter M., Cuoco E., 2009. *Strategic Research Agenda for Organic Food and Farming*. Brussels, Ifoam-EU Group.

Schmid O., Padel S., Levidow L., 2012. The Bio-Economy Concept and Knowledge Base in a Public Goods and Farmer Perspective. *Bio-Based and Applied Economics (BAE)*, 1(1), 47–63. http://orgprints.org/20942/, http://www.fupress.net/index.php/bae/article/view/10770.

TP Organics, 2011. *TP Organics Response to the Consultation on the "Green Paper on a Common Strategic Framework for Future EU Research and Innovation Funding"*. BrussB Brussels, Technology Platform Organics.

TP Organics, 2017. *Innovating for Organics: Organics in EIP-AGRI Operational Groups*. Brussels, Technology Platform Organics. http://tporganics.eu/wp-content/uploads/2017/09/TPO_brochure_EIP_AGRI.pdf.

Van Dam D., Streith M., Nizet J., Stassart P., 2012. *Agroécologie: entre pratiques et sciences sociales*. Dijon, Educagri éditions.

Vanloqueren G., Baret P.V., 2009. How Agricultural Research Systems Shape a Technological Regime that Develops Genetic Engineering but Locks out Agro-ecological Innovations. *Research Policy*, 38(6), 971–983.

Vivien F.-D., 1998. Bioeconomic Conceptions and the Concept of Sustainable Development. In Faucheux S., O'Connor M. and Van Der Straaten J. (Eds.), *Sustainable Development: Concepts, Rationalities, Strategies*. Alphen-sur-le-Rhin, Kluwer Academic Publishers, pp. 57–68.

Vivien F.-D., 1999. From Agrarianism to Entropy: Georgescu-Roegen's Bioeconomics from a Malthusian Viewpoint. In Mayumi K. and Gowdy J.M. (Eds.), *Bioeconomics and Sustainability: Essays in Honour of Nicholas Georgescu-Roegen*. Cheltenham, Edward Elgar, pp. 155–172.

Wezel A., Bellon S., Doré T., Francis C., Vallod D., David C., 2009. Agroecology as a Science, a Movement and a Practice. *Agronomy for Sustainable Development*, 29(4), 503–515.

White House, 2012. *National Bioeconomy Blueprint*. Washington, DC, The White House.

World Economic Forum, 2010. *The Future of Industrial Biorefineries*. Geneva, World Economic Forum.

Part III

Cases studies

Competition in markets and policies

10 Organic farming in France

An alternative project or conventionalisation?

Thomas Poméon, Allison Loconto, Eve Fouilleux and Sylvaine Lemeilleur

Introduction

While organic farming is widely recognised by markets and public institutions alike, this does not prevent heated debates over its definition and the boundaries of its legitimacy. Indeed, if it is subject to public regulation and labelling (European Regulation (EC) No 834/2007), there are still divergences as to its nature, its objectives and the representations and values associated with it. There is a continuum of positions and practices between the proponents of organic as an alternative social project and those who see it more simply as a technical model of agriculture with a market that is complementary to the dominant model (Verhaegen, 2012). These divergences are emblematic of the different positions actors have taken with regards to the initial political project of organic farming and its evolution.

The practices of a type of agriculture called 'organic' have been referred to as such since the organic pioneers in the 1930s, to distinguish theirs as an alternative model to the then emerging industrial model of agriculture. This alternative model integrated agronomic, economic, social, political and philosophical dimensions (Besson, 2011; Leroux, 2014). But the institutionalisation of organic farming did not formalise until the early 1990s. Organic became an official and certified market standard, before becoming enrolled into a broader field of sustainable development (Fouilleux and Loconto, 2017). Despite the small size of the organic sector if compared to the conventional agriculture sector (5.7% of the agricultural land and 3.5% of the food market in France, according to Agence Bio in 2016), it has shown much stronger growth. In France in 2016, there was a 17% increase in producers and 21.7% increase in consumption, and consolidation and integration of its sectors and professional networks. This process of institutionalisation has at the same time distanced organic from many of its founding principles (Freyer and Bingen, 2014). This 'conventionalisation' critique of the organic movement was introduced by Buck et al. (1997) and has been taken up widely in scientific, professional and activist circles (Darnhofer et al., 2010; Baqué, 2012). These criticisms have led to the (re)emergence of cognitive and organisational devices that reaffirm the social movement dimension of organic that critiques capitalism. Two movements can be traced, one in the promotion of participatory guarantee systems (PGS) as alternatives to third-party certification, and a second in the emergence and strengthening of private, 'bio +' standards that go beyond the regulated organic requirements.

In this chapter, we trace the evolutionary trajectory of organic farming as it has moved from representing a political critique to a market segment based on an official label. We explore the tensions that these dynamics created among the actors in the organic field. The analysis is based on three separate studies of governance, certification and private standards for organic farming (Fouilleux and Loconto, 2014; Lemeilleur and Allaire, 2014a; Poméon et al., 2014) that were presented at the 2014 Symposium. Primary data were collected from interviews with producers, representatives of professional organisations and management bodies, various experiences of participating observation, and analysis of different written sources (websites, specifications, etc.). In the first section, we explain the initial project of organic agriculture and its transcription into a set of practices and devices. Here, we focus on the tensions and debates arising from its progressive institutionalisation in France. Then we will explain how this regulatory and market institutionalisation of organic agriculture has profoundly modified the rules of the field, the type of actors involved, and their practices. This leads into the third section, which focuses on the contemporary movement to reactivate the critical and alternative dimension of organic through the revival of participatory certification and the multiplication of private standards aimed at going beyond the public standard. Despite a focus here on the French case, our analyses are representative of what happened in the worldwide organic field in terms of conventionalisation dynamics and counter dynamics.

From a multifaceted critique of capitalism to conventionalisation trends

While scholars point to a progressive conventionalisation of organic, the question of coherence between the practices, principles and values associated with organic farming has also been raised. To address these concerns, we analyse the evolution of this field, both from the point of view of the ideas carried by the historical organic actors and the devices they have developed. We borrow a conceptual framework that positions the dynamics of these ideas within critiques of capitalism, specifically, how organic agriculture emerged in an environment that challenges the inclusion of agriculture into a capitalist logic. The concepts and critical grammar developed by Boltanski and Chiapello (2005) and Chiapello (2009) allow us to describe and characterise the critiques of capitalism, its forms, its effects and its appropriation by capitalist actors since the mid-19th century. This framework enables us to describe and analyse the forms of capitalist critique that are present in the organic movement and how these have changed over time. This is, of course, a critique that is situated in time and space, which evolves, and which is either advanced or side-lined. To characterise the forms taken by the critique, two questions adapted from Chiapello's (2009) conceptual framework (see Table 10.1) must be analysed: *What is denounced in 'conventional' agriculture by organic advocates*? Based *on which values?*

Table 10.1 The four types of criticism of capitalism.

	Conservative criticism	Social criticism	Artistic criticism	Ecological criticism
Causes of indignation	Poverty/insecurity, moral disorder, destruction of solidarity, class struggle	Poverty/inequalities, wage relations, exploitation, command of capital, class domination	Mediocrity, stupidity; uniformisation, massification, commodification, conditioning; alienation	Destruction of ecosystem, species, and human habitats
Underlying values	Shared dignity common to all human beings, class interdependence, moral duty of the elite	Labour, equality (in economic terms and in decision-making) as the necessary condition for a true freedom	Personal autonomy (internal and external), taste and refined existence (art, philosophy, truth, etc.)	Shared dignity common to all living beings, life of future generations

Source: Chiapello (2009: 65)

In France, and around the world, organic pioneers denounce the artificialisation of agriculture based on three core arguments: the link between production methods, food and health; the question of farmers' autonomy vis-à-vis the purchase of inputs and the sale of products; and, a concern for the preservation of natural resources (Piriou, 2002). Leroux (2014) proposes that the socio-genesis of French organic farming is linked to a critique of industrial capitalism and its control of the food and agriculture sectors. With significant involvement from non-farmers (doctors, consumers, etc.), organic first emerged in France more as a social movement than a professional farmer identity. Scholars cite a historical link between critiques of capitalism and those of the transformations of agriculture and the agro-food systems that began at the end of the 19th century (Viel, 1979; Piriou, 2002; Leroux, 2014). These transformations produced a pluralistic critique from the organic movement as they resisted what the early actors saw as processes of artificialisation, industrialisation, commodification, and market dominance by agribusiness (seed and input suppliers), large-scale processing and distribution groups. GABO (Association of organic farmers in the West), the first organic farmers group in France, was founded in 1959, and formed 'in reaction to this modernity (referred to as "progress"), which challenges a set of traditional peasant values for which and through which they view the future' (Leroux, 2011: 42–43). Thus, in France as in other countries, the roots of organic are marked by a rejection of capitalism in its technical and industrial forms, rather than by the social inequalities it generates. It is the physical and moral decay brought about by industrialisation and commodification that is pointed out first and foremost, with an aspirational return to an idealised peasant society. For Viel (1979), this early agrarian conservatism, which he calls 'reactionary', would merge with other critiques of capitalism that eventually give rise to values specific to organic agriculture.

In 1962, the French Association of Organic Agriculture (AFAB), was founded as the national version of GABO. But quickly two visions clashed and gave rise to a split in the movement. On one side, the Lemaire-Boucher group promoted a commercial orientation with conservative values (Viel, 1979; Leroux, 2011). In reaction to this commercial positioning, Nature et Progrès (N&P) was created in 1964 as an association that brings together farmers, processors, suppliers, distributors and especially consumers. Structured around relatively autonomous local groups, it simultaneously promotes a set of production practices and a lifestyle. N&P's mission developed a more societal and transversal character, in opposition to the productivist and consumerist capitalist society. Progressively, N&P aligned itself more closely to the protest movements linked to May 1968, anti-capitalism, anti-centralism and environmentalism. The association thus gradually detached itself from the original conservative matrix, focusing instead on values such as autonomy, energy and technological conservation and local economies (Viel, 1979; Leroux, 2014). These elements echo the artistic, social and ecological criticisms (see Table 10.1) that characterise the protest movements of the 1960s and 1970s. These movements would thus play a significant role in the views about and development of organic agriculture in the 1980s. During this period, the rejection of consumerism and its conventional distribution channels became a core value reinforced with a collectivist, commune character that was found in many of the neo-rural projects of the time. Beyond the networks of actors involved, the critical dimension of organic farming can also be found in the model advocated and practised: small, diversified and autonomous farms with a preference for a system of direct sales of raw or lightly processed products (Viel, 1979).

The vision advocated by N&P gradually overtook that of Lemaire-Boucher (Leroux, 2011), and in the late 1970s became the driving force of the organic movement. It is therefore this vision, criteria and standards that served as the main reference point in the process of institutionalisation. As described by Piriou (2002), this began in 1980 with the Loi d'Orientation Agricole. This process, demanded by the organic movement and progressed by N&P, led to the official recognition and regulation, in France and then in Europe, of organic agriculture. The progress that we can trace is from 11 private standards recognised by the AB label in France in 1989, to a single standard and label controlled by the European Union from 1991 onwards. Organic agriculture thus gradually became a unified reality, at least from an institutional point of view.

In this process of regulatory institutionalisation, the ecological critique became central to the form of capitalist critique that we currently see in organic agriculture (Chiapello, 2009). While it became a promising, unifying element (around the need to reduce input costs and promote the multifunctionality of agriculture) of the organic movement, thus making space for organic practices in different public and private arena, the ecological critique marginalised the socio-economic dimensions of the movement. Organic products were brought to market and, as such, provided the justification to the State to protect consumers via an official label. In addition, as organic gradually emerged as a credible alternative, a growing number of actors who were not initially present in the movement seized the opportunity to respond to the critiques of the environmental impacts of chemical agriculture and to the successive food crises (mad cow, dioxin, etc.). But this appropriation by external actors is accompanied by compromises that result in a selection and limitation of criticism and its transformative potential. This dilution of the critical stance is similar to that shown by Boltanski and Chiapello (2005) on the integration of artistic criticism into management practices following the May 1968 movement. In the case of organic, it has led to the creation of a market standard, supported by Europe in its role as a provider of public environmental goods, and as a niche market.

This position is condemned by many actors, who view the market mainstreaming as a dilution of the transformative social mission of the organic project (Baqué, 2012). From the outset, organic farming has always been plagued by ambiguities about its relationship to the market. Yet despite the critique that this market position threatens the agricultural practices and the quality of organic products, most organic farmers have been and still are included in the market economy. Although organic farmers often prefer short supply chains, long supply chains have developed relatively quickly, with their own wholesalers, processors and distributors. These debates on the positioning of organic within conventional markets are still very present in France, particularly around the role of large retailers, the increasingly transactional nature of certification, or the market segmentation of organic (the niche market vs the overall transition from agriculture to a new socio-technical model). These debates testify both to the critical position of organic as a social movement in opposition to liberal capitalist regulation of trade and the difficulties that actors face in balancing their activities between ideas and practices.

It is thus in the context of renewed power struggles that organic farming has evolved and is currently defined (Piriou, 2002). Those actors linked to a more radical and holistic critique of the capitalist system and the agro-industrial complex must deal with new players, attracted by the opportunities of the organic market and who do not share the same visions and objectives. The domination of ecological criticism has not, however, evacuated other forms of criticism, to which it may even be associated (Chiapello, 2009). The challenge is for the different actors to build the political value of their vision, such as the one that combines ecology and autonomy, against a technocratic vision of ecology is often antagonistic to it.[2]

The success of organic agriculture, attested by the development of a specific market (Allaire, 2016), is balanced by debates around the weakening of its critical scope, expressed in the concept of conventionalisation (Coombes and Campbell, 1998; Darnhofer et al., 2010). The notion of conventionalisation appeared for the first time in the Buck et al. (1997) study of California organic and refers to a variety of processes. A first definition is linked to the evolution of the motivations of the practitioners of organic farming, among which the economic opportunity would play an increasing role. Here, the structures and management tend to resemble those of conventional agriculture, including a substitution of land and labour for capital (Guthman, 2004). Conventionalisation can also be understood as a standardisation of agronomic practices that would reduce its ecological and ethical significance, attributed to the reduction of principles that are used in certification or the dynamics of knowledge systems (Stassart and Jamar, 2009). The logic of input substitution (using organic inputs instead of synthetic inputs) prevails over systems redesign that requires integrated crop and livestock systems (Rosset and Altieri, 1997; Lamine and Bellon, 2009). Another definition refers to the development of market channels for organic products within conventional and global value chains driven by multinational agribusinesses and supermarkets (Jaffee and Howard, 2009), associated with a diversification of buyers and purchasing practices of organic products.

Generally, conventionalisation refers simultaneously to the changes made by the historical actors of the organic sector and to the characteristics of the new arrivals, who often respect the regulatory requirements of organic but not its historical principles and values. However, a key issue is the capacity of standards, and their social and political dynamics, to take these principles and values into account and to implement them (De Wit and Verhoog, 2007; Darnhofer et al., 2010).

The effects of institutionalisation centred on a market standard

Can the initial organic agriculture project survive its institutionalisation in the form of a standard? For Piriou (2002: 409), 'as long as organic agriculture is recognised only as an ensemble of technical or market requirements and not as an innovative production system it cannot question the dominant agricultural model'. Thus, we argue that we must look beyond what happens to the critical capacity of organic when it shifted from a political agenda to a set of technical standards. We must also consider the institutional effects associated with standard-setting in the analysis of the processes of conventionalisation. There are several insights to be gained from examining the development of service markets for certification and accreditation activities.

Indeed, the 'voluntary standard' instrument that regulates the organic field has its own institutional effects, linked to the fact that it is based on a set of imbricated markets. It is based first and foremost on standards (a list of specific practices to be implemented by the producer or manufacturer), to which an associated on-product label can be used by producers to distinguish themselves from other products that look similar. The label signals to the consumer that the conditions of production that are invisible to the naked eye (no chemicals, etc.) have been used. To achieve this, operators must undergo a process called 'third-party certification', which guarantees the credibility of the system. In other words, they must accept that their production practices are controlled by an independent operator, the certifier, who, if he considers that they comply with the specifications, issues them a certificate of conformity authorising the use of the label. The producer pays the certifier for this inspection service. The last element to ensure trust and credibility in this system is that the certifier is itself controlled by an 'accrediting' body, which he pays, who oversees the certifier's competence to certify compliance with a given voluntary standard. The voluntary standard thus refers to a complex regulatory system based on multiple market transactions, associating standards (with or without a label), certification and accreditation. This interweaving of markets for products and services can be described as a '*tripartite standards* regime' (TSR) (Loconto, Stone and Busch, 2012; Fouilleux and Loconto, 2017), which is an internationalised institutional system that combines public and private actors (see Chapter 4). The establishment of a TSR is justified by the desire to legitimise a standard and to guarantee the effectiveness of the practices that it covers. It relies on a set of principles: harmonisation and readability (simplicity) of rules and criteria, independence, and impartiality of control. It is the effectiveness of these principles and the effects of the implementation of an organic TSR that can shed light on the past and present dynamics in this field.

In the European Union, public authorities imposed third-party certification after the organic regulation was introduced in 1991. It involves standardised procedures, mainly based on document control (accounting, invoices, etc.) and a visit to the headquarters of the farm and sometimes a visit to the fields. No interaction other than control-related interaction should take place between the farmer and the certifier – advice or tips on how to improve the farming practices are strictly forbidden. Accreditation is also organised at European level and governed by Regulation (EC) No 765/2008, which implies that a body specifically designated by the State controls and accredits the capacity of the certifier to certify in general (in conformity with the European standard ISO 17065/NE 45011), and on the other hand that the certifier is accredited to certify the European standard of organic farming. Within the European Union, these two procedures are delegated to the Member States, which themselves delegate to private or public actors (depending on the country). In France, the responsible organisations are the French Accreditation Committee (COFRAC) and the National Institute of Origin and Quality (INAO).

The European regulation was quickly followed by the emergence of public organic standards around the world, all following the TSR model (87 countries in 2016) (Willer and Lernoud, 2017). From this point forward, discussions focused mainly on the means of facilitating and increasing trade as much as possible at both European and international levels. While the harmonisation efforts made it possible to extend the markets for organically grown products, including the market for auditable standards, they have also made it possible to transfer to the Global South a vision of organic agriculture. In developing countries, organic was mainly developed by international and bilateral cooperation agencies and importers, and not as a pluralistic social movement driven criticism anchored in local environments, as was the case in Europe.[3] Rather, organic emerges here quasi-exclusively around export products and according to the rules of the TSR, mainly embodied by US and European certifiers (Willer and Lernoud, 2017). The aim was to encourage the development of organic products by structuring a differentiated and remunerative market capable of attracting new players. At the level of the organic standard, this resulted in a desire to harmonise and simplify the rules, to make them more accessible to producers and more intelligible to consumers. Harmonisation of rules at the national, European (Gibbon, 2008) and international levels (through the projects financed by UNCTAD, FAO, IFOAM, SIDA, etc.) has effectively excluded the idiosyncratic dimensions of organic and the particularities of each agro-ecosystem. The need to simplify, so as to put in place auditable criteria, has forced the shift from an organic based on a set of global principles and objectives shared and translated to each local context to a set of reductionist technical criteria. We see emerging from this experience an organic standard that aims mainly to ensure the circulation of organic products and the extension of its market (Fouilleux and Loconto, 2017).

The functions and organisation of certification systems have also changed considerably. Prior to the introduction of third-party certification, the control system was based on farmer exchanges and support with other organic stake-holders, especially consumers, being highly involved in the control mechanisms. With the introduction of the TSR, certifiers have become key actors in the institutionalisation of the organic field, even as their interests and strategies have become increasingly estranged from the original values and principles of organic. What began as informal certification groups and associations have become auton-omous enterprises. Specialised multinational certification companies (e.g., Bureau Veritas), seeing the market opportunity, have purchased small organic certifiers and begun to compete with the mission-driven certifiers (Garcia-Parpet, 2012). These market dynamics have in fact shifted the practice of certification away from the original intent of certification by committed members of the organic movement. Instead, the current focus of certifiers is to provide internationally competitive prices for a broad portfolio of services, including the creation of new standards that go beyond the public organic regulation and even create competition for it. Organic agriculture finds itself competing with other labels and standards, developed by certifiers or other private firms or corporations, both in terms of market share and political and social legitimacy. This pressure has led IFOAM to consider aligning their standards with 'weaker' ones that dominate some trading systems such as GlobalGAP (Fouilleux and Loconto, 2017). Certification within the TSR has become a market in and of itself (certification market) that has a value (price premium or market access) and is exclusive from other forms of certification. It may exclude legitimate actors, for example producers who cannot afford to pay a certifier, and it can block peasant innovations (Lemeilleur et al., 2015). Its homogenising character ignores the diversity of contexts that were fundamental to the origins of organic standards and control systems.

The accreditation system, which underpins the credibility and coherence of the entire TSR, has also undergone a series of changes. The European Union defines accreditation as a non-profit activity entrusted to State-sanctioned national monopolies (either public or private) so as to privilege the credibility mission rather than the market mission of competition and profit. Yet if competition is absent at the European level, we find accreditation bodies competing internationally for markets, which is a direct result of the European regulations for conformity assessment that have been adopted by the organic regulation. Indeed, for countries outside the EU that do not benefit from equivalency agreements,[4] EU-approved certifiers must monitor the implementation of EU rules. These EU-approved certification bodies are thus transformed into standard-setters who must adapt EU public standards for a global private organic market, which amounts to a parallel accreditation system controlled by the European Commission. Furthermore, accreditation bodies use the public authority allocated to them at European level to become 'accreditors for the world' and to expand their markets in countries in the South. For example, the German National Accreditation Body (DakkS) accredits Biolatina (Peru), COAE and ECOA (Egypt), CERTIMEX (Mexico), Argencert (Argentina), and INDOCERT (India). Finally, EU accredited multinational certification bodies, who do not have local branches in all countries, tend to outsource the inspection services to non-accredited local inspectors, which in turn induces the emergence of a parallel local market for auditing.

More generally, this focuses the public and private organic standards of different countries towards a hegemonic global model that promotes the EU organic standard, with its rules, its dominant players and its logics of market development.[5] This model contributes to the conventionalisation of organic for two reasons: first, the watchdog approach leaves little room for political, ethical and critical stances by actors in the system; and second, its hegemony prohibits the original intent of the founders that general organic principles should be translated and adapted by the local groups (including players in the industry but also consumers, doctors, etc.) to their contexts. This points the movement in the direction of a dispossession of organic agriculture, which grows in response to some questioning of the different dimensions of the organic TSR that we return to in Section 3 of this chapter.

In summary, the organisation of the organic field as a TSR has been a key 'hidden' force in the conventionalisation of organic. It has strengthened and legitimised the concept of organic by guaranteeing a market demand. Nonetheless, these same mechanisms have kept organic as a niche market where its products must compete not only against conventional products but also against other 'sustainable' products. The consequence of this form of conventionalisation is that the original leaders of the organic movement (farmers, doctors, consumers, nutritionist, and technicians) have been disenfranchised of their ability to define organic and have been replaced by the State, certifiers and accreditors. We can trace through this movement an impoverishment of organic's political project via the transformation of principles into technical criteria and the commodification of relationships within the industry (between producers, certifiers and accreditors). The constraints related to the internationalisation of the organic standard (and related markets) have further reinforced this trend. Faced with the reality of its progressive conventionalisation and for the sake of re-appropriation and reactivation of the social and political dimensions of organic farming, some organic actors in France and elsewhere have created or reactivated different devices existing in the margins of the TSR to regain the initial project; these include the development of private standards and a revival of PGS.

The proliferation of standards and the return of participatory certification

Even though the European Regulation replaced the existing national specifications beginning in 1991 (1999 for animal products), this does not mean that private standards disappeared. There are currently about 15 private standards in France that are legally registered as private brands. If the brand is not used as a complement to the EU organic label – put differently, if the producer is not also certified against the European regulation – the users of the brand cannot use the term organic to describe their production practices or products. Of the 11 private standards that were approved by the French state in 1991, four still existed in 2013:

- **Nature and Progrès** (N&P), as mentioned above, a pioneer in the French organic movement;
- **Demeter**, a biodynamic standard and the first organic standard in the world, established in Germany in 1928;
- **Biobourgogne**, an association and regional brand, created in 1981 and linked to N&P;
- **SIMPLES** (InterMassif Union for a Simple Production and Economy), focused on plant production and collection, linked to N&P.

Since 2002, 13 new private standards have emerged:

- **Biobreizh**, created in 2002 by the Association of Organic Fruit and Vegetable Producers of Brittany, which is a sector and regionally specific brand and **Bio Loire Océan** (2005), its counterpart in the Loire Valley;
- **Biodyvin** (2002), a splinter group from Demeter, this standard focuses on biodynamic wine producers;
- **Bio Solidaire,** created in 2009 by the Bio Partenaire association as a means to promote the 'North–North' exchanges of the BioEquitable brand, which combines organic and fair trade;
- **Bio Cohérence**, led by FNAB and established in 2010 as a reaction against the weakening of the European regulation during its 2007 revision[6];
- Regional brands such **Alsace Bio** (2004), **Paysan Bio Lorrain** (2005), **Mon Bio Pays de la Loire** (2012), **Saveurs Bio Paris Ile de France** (2011), **Bio di Corsica** (2013), **Bio Sud Ouest** (2013), or the charter **Bio Rhône Alpes** (2010).

These private standards represent about 2,285 producers, which is 9% of organic farmers, who themselves represent 4.9% of French farmers in 2013 (see Table 10.2).[7] Some private standards also combine processors, distributors, and even consumers (in N&P and Bio Cohérence).

In 1989, N&P and Demeter, with 1,161 and 222 producers, respectively, represented 42% and 8% of the 2,768 organic producers (0.27% of farmers at the time) using private standards (Robidel, 2014). The remaining producers were using the now extinct standards. In the 1990s, we witnessed a relative and absolute decline in the use of private standards as the French organic movement was restructured and realigned around a public standard. For example, N&P experienced a mass exodus of producers from its groups during this period. But a differentiation process was reactivated in the early 2000s and private organic standards regained ground.

One explanation for this resurgence lies in the use of differentiation strategies to enhance value, in an increasingly competitive organic market. But we also observe new memberships and new standards as an expression of dissatisfaction with the public standard. We can identify attempts to regain the demanding and holistic approach to organic, as illustrated by the example of Bio Cohérence.[8] N&P's standard for crop production also illustrates this attitude:

> In light of the preamble to the EEC regulation for organic agriculture which speaks only of 'markets', Nature & Progrès considers 'Organic' in its global dimension, is still not recognised, thereby risking its abuse. This single preamble of 'official organic' fully justifies the keeping the Nature & Progrès *label*.
>
> (Nature et Progrès 2010)

From this point of view, the existence and the creation of private standards that are more stringent than the official organic label can be understood as a part of

Table 10.2 Number of farmers under private standard in 2013.

	Nature & Progrès		Demeter	Biodyvin	Bio Cohérence	Bio Solidaire	Biobreizh	Bio Loire Océan	Others[1]	Total
	Off label AB	In AB label								
Number of producers	350	350	430	90	300	300	65	50	350	**2,285**

1 Includes BioBourgogne, SIMPLE and the other regional brands.

Source: Producers' numbers are approximate and were obtained during interviews and/or via publicly available information about the organisations (brochures, website, etc.).

the negotiation process among actors in the organic field. Indeed, the groups who created these private labels are also involved, in different ways, in the negotiations on the revision of the European standard.

Private standards differ in their purposes, their visions of agriculture, operating procedures (organisational form, reference materials, control practices, financing, etc.) and their relationship to the rest of the industry (Espagne, 2014). They are part of a trend of 'bio+' labels that add requirements to the public standard without opposing it outright. Indeed, apart from SIMPLES union and N&P,[9] the private standards make compliance with the European standard a baseline requirement. Even if these private standards are marginal in terms of product sales, their promoters are often active in the dissemination of alternative ideas and practices within the organic movement. An analysis of their historical dynamics reveals three typical logics driving their development:

- Segment and differentiate supply to meet demand, increase competitiveness and create a niche;
- Better coordinate and structure the supply, especially to create economies of scale in logistics and marketing;
- Compensate the deficiencies of the public organic standard by proposing a promising alternative standard of what organic should be.

While the first two logics are related to the organisation of the sector and its market, the third is a socio-political logic that questions the collective future of organic according to the social change that it promotes. These three logics are found in all the private standards, but they differ according to which logic dominates their messaging efforts.

As part of this last logic, we can identify diverging visions among these different private standards. First, many regional brands impose rules on the origin of products, sometimes with a set of guiding principles, but without much else restricting production. Promoted by a range of actors (e.g., producers' organisations, processors' associations, and local authorities), their main goal is to organise and promote a local organic supply. We can find other organising activities associated with these initiatives such as common pooling of resources, using marketing instruments (logos, trade fairs, publicity) or market placement (aggregating supplies through platforms), and knowledge exchanges.

Other private standards adopt an agroecological positioning (similar to that defined by Guthman, 2004) that goes beyond the public standard on several criteria where the EU standard is found lacking, e.g., production diversity, autonomous fodder, restricted list of authorised inputs, etc. They also establish stricter composition requirements for processed products. These standards introduce not only additional environmental requirements, but also requirements for autonomy, local sourcing and the authenticity and naturalness of the products. Products carrying these standards' labels are almost exclusively sold in specialist organic boutiques or via direct sales. At one end of the spectrum, Bio Solidaire and Bio Cohérence explicitly prohibit the sale of their products in conventional supermarkets and Bio Cohérence offers a certification for retailers. At the other end, Biobreizh is traded in long supply chains and does not exclude supermarkets.

In general, the standards' social, ethical and economic criteria (e.g., farm size, economic relations, price, market, employment, etc.) are designed and evaluated with a 'non-industrial' or 'traditional' form of agriculture in mind. Measurable criteria are often supplemented by principles with varying levels of sanction (including the possibility of expulsion) that are contained in an associated charter signed by members. N&P and Bio Cohérence are the strictest standards and Biobreizh is working to improve its requirements on these dimensions. Bio Solidaire differs from other standards by its collective approach of including supply chain actors in its system (producers, buyers, processors). By promoting North–North fair trade, Bio Solidaire imposes strict rules on contractual exchanges, guaranteed minimum prices and sustainable partnerships between producers and buyers. At N&P, issues of equity and the nature of trade are included in a separate charter that lists their social, economic and environmental principles. Demeter refers to the principles of anthroposophy in its reference values while Bio Cohérence explicitly recommends criteria for certified distributors (e.g., wage differentials, share of permanent employees) and prohibits the hiring of workers through contracts that apply a foreign, rather than the French national, labour law.

In terms of certification, Demeter, Biobreizh, Bio Cohérence and Bio Solidaire delegate their audits to third-party certifiers, who conduct joint audits (of both the private and public standards) so to reduce costs. N&P, BioBourgogne and SIMPLES use the participatory guarantee systems (PGS)[10] that they had developed in the 1970s. PGS are local groups of producers, processors and consumers who assume the audit functions (peer control procedures) and conduct social controls. While they used only marginally in France, PGS have been on the international organic scene since the early 2000s (Lemeilleur and Allaire, 2014a, Loconto et al., 2016; Fouilleux and Loconto, 2017). In 2016, PGS were active or in development in 72 countries and are recognised in the public organic standards of Brazil, Bolivia, Chile and India.[11] IFOAM defines PGS as 'locally focused quality assurance systems. They certify producers based on active participation of stakeholders and are built on a foundation of trust, social networks and knowledge exchange'.[12] In some cases they are close to the internal control systems[13] set up for farmer groups in the South to reduce the costs of certification (Van der Akker, 2009). Yet there are several key differences, such as the inclusion of different types of actors in the control system (consumers, municipal officials, researchers), the ability to remain organised as a network without a mandatory requirement to form a cooperative, and the reliance on farmer-led 'peer-reviews' without a third-party control (Lemeilleur and Allaire, 2014b; Loconto, 2017).

The recent renewed interest in PGS is motivated by a critique of the logics and effects of the hegemonic TSR explained in the last section. This critique has inspired debates within IFOAM and other organisations (IFOAM, 2014). Indeed, third-party certification is based on a privileging of distance (lack of communication between auditor and producer, no advisory function) and organisational independence to ensure compliance (Jahn et al., 2005; Hatanaka and Busch, 2008). Rather than being accompanied,[14] the producer is subjected to an audit that serves to prove their innocence (of not using banned products) (IFOAM, 2005). Moreover, the cost of third-party certification is considered too high, particularly for small producers and highly diversified production systems.[15] This system can create exclusion and thus unfairly apply the right to use the term 'organic' (Lemeilleur and Allaire, 2014b). Finally, it reduces organic to a set of control points and predefined indicators. It does not consider local specificities and production systems or the dimension of making progress over time. The desire for measurability and simplified audits has increased traceability, but there is no clear record of effective environmental and social performance of certified production (Lemeilleur et al., 2015). This contradicts the founding principles of organic, which reject the privileging of generic technical solutions over adaptation to the natural environment (including both its potential and its limitations). Third-party certification thereby decouples the organic standard from a holistic vision of organic as a set of values and practices.

These values and practices of organic agriculture could be considered as intellectual common-pool resources threatened by privatisation of the resource, and in this sense PGS appears as the re-conquest of the common (Lemeilleur and Allaire, 2016). Indeed, PGS are collective management and audit devices that are flexible and participatory as they are entrusted to local groups composed of producers, consumers and other stakeholders (Nelson et al., 2010). Each participant, auditor or audited, is involved at an equal level of responsibility. The PGS is first and foremost seen as an advisory tool for continuous improvement of the producer, in pace with their specific environmental and socio-economic context and ambitions (Hochreiter, 2011). The standard-setting, verification and decision-making procedures are based on the principles of participation and horizontality to develop a 'shared compromise'. Local groups are typically federated into a larger association that ensures coordination, guidance and support (e.g., document templates, training) and, if necessary, arbitration in disputes. The audit reports and decisions are usually made public (Nelson et al., 2010), ensuring a full responsibility of the group who conducts the audit. Experience shows that non-compliance is relatively low and often linked to practical registration problems (Van der Akker, 2009; Hochreiter, 2011). In practice, those PGS that include reinforce responsibility, gender equity and social control, are generally as effective as third-party certification (Lemeilleur and Allaire, 2014a). PGS offer a solution to the impasses linked to the need for detailed global standards and diverse local conditions (Vogl et al., 2005). Consistency between principles and practices is prioritised in PGS and can counter the homogeneity imposed by the public standard.

While generally positively received, PGS is critiqued by movement actors. The local adaptation of the standard and audit rules could lead to heterogeneity, which if poorly managed could increase confusion, asymmetric information, and eventually reduce confidence in organic production. However, this criticism is likewise levelled against third-party certification systems (Baqué, 2012). Another critique refers to the inherent difficulties related to managing participatory processes that can generate tensions and hinder access for some producers. Finally, the effectiveness of participatory approaches depends heavily on the motivations and capacities (material, institutional and human) of different actors. There is no escaping knowledge asymmetries and balances of power (Lemeilleur and Allaire, 2014a). The current political and institutional activities on PGS carried out by IFOAM and international organisations recognise these challenges.

Overall proponents of PGS, N&P in France, view it as a certification mechanism that is better suited to the holistic vision of organic. It is a work in progress rather than a recipe. However, unlike the TSR, PGS cannot claim exclusive legitimacy: some PGS are experimenting with hybrid systems, for example with the introduction of a third-party certification of the PGS groups (New Zealand, Bio Caledonia) or by requesting equivalence recognition with internal control systems (Van der Akker, 2009) in group certification. To Teil (2013), the coexistence between different certification mechanisms is desirable based on the principle of complementarity – i.e., there is a role for everyone and diverse logics in contributing to the development of organic. Nonetheless, we must be able to facilitate the dissemination and circulation of organic products with stable and comprehensible systems and to simultaneously allow organic agriculture to evolve in its transversal project.

Conclusion

In the history of critiques of capitalism, the proposed alternatives have typically looked for transformative change either by replacing the existing system or building a parallel system that could push whole system change. Organic agriculture was originally part of this history; its identity was first built in parallel to the dominant paradigm of 'conventional' agriculture, and then in interaction with it. This interaction led to a compromise of coexistence where organic became an institutionalised niche market and just one sustainability standard among many others. Radicals and reformers faced off so as to reclaim the classic terminology. On one side was those who believed that the compromise would weaken the political and systemic project, and on the other side was those who feared that the inability to reach a compromise would confine the movement to a comfortable and unthreatening margin of the dominant system. Organic agriculture, when considered as a public good and a political pathway for greening practices, remains the object of permanent revision and renewal among different societal actors (Allaire, 2016).

The history of organic tells the story first of a search for revendicated pluralism, then for homogeneity. In both stories the identity of organic is the plotline: how to define and implement this identity and how to audit it are what merge the political and market projects of the movement. Under the influence of the TSR, organic's critical and confrontational character has been marginalised. Nonetheless, organic actors (producers, processors, consumers, and others) mobilise alternative arrangements, such as PGS and private standards, to reclaim the 'true' organic and revendicate their pluralist critique. This plurality is both a strength and a source of tension, which in turn is used by the actors to position themselves between the logic of the market and sustainability standards, and the questioning of the current agri-food system paradigm. Organic is thus caught between the guarantee of predefined, objective, controlled criteria and an inclusive but unique political philosophy. In this sense organic has become a quality revealed by subjective judgement, but which wants to be 'a global action framework that subordinates other economic and social goals' (Teil, 2013: 217).

The analysis presented in this chapter on the organic TSR, the rise of private standards and PGS has traced the weakening and reactivation of the critical dimension of the organic movement. Beyond simply coordinating supply chains and market segmentation, private standards might be vehicles for both defending the pluralist message and delivering a structural critique of current agri-food systems. Along with PGS, which are built into some of these standards, they represent a critical structural element of the current movement that could enable actors to reclaim the organic identity. PGS put forward a vision that is coherent with the founding principles of the organic movement that revelled in the diversity of contexts, rather than the reductive homogeneity that is mostly linked to the governing model of the public organic standard. Yet while the attractiveness of PGS as a regulatory tool controlled by the movement itself is enticing as a solution, it is important to not forget that this tool itself cannot influence system change. The regulatory regimes and agricultural policy in general are just as important. Even though the percentage of organic farmers in France increased from 0.27% in 1989 to 4% in 2014 because of specific organic regulations, the long-term success and sustainability cannot be achieved without a profound reform of the European and national agricultural policies and the creation of a system of incentives that are far more favourable to agriculture, and food practices that are more respectful of balancing both ecological and socio-economic needs of future generations.

Notes

1 This chapter was translated from French by Allison Loconto.
2 To illustrate this vision, the arguments developed include GMOs as environmental services technology for restricting the use of pesticides.
3 This doesn't mean that critiques and alternative forms of agriculture have not emerged in these countries, but rather they have been focused on different issues such as agroecology or land reform.
4 In third countries benefiting from an equivalency agreement with the EU, the products certified organic by accredited certification bodies in their countries of origin per the procedures of that country can be exported to the European market without the need for additional EU accreditation of the certification bodies.
5 The European hegemony in the field of organic agriculture can be explained by several factors, especially the late development of a national standard in the US (2000) and the importance of early European movements (i.e. Nature & Progrès, Soil Association, Demeter) in the international organic movement, led by IFOAM.
6 These relaxations are mainly the authorisation of 'parallel production' (organic and non-organic on the same farm), a weakening of the need for agriculture to be linked to the soil, a reduction in the minimum slaughter age for chickens, and increased tolerance for GMO traces from 0.1 to 0.9%.

7 We find private organic standards in all European countries, standards that generally add stricter requirements to those included in the EU organic label. In some cases, their influence in the national organic movements is more significant than in France. For example, the Soil Association is used by more than 50% of the organic producers in the UK, and in Germany more than half of all organic producers are also certified for a private standard (e.g. Naturland, Bioland or Demeter).

8 The National Federation of Organic Agriculture (FNAB), a national union that counts half of the country's organic producers as members, raises the dual question when introducing its Bio Cohérence standard: 'Can and should the development of organic escape the "conventionalization" trap?' See: www.fnab.org/index.php?option=com_content&view=article&id=655:bio-coherence&catid=22:news-of-partners.

9 Half of the N&P producers are nonetheless also certified AB.

10 Demeter is also working on developing a PGS.

11 The implications of alternative certification systems is also valid for other sustainability standards.

12 www.ifoam.bio/sites/default/files/pgs_definition_in_different_languages.pdf

13 This type of certification is based on two stages: an internal control carried out by the producers or the federating structure and an external audit of the system carried out by an independent certifier. This systems has existed in southern countries since the 1990s and in Europe since 2008 for the protected geographical indications.

14 Which was originally one of the objectives of CABO (Association of Independent Advisors in Organic Agriculture) – one of the originators of the organic control system in France and which gave birth to Ecocert.

15 Since each product must be certified, the cost of the audit is generally based on the number of products and activities (crop production, breeding, processing and/or sales) to be controlled.

References

Allaire G., 2016. Que signifie le 'développement' de l'Agriculture Biologique (AB)? *Innovations Agronomiques*, 51(2016), 1–17.

Baqué P. (ed.), 2012. *La bio entre business et projet de société*. Marseille, Ed. Agone.

Besson Y., 2011. *Les fondateurs de l'agriculture biologique. Albert Howard, Rudolf Steiner, Maria & Hans Müller, Hans Peter Rush, Masanobu Fukuoka*. Paris, Le Sang de la Terre. p. 775.

Boltanski L., Chiapello E., 2005. *The New Spirit of Capitalism*. London, Verso.

Buck D., Getz C., Guthman J., 1997. From Farm to Table: The Organic Vegetable Commodity Chain of Northern California. *Sociologia Ruralis*, 37, 3–20.

Chiapello E., 2009. Le capitalisme et ses critiques. 4ème congrès du RIODD : la RSE, une nouvelle régulation du capitalisme, Lille.

Coombes B., Campbell H., 1998. Dependent Reproduction of Alternative Modes of Agriculture: Organic Farming in New Zealand. *Sociologia Ruralis*, 38, 127–145.

Darnhofer I., Lindenthal T., Bartel-Kratochvil R., Zollitsch W., 2010. Conventionalisation of Organic Farming Practices: From Structural Criteria Towards an Assessment Based on Organic Principles: A Review. *Agronomy for Sustainable Development*, 30, 67–81.

De Wit J., Verhoog H., 2007. Organic Values and the Conventionalization of Organic Agriculture. *NJAS- Wageningen Journal of Life Sciences*, 54, 449–462.

Espagne C., 2014. La différenciation au sein des produits issus de l'agriculture biologique en France: standard public et standards privés. Rapport de stage AgroParisTech. https://odr.inra.fr/intranet/carto_joomla/index.php/l-equipe-odr/menu-odr-theses.

Fouilleux E., Loconto A., 2014. Du projet politique alternatif à la multiplication de services et de marchés. Les trajectoires régulatrices de l'agriculture biologique. In Colloque *Renouveler les approches institutionnalistes sur l'agriculture et l'alimentation: La "grande transformation" 20 ans après*, Montpellier, 16 et 17 Juin 2014.

Fouilleux E., Loconto A., 2017. Voluntary Standards, Certification and Accreditation in the Global Organic Agriculture Field. A Tripartite Model of Techno-Politics. *Agriculture and Human Values*, 34(1), 1–14. doi: 10.1007/s10460-016-9686-3.

Freyer B., Bingen J., 2014. *Re-Thinking Organic Food and Farming in a Changing World*. Dordrecht, The Netherlands, Springer.

Garcia-Parpet M.-F., 2012. Le marché des certificateurs de l'agriculture biologique. In Bonnaud L. and Joly N. (Eds.), *L'alimentation sous contrôle. Tracer, auditer, conseiller*. Dijon, Editions Educagri et QUAE, QUAE, pp.109–123.

Gibbon P., 2008. An Analysis of Standards-Based Regulation in the EU Organic Sector, 1991–2007. *Journal of Agrarian Change*, 8, 553–582.

Guthman J., 2004. *Agrarian Dreams. The Paradox of Organic Farming in California*. Berkeley, University of California Press.

Hatanaka M., Busch L., 2008. Third-Party Certification in the Global Agrifood System: An Objective or Socially Mediated Governance Mechanism? *Sociologia ruralis*, 48(1), 73–91.

Hochreiter C., 2011. *Certified with Trust and Solidarity? Attitude, Benefits and Challenges of Organic Farmers in Participatory Guarantee Systems, Cacahoatán, Mexico*. University of Natural Resources and Life Sciences, Vienna.

IFOAM, 2005. *PGS Case Studies 2005 in Brazil, France, India, New Zealand, USA*. Germany, IFOAM Studies.

IFOAM, 2014. Membership Vote on Motions to the IFOAM Norms. www.ifoam.org/sites/default/files/ifoam_norms_motions_membership_vote_2014.pdf.

Jaffee D., Howard P., 2009. Corporate Cooptation of Organic and Fair Trade Standards. *Agriculture and Human Values*, 27, 387–399.

Jahn G., Schramm M., Spiller A., 2005. The Reliability of Certification: Quality Labels as a Consumer Policy Tool. *Journal of Consumer Policy*, 28(1), 53–73.

Lamine C., Bellon S., 2009. Conversion to Organic Farming: A Multidimensional Research Object at the Crossroads of Agricultural and Social Sciences. A Review. *Agronomy for Sustainable Development*, 29, 97–112.

Lemeilleur S., Allaire G., 2014a. Peut-on faire entrer le Développement Durable dans des labels? In Colloque *Renouveler les approches institutionnalistes sur l'agriculture et l'alimentation: La "grande transformation" 20 ans après*, Montpellier, 16 et 17 Juin 2014.

Lemeilleur S., Allaire G., 2014b. Standardisation and Guarantee Systems: What Can Participatory Certification Offer? In Paper presented at the Conference IIPPE 2014: Fifth Annual Conference in Political Economy, *The Crisis: Scholarship, Policies, Conflicts and Alternatives*.

Lemeilleur S., Allaire G., 2016. Certification participative des labels du mouvement de l'agriculture biologique: Une réappropriation des communs intellectuels. In 12th AFD International Conference, *Commun et Développement*, 1–2 décembre 2016, Paris, France.

Lemeilleur S., N'Dao Y., Ruf F., 2015. The Productivist Rationality behind a Sustainable Certification Process: Evidence from the Rainforest Alliance in the Ivorian Cocoa Sector. *International Journal of Sustainable Development (IJSD)*, 18(4), 310-328.

Leroux B., 2011. *Les agriculteurs biologiques et l'alternative. Contribution à l'anthropologie politique d'un monde paysan en devenir*. PhD Thesis in Sociology, EHESS.

Leroux B., 2014. Une perspective sociohistorique des agricultures biologiques en France. In Leroux B. (Ed.), *Dynamiques des agricultures biologiques. Effets de contexte et appropriations*, Versailles (FRA), Editions Quae, pp. 19–43.

Loconto A., 2017. Models of Assurance: Diversity and Standardization of Modes of Intermediation. *The Annals of the American Academy of Political and Social Science*, 670(1), 112–132.

Loconto A., Poisot A.S., Santacoloma P., 2016. *Innovative Markets for Sustainable Agriculture: How Innovations in Market Institutions Encourage Sustainable Agriculture in Developing Countries*. Rome, Food and Agriculture Organization of the United Nations and Institut National de la Recherche Agronomique.

Loconto A., Stone J.V., Busch L., 2012. Tripartite Standards Regime. In Rtizer G. (Ed.), *The Wiley-Blackwell Encyclopedia of Globalization*, Malden, MA, Blackwell Publishing Ltd, pp. 2044–2051.

Nelson E., Tovar L.G., Rindermann R.S., Cruz M.A.G., 2010. Participatory Organic Certification in Mexico: An Alternative Approach to Maintaining the Integrity of the Organic Label. *Agriculture and Human Values*, 27(2), 227–237.

Piriou S., 2002. *L'institutionnalisation de l'agriculture biologique (1980–2000)*. Thèse de l'Ecole Nationale Supérieure Agronomique de Rennes, l'Ecole Nationale Supérieure Agronomique in Rennes.

Poméon T., Desquilbet M., Monier-Dilhan S., 2014. Les standards privés dans le développement de l'agriculture biologique. In Colloque *Renouveler les approches institutionnalistes sur l'agriculture et l'alimentation: La "grande transformation" 20 ans après*, Montpellier, 16 et 17 Juin 2014.

Robidel J.P., 2014. Quantifier et cartographier l'agriculture biologique des années 1980–90 en France. In Cardona A. et al. (coord.) *Dynamiques des agricultures biologiques : effets de contexte et appropriations*. Versailles, Ed. Quae – Educagri, pp. 45–56.

Rosset P.M., Altieri M.A., 1997. Agroecology Versus Input Substitution: A Fundamental Contradiction of Sustainable Agriculture. *Society & Natural Resources*, 10, 283–295.

Stassart P.M., Jamar D., 2009. Agriculture biologique et verrouillage des systèmes de connaissances. Conventionnalisation des filières agroalimentaires bio. *Innovations Agronomiques*, 4, 313–328.

Teil G., 2013. Le label AB, dispositif de promesse ou de jugement? *Natures Sciences Sociétés*, 21, 213–222.

Van der Akker J., 2009. Convergence entre les Systèmes Participatifs de Garantie et les Systèmes de Contrôle Interne dans un projet pilote européen d'IFOAM. *Innovations Agronomiques*, 4, 441–446.

Verhaegen E., 2012. Les réseaux agroalimentaires alternatifs: Transformations globales ou nouvelle segmentation du marché. In Van Dam Denise, Nizet Jean, Streith Michel (eds.), *Agroécologie entre pratiques et sciences sociales*. Dijon, Educagri editions, pp. 265–279.

Viel, 1979. *L'agriculture biologique, une réponse?* Paris, Editions Entente, p. 96.

Vogl C.R., Kilcher L., Schmidt H., 2005. Are Standards and Regulations of Organic Farming Moving Away from Small Farmers' Knowledge? *Journal of Sustainable Agriculture*, 26(1), 5–26.

Willer H., Lernoud J., 2017. *The World of Organic Agriculture – Statistics and Emerging Trends 2017*. Bonn, Germany, Research Institute of Organic Agriculture (FiBL) and IFOAM – Organics International.

11 Japanese agri-food in transition

D. Hugh Whittaker and Robert Scollay

Introduction

This chapter explores continuity and change in Japan's agriculture system, and tentatively places it in a broader agri-food framework that extends to consumers and agri-food importers and their overseas interests.[1] Without this framing we run the risk of reinforcing Japanese exceptionalism, but it has to be done sensitively, and not by simply imposing a framework developed in other contexts like a suit of ill-fitting clothes. Although there is a voluminous literature about Japanese domestic agriculture, little attention has been paid to how imported food and food inputs reach Japanese plates. The chapter draws inspiration from the Régulation and Convention approaches in other chapters and earlier literature, such as Allaire and Boyer (1995), but it should be seen as the beginning rather than the conclusion of a dialogue.

Recent upheavals in Japanese agriculture were triggered by an accumulation of contradictions in the domestic system, the global food crisis of 2007–08, and Japan's decision to enter into negotiations over the TPP (Trans Pacific Partnership). Those opposed to the TPP soon mobilised, notably the powerful agricultural cooperatives (collectively referred to as JA), but they in turn came to be portrayed as entrenching vested interests and holding the country hostage to the past, giving incoming Prime Minister Abe the opportunity to introduce a package of reforms, including reform of JA itself.[2]

Change has come to Japanese agriculture. There are now two broad streams, one moving in the agro-industrial direction, and the other towards 'differentiated quality'. (In fact, there are several currents within each stream.) They represent an embrace and rejection, respectively, of market allocative efficiency, with some middle ground. This middle ground, as we shall see, has historical roots.

The chapter begins with a brief consideration of pre-World War II agriculture and its role in Japanese economic development, without which we cannot understand the present. We then consider post-World War II economic reconstruction and the role of domestic agriculture in the high-growth model, followed by its mounting problems and contradictions. The system withstood pressures for change during the Uruguay Round, into the 2000s. From the late 2000s, however, internal problems were compounded by the global food crisis and then the prospect of TPP. Here we attempt to assess the changes that have followed.

Agriculture in Japan's economic development: fly in theoretical ointments

Wilkinson and Goodman (Chapter 7) rightly cite Japan's challenge to Food Regime Analysis. In fact, the challenge posed by Japan extends to basic Anglo- or Euro-centric models of industrialisation, economic development, and world systems. For this reason, as well as to understand contemporary agriculture, we begin with the Tokugawa period (1600–1868).

In *The Great Divergence* (2000), Pomeranz challenges the notion that the preconditions for industrialisation in the 18th century were only present in Europe, and therefore proposes that the divergence brought about by the Industrial Revolution must be based on more contingent factors. While China is Pomeranz's main counter-example, Francks (2016) extends this analysis to Japan. Far from Japan being closed, static and feudal, she argues, there was considerable international trade, especially in the first half of the Tokugawa period, and subsequent import substitution when the central authorities attempted to stem the outflow of silver to China. Limited market relations penetrated into the countryside, and small-scale rural industry was both vibrant and innovative. By the 1850s, non- primary sectors accounted for around two-fifths of GDP, stimulated by rising domestic consumption which sumptuary legislation failed to stem. Francks suggests that comparative data which suggests a significant gap with Europe by 1700 exaggerates the differences. Real wage data is problematic because many rural Japanese households relied on multiple sources of income. And income was comparatively equally distributed, implying higher *average* levels than in less equal Europe.[3]

Francks' argument is not simply concerned with preconditions of industrialisation. She argues that the rural, small landholding, mixed income, semi-commercial and innovative economy of the Tokugawa period was carried over into the Meiji period (1868–1912). The Meiji 'Restoration' may have unblocked impediments to change, and opened the country to the importation of European institutions and technologies, but the fundamental growth pattern did not change. This contradicts views derived from Gerschenkron's late development thesis, which see Japan as advancing primarily by the rapid adoption of capital-intensive, urban-based technologies and institutions. Further, and contrary to the common view that Japanese imperialism was motivated by resource shortages,

> Japan continued to be a net exporter of land-using primary produce (if this includes raw silk, alongside such items as tea and rice) through much of the Meiji period, and it was only the drive towards heavy industrialization itself, in the interests of the militarist state of the 1930s, that caused the shortages of mineral and energy resources that imperial expansion was supposed to resolve
>
> (Francks, 2016: 52; cf. also Yasuba, 1996).

Ownership rights and social relations in the countryside also shaped Japan's path. Nakabayashi (2017) points out that feudal lords of the Tokugawa period were entitled to rents, but that the property rights belonged to farmers, and by the 1670s were legally protected. Social relations based on small landholdings, mechanisms to alleviate risk and pool fiscal responsibility, evolved in this context.[4] Although farmers retained their property rights, after the Meiji Restoration they were obliged to pay their taxes to the government in cash. They were no longer allowed dispensations for bad harvests, and the collective (village) payment system was abolished, exposing them to greater risk and volatility. As a result, tenanted farmland rose from 30% in the 1870s to 45% in the early 1900s. Many tenants continued to work on their former land, however, so (moderate) concentration of ownership did not equate with larger units.

And while the landlords became more prominent, many were still involved in agriculture themselves. Dore (1965) distinguishes between two types of landlords – those who acquire control over a territory by military conquest or infeudation, and those who achieve their position by economic means within the framework of the established political order. With the removal of the *daimyō* lords, only the second type were left in Japan during the Meiji period. And on balance, he argues, they were a force for development, at least until around 1920, as they were often responsible for the introduction of new ideas and farming methods to their villages (cf. also Waswo, 1977).

Although a new land tax accounted for almost 80% of government revenues in the first half of Meiji, the rural tax burden was no higher than during the Tokugawa period. Rural output and productivity increased, and living standards in the countryside were maintained (Smith, 1959). New frontiers for agriculture were opened up in Hokkaido, and later in Japan's overseas colonies, but increased output came more from greater productivity of existing resources in small-scale units and the expansion of side income than in such expansion.

From around 1920, however, things began to change. Landlords had become less involved in agriculture, and tenants had become more educated and less deferential, sparking a rise in tenancy disputes. This was fuelled by falling silk, rice and other primary goods prices from the mid-1920s, leading to an 'agrarian crisis' of generalised rural hardship, providing grist for militarists and anti-capitalists.[5] Although it took an extreme form in the 1930s, in general one can say that the case of Japan confirms Davis (2004) argument that rural 'middle classes' often play an unacknowledged role in economic development by supporting government attempts to discipline both capital and labour.[6]

The rise and decline of the post-war system

The 'Japanese model' of economic development was a product of pre-war and wartime continuities and post-World War II reforms. Originally intent on demo-cratising Japan, Occupation policy shifted to economic rehabilitation, and aligning Japan with the US in the Cold War. Beginning with provisioning of US-led forces in the Korean War (1950–53), Japanese manufacturing came to enjoy a remarkable resurgence and productivity growth, with the spread of mass production generating high growth until the late 1960s. Middle school and then high school graduates from rural Japan boarded trains to swell the ranks of factory workers in the key industrial belts.

But in the bleak immediate aftermath of the war, the government's first priority for agriculture was to stabilise prices, and to ensure enough food was produced domestically to feed the population, with colonies now lost and their million Japanese citizens repatriated. The Occupation authorities demanded a more radical land reform than that contemplated by the Japanese government; General MacArthur was to hail it as the most far-reaching – and successful – land reform since the Roman Gracchus brothers in the 2nd century BC. Almost 2 million hectares were released to tenants. The corollary, however, was a perpetuation of the system of small land holdings, typically of one hectare or less. The 1952 Agriculture Land Law consolidated this structure by creating stringent restrictions on farmland ownership.

The Occupation authorities saw agricultural cooperatives as a vehicle for democratisation in the countryside, along with labour unions in the cities. But in the war's aftermath the collection and distribution of food was the more urgent priority, and wartime agricultural associations were resurrected. Almost 14,000 'cooperatives' were rapidly formed, along with prefectural and national organisations. 'Bottom-up' in principle, many ran into financial trouble, and in the ensuing restructuring, 'top-down' processes were strengthened. They became a bastion of the post-war system.

The 1961 Basic Law on Agriculture sought to promote modernisation and mechanisation, as well as greater efficiencies through amalgamation, but there was little political appetite for reversing the results of the land reform.[7] With the path of amalgamation foreclosed, land and labour productivity improvements advanced on the basis of small holdings, which became highly mechanised, enabled by generous subsidies to buy equipment and fertiliser from manufacturing. By the mid-1960s, Japan had 'reached the top of the world league table for the horsepower available per acre of cultivated land' (Fukutake, 1989: 93). Ironically, however, 'the cost of this mechanisation, together with increases in the cost of chemicals for pest control and of fertilisers, provided a brake on farm incomes in spite of increases in yields' (ibid.).

However, and reminiscent of the Tokugawa and Meiji eras, income from farming was supplemented with other sources of income, notably work in factories that were being relocated from the major urban centres to their hinterlands. By that time the the Basic Law had been passed, and full-time farmers were a minority, with the remainder split between 'type 1' (most income from farming) and 'type 2' (minority of income from farming) farmers. By 1970, the ratios were, respectively, 16%, 34% and 51%. Government subsidies for rice, tax advantages and rising land prices further discouraged farmers from cutting ties with their land (Rothacher, 1989).

By 1970, a turning point had been reached. Rice production exceeded consumption, which was declining as diets diversified, but the excess could not be exported competitively. A new policy of rice diversion (acreage control) was implemented, in which farmers were paid subsidies to produce crops other than rice. The policy was expanded over time.[8] Still, rising costs of supporting the rural sector were borne by urban middle-class families, who valued improved quality, and enjoyed a wider range of products that their rising incomes enabled them to purchase. Government policy and the JA structure had promoted 'standardisation' (Allaire and Wolf, 2004), but there were extensive markets for specialist quality produce, displayed, for example, in the lower floors of department stores, or served in restaurants on the upper floors.

As diets diversified, Japan's food self-sufficiency ratio steadily declined, from just under 80% in 1960 to around 60% in 1970 (and 40% by the mid-1990s on a calorie basis: Godo, 2013). As livestock numbers increased, feed grain had to be imported, making livestock farmers susceptible to volatility in international markets. Such vulnerability was highlighted by the temporary US embargo on soybeans in 1973, which compounded the 'oil shock' waves from the Arab-Israeli War. Japan's large general trading companies (sōgō shōsha), which had expanded hugely with Japan's post-war growth and integration into the international economy, were hit hard by the transition to slow growth and structural changes in the Japanese economy. Forced to change their business models, they became more involved in domestic value chain integration, and began to invest abroad to secure their own upstream inputs (Yoshino and Lifson, 1986; Ryan, 2017).

In the early 1970s, with ODA support, three of Japan's trading companies acquired almost 17,000 hectares in Lampung, Indonesia, to grow corn. The ventures failed, and the farms were sold to the Indonesian government. In 1979, again with ODA support, a 350,000-hectare soybean project was launched in Cerrado, Brazil. From 230,000 tons in 1975, the region grew to produce 17.6 million tons by 1999–2000, and 15% of Japan's soybean imports by 2008 (MAFF, 2013). By and large, however, the trading companies were cautious about investing directly in overseas farming, at least by themselves. Their interest was primarily in securing stable, quality supplies and diversifying risk. Over time this led them to strategies of value chain integration, with investments or downstream as well as upstream.

Japan entered the 1980s in much better shape than most industrialised economies. Compromises by labour unions over wage increases had brought inflation under control, and Japanese manufactured products were sweeping world markets, generating intense trade friction. Murmurings began to be expressed of discontent with Japan's 'triple burden' of food surpluses, high consumer prices and rising government expenditure, however, and in 1980 the peak business federation Keidanren began to call for agriculture reform and liberalisation. US trade negotiators had pressed Japan to liberalise its beef and citrus during the GATT Tokyo Round in 1978–79, and Keidanren feared retaliation against the manufacturing sector (Yoshimatsu, 1998).[9]

JA retaliated with boycotts against the companies of prominent reform advocates, and Ministry of Agriculture, Forestry and Fisheries (MAFF) exerted pressure on Keidanren to moderate its demands. US pressure was harder to deflect. As the Uruguay Round got under way, Japan was forced to end quotas on beef and oranges (in 1991). In the 1989 election after this was announced, the Liberal Democratic Party (LDP) lost its upper house majority as a result of rural protest votes.[10] The nerve point was rice. Japan was prepared to consider a reduction in total domestic support for agriculture, but it strongly resisted pressures for tariffs on rice. Officials declared, 'we will never allow even a single grain of rice to enter Japan'. When rice was eventually 'tariffied', the rate was set at ¥341 per kg, which equated to an *ad valorem* tariff of several hundred percent on US, Thai and Vietnamese rice (with small specified quantities imported by MAFF's Food Agency under a tariff rate quota scheme at zero tariff rates).

A sum of ¥6 trillion was budgeted to prepare the countryside for trade liberalisation. Much of this money (plus ¥1.2 trillion for rural 'hometown' works) in fact went into public works, with tenuous links to agriculture. Although the level of market price support for Japanese agricultural products did decline by over a third between 1986 and 2006, when measured by the OECD's producer support estimate which includes budget transfers as a percentage of farm receipts (%PSE), the decline was more modest. OECD (2006: 34–35) figures suggest a decline from around 66% in 1986 to just under 50% in 2006, making it the fifth highest level of support in the OECD. Others place the figure higher (cf. Yamashita, 2008b; Shogenji, 2010).

Internal erosion and disjointed policy

Transfers and stopgap measures taken to revive Japanese agriculture were not able to arrest, or even mask, the decline of Japanese agriculture, in which the total value of agricultural production declined by almost a third from the mid-1980s to 2009. The ratio of 'commercial' farms (which earned at least ¥500,000 per year from their produce, from farm land of 0.3 hectares or more) to 'non-commercial' farms (under the commercial threshold, or having agricultural land but not cultivating it) was almost 2:1 in 1990; by 2010 it had reversed to 1:1.4 (MAFF, 2010). Underlying relative stability of the number of full-time commercial farms over this period, moreover, was a shift from part-time to full-time farming with retirement from non-farming jobs.

This in turn signalled the ageing of full-time farmers. In 2009, not only were 48% aged 70 and over, as would be expected with an average age of 68.6, but 31% were aged 75 and over. Only 7% were aged between 20 and 39. In addition, non-farm jobs were disappearing, with factories relocating offshore, and administrative amalgamations and cuts squeezing rural public sector jobs. Income-supplementing jobs for successor children thus became limited. Indeed, in many cases there was no successor (MAFF, 2011a). Land under cultivation declined.

> [S]ome 2.60 million hectares—more than 40% of the 6.09 million hectares that existed in 1961—have disappeared due to abandonment and conversion for residential or other purposes. This is about the same as the total area of rice paddies in Japan today
>
> (Yamashita, 2008b).

Of course there are differences by agricultural sector. Around 70% of commercial farms plant paddy rice, although the share of rice in the value of Japanese agricultural production fell from over 50% in 1955 to little more than 20% by 2010, with per capita consumption halving. Other sectors are seen to be in better shape. Although dairy appears to be a relative success story, it too has many problems. National output of milk declined from around 8.5 million tonnes in the mid-1990s to 7.5 million tonnes in the early 2010s, and the number of dairying units has declined. This has been accompanied by a steady increase in the number of cows per farm, as well as yield per cow. Average incomes are also relatively high, at least in the northern island of Hokkaido. But risk rises with increasing farm scales, for example risks from volatile imported feed prices and capital costs.[11] And many of the ageing and succession problems found in rice farming are also found in dairy, even in bigger Hokkaido farms.[12]

The decline of Japanese agriculture has been such that few oppose reform in principle, openly at least. But there are degrees of reform, and different timelines. Minimalist, gradualist and radical reformers have very different visions. Until 2013, the former two were the dominant force. Under a gradual reform vision, the 1999 Basic Law on Agriculture recognised that policy needed to be nuanced according to the local needs, and envisaged a role for local bodies to take part in both policy formulation and implementation. However, this was disrupted by the amalgamation of local bodies, also beginning in 1999. The ability of local bodies to engage in more nuanced policy formation was undermined, just as new expectations were being thrust on them (Shogenji, 2010).

The 2000 Basic Plan of Agriculture reiterated the goal of raising food self-sufficiency from 40% to 45% (on a calorific basis) by 2010, without realistic measures to achieve the goal, and while maintaining diversion policies that themselves had contributed to a decline. A proposal for 'sextiary industry' (*rokuji sangyō*) sought to link agricultural production (primary industry), processing (secondary industry) and sales and marketing, extending into export markets (tertiary industry), but from the logic of upstream growers moving downstream, rather than the reverse. Proposed as a way forward for farming villages, it was congruent with the multiple income stream concept, which led to the growth of part-time farming and away from consolidation. Consolidation proposals envisaged local farming organisations taking the lead. The revised Agriculture Land Law of 2009 began the process of deregulating who can participate in agriculture, but local farming organisations that could reinforce the positions of local stakeholders were allowed to play a key role (Jentsch, 2017).

Overall, the gradualist approach did little to arrest the decline of Japanese agriculture. Indeed, Shogenji (2010) describes a 'triple deterioration' of agriculture policy, agriculture production, and rural villages. Although it introduced an income stablilisation programme for farmers in 1998, in 2008 the LDP indicated that it would begin to focus support on 'core' farmers. This opened a wedge for the opposition Democratic Party of Japan (DPJ) to exploit, by offering income support payments to all farmers, regardless of size, or whether full-time or part-time. The DPJ swept into power in 2009, and introduced the scheme initially as a pilot programme for rice farmers, to be subsequently extended to producers of crops such as wheat, barley, soybeans, sugar, and starch potatoes.[13] In 2010, the core of Japanese agriculture was highly protected, and very fragile.

New external threats: food crisis and TPP

Two new external threats added to the internal problems of Japanese agriculture, and tipped the scales towards more radical reform. The first was the 'food crisis' of 2007–08, which rekindled fears about food security. The fears were compounded by the growing appetite for food and land by Japan's giant neighbour China, and other emerging economies. How was Japan to make its way on a planet with growing competition for finite resources? The second was a sense that Japan was being left behind in the emerging 'noodle bowl' of bilateral and multilateral free trade agreements, largely because of its insistence on agriculture protection. The costs of this stance were perceived to be growing. The TPP train was leaving the station; would Japan be on it or not?[14]

The response to the 'food crisis' echoed the 1970s response. Mitsui & Co. purchased 100,000 hectares of land in Brazil in 2007 and Sojitz established a farming unit and purchased 11,000 hectares in southern Argentina. But as Hall (2015) points out, the *sōgō shōsha* were generally cautions about becoming directly involved in farming; more significant were their downstream investments aimed at giving them greater control over food chains. These included Mitsui & Co.'s 25% stake in Multigrain AG (2007, subsequently raised to 45% in 2010), and Marubeni's purchase of US grain trader Gavilon in 2011. Those investments had China's expanding market in mind, moreover, and included Itochu's alliance with COFCO and Marubeni's investment in Sinograin.

Such investments were probably more defensive than offensive. As Hall notes, Japan's trading companies feel increasingly squeezed between the US ABCD majors and China's emerging giants like COFCO. As China currently imports 60% of the world's exported soybeans, to stay out of that market means being side-lined. And as trading company officials often assert, the scale economies generated from Asian business bring benefits to Japan in the form of cheaper import prices. They thereby claim to be preserving their historic role in Japan's agri-food system rather than abandoning it.[15]

The Japanese government was also constrained. On the one hand, it started a massive triangular project in 2009 in Mozambique, also involving the Brazilian Cooperation Agency, to convert 11 million hectares of land primarily used for peasant agriculture into commercial agricultural production. (The project soon ran into problems; attempts were being made to revive it in 2017.) On the other hand, it was alarmed at the scale of 'land grab' acquisitions by other countries, and strongly supported the Responsible Agricultural Investment initiative and its seven principles at the G8 Summit in 2009.

The second threat was the TPP. In October 2010, Prime Minister Kan announced that Japan would consider participating in the TPP negotiations. Opponents sprang into action. Centring on JA and indeed the MAFF, they produced a number of publications claiming that participation in TPP would destroy not just Japanese agriculture, but Japan's very identity, rooted in farming. An anti-TPP petition collected 11 million signatures. MAFF's (2010) GTAP (Global Trade Analysis Project) model suggested that half of Japanese agricultural output and 90% of rice production would be destroyed through TPP participation, based on questionable assumptions about the price, source and amount of rice which would be imported (Honma, 2012).

Although slower to mobilise, proponents of deep, liberalising reform gained a public platform. Their GTAP models produced very different results. They argued that TPP participation need not be negative for agriculture. Indeed, they argued that it was the policy framework supported by the agriculture lobby that was trapping Japanese agriculture in a downward spiral, and that if agricultural reforms were carried out, Japanese agriculture could be transformed into a vibrant sector with significant exports – the most effective way of bringing about reform would be through participation in TPP.

Crucially, agriculture moved from the fringes of public debate into the mainstream media. One factor was the realisation that it was not just the interests of the agriculture sector that were at stake in the debate but Japan's economic, diplomatic and political interests as well. Another, conversely, was the rediscovery in the public mind of the issues facing the agricultural sector, and the possibility that it could be much more than a subsidy-absorbing 'traditional' sector, and might actually become a dynamic part of Japan's future.

This was the line taken by Prime Minister Abe. Following his victory in the December 2012 lower house election, Abe took the decisive step of entering into the TPP negotiations, presenting this as a key part of the 'third arrow' of 'Abenomics', aimed at restoring Japan's economic vitality and power.[16] By also placing agriculture within the third arrow, the Abe government sought to shift the ground of the debate over the implications of TPP for Japanese agriculture. A 'Vitality Creation Headquarters of Agriculture, Forestry, Fisheries and Rural Regions' was set up in 2013, tasked with mapping out a strategy to double agricultural/rural income over ten years. This including a doubling of agricultural exports to ¥1 trillion by 2020; reducing production costs by 40% over ten years through consolidation with the help of a Farmland Mediation Organization; doubling the number of people entering farming, and increasing corporate farming units to 50,000 in ten years; and so on (Sakuyama, 2014). A Food Industry Division was established within MAFF. Food had become an 'industry', reflecting a pendulum swing towards the reformers within the Ministry.

This shift in the Ministry and defection of key politicians (*zoku giin*) also weakened Japan's agricultural 'iron triangle' (George Mulgan, 2017), allowing Abe to begin an assault on 'fortress JA' in 2015. With almost 10 million members – more of them 'associate members' (not engaged in agriculture) than regular farmer members – and, in the past at least, a crucial base for electoral success, many had predicted that this would not happen.[17] But it did. In principle, the reforms will weaken the peak organisation of JA and shift the weight towards individual cooperatives (and towards agriculture), give them new choices through competition, and encourage local independence and creativity. The immediate impact was not dramatic; inevitable political compromise watered down the impact although not the direction of reform. Similarly, although subsidies for rice diversion (acreage reduction) would be stopped in 2018, 'voluntary' targets would still be set in most prefectures.[18] In the TPP negotiations Japan fought hard, and for the most part successfully, to preserve protection for the 'five sacred items' of rice, wheat, beef and pork, dairy products, and sugar, and to minimise the overall impact on Japanese agriculture (cf. Yamashita, 2015).

Irrespective of the eventual TPP outcome, however, the possibility – indeed likelihood – of such an agreement has focused attention on Japanese agriculture, and accelerated and deepened proposals for change. What started as an all-out offensive against TPP ironically opened the door to change more quickly than anyone envisaged at the start. From 'reactive' agriculture, a new future narrative is being constructed around a discourse of 'strong agriculture', or 'agriculture on the front foot', buoyed by new entrants and investors. It is to these that we now turn.

What kind of rebirth?

Corporate participation in agriculture was accelerated when in 2014 the Farmland Mediation Organization (commonly called the Farmland Bank) was set up as an intermediary for those wishing to lease farmland, and those wishing to make it available.[19] The lease period was simultaneously extended from 20 to 50 years. Retailers had been preparing for this. Aeon, with 1,500 retail outlets nationwide, announced that by 2015 it would be raising its own-brand vegetable share from 4% to 20%, or ¥100 billion annual sales, through a combination of direct farm management and contract farming. Its own farms would average 15 hectares, and it would secure at least 500 hectares.[20] It may become the biggest farm cultivator in Japan. Seven & i Holdings Co also announced it would triple its private brand fruit and vegetables through a combination of direct farm management and contract farming. Convenience store chain Lawson, by contrast, announced tie-ups with larger growers through minority stakeholdings, mindful that non-agriculturalists have a poor record of success at farming in Japan. For its part, JA started entering into tie-ups with non-agriculture companies for production, processing, logistics and dining.[21]

Financial institutions, including the main city banks as well as regional banks, rushed to set up new funds for agriculture. Seeking higher returns than from Japan Government Bonds, life insurance companies also announced plans to invest in 'growth' sectors, including agriculture. Manufacturers also set up farm divisions, including almost all the major electronics companies who see real-time data management and automated control systems as a new business frontier, and hope to turn such 'IT farming' into an export industry.[22] Around 50 major IT and related companies joined a joint project by Keidanren and JA on 'next generation agriculture'. Toyota has been honing its *kaizen* practices for rice, aiming to reduce material costs by 25%, labour costs by at least 5% (while securing wages at manufacturing levels), and to bring the cost of producing rice down to ¥100 per kg, or ¥6,000 per60 kg. It has experimented on its own farmland, and created a subsidiary to commercialise its methods.[23]

Nor is this just being done in Japan. In Dà Lat, Vietnam, Japanese companies are expanding their presence through local tie-ups to raise productivity and quality, selling the produce with a Japanese quality label through the Aeon Mall in Ho Chi Minh City, or exporting it to Hong Kong and Japan.[24] There are growing opportunities for this kind of business in view of numerous food safety scares in Asia.

Finally, while predating the Abe administration's reforms, the government introduced loans and tax incentives to entice young novices into farming. The schemes initially resulted in considerable churning, the key stumbling block being skill acquisition (Kobari, 2012). But support structures have started to deepen, and new opportunities for both on-job and off-job training have sprung up, providing a stepping stone to eventual owner-cultivator farming, or other agricultural employment. Many seeking to make the transition to owner-cultivator farming are university-educated married couples, seeking a new way of life for themselves and their children. Applicants for agriculture studies in universities have also risen sharply, prompting some universities to strengthen their agriculture departments, and to acquire new farmland.[25]

As this suggests, agriculture is now being positioned as a growing 'new industry'. Value chains along which agricultural produce moves are being reformed, with and without JA, and expanded internationally. Some inspiration has been gleaned from abroad. France has attracted Japanese interest in the area of land consolidation while maintaining the spirit of family farming. Indeed, between 2005 and 2015 the land share of agriculture units of 10 hectares or more outside Hokkaido increased from 11% to 27%, marking a significant step towards concentrating land use in the hands of 'willing and able' farmers (*ninaite*).[26] Holland, too, has provoked considerable interest, especially outside of farming circles. Harada (2011) points out that despite its small size, Holland exports US$80 billion in agricultural products (cf. also Kawashima, 2011). If Holland can do this, why not Japan?

This new narrative is one of 'strong agriculture' – agriculture on the front foot, and not on the defensive. Buds of new agriculture supporting the discourse are showcased. They point to a new alignment of agriculture with Japan's manufacturing strengths, as well as internationalisation, capitalising on the popularity of Japanese cuisine abroad, a new geographic indication system, and reputation for quality and safety derived from manufacturing. Agricultural exports may not reach the targeted ¥1 trillion by 2019, but they rose from ¥400 billion in 2005 to ¥750 billion in 2015, while agricultural output overall increased by 5% alone in 2015 (MAFF, 2017).

But there is another narrative as well, especially among those looking to agriculture for a new, eco-friendly, local, personal and non-industrial way forward. It surfaced in the mainstream media in the aftermath of the 11 March 2011 triple (earthquake, tsunami and nuclear) disaster, before being swamped by the 'strong agriculture' narrative now associated with Abenomics. It flourishes, however, in interpersonal networks that link regional growers and 'cool' restaurants and consumers in large cities, along which flow speciality goods and pesticide-free or low-pesticide produce, as well as among those seeking to ignite individualised rural regeneration. In part, it is shaped by a rejection of the former narrative.

Young people are seldom attracted into agriculture by the prospect of efficiency gains; Japan's tightening labour market would give such people less arduous opportunities elsewhere. Many are seeking a different way of life, described by Rosenberger (2017) as 'edgework':

> With their eyes on making an ethical lifeworld within the neoliberal situation in which people seek freedom, they take the risk of being hybrid and partial as they link with consumers and use market and government initiatives to their advantage
>
> (p. 28).

In this they are different from the generation that founded the Japan Organic Agricultural Association in 1971, who opposed direct participation in market and government (ibid., p. 17).

Thus it appears that two broad narratives, or streams, have emerged in Japanese agriculture, with broader changes in Japanese capitalism and society as a background. Superficially, the first stream resonates with the agro-industrial model (cf. Chapter 3 in this volume), which was always compromised in Japan by small landholdings. This stream links with changes on the consumption side, in which corporate restructuring and the growth of non-regular employment have heightened price consciousness by many consumers, exerting significant downward pressure on prices, seen in the price of milk and yogurt in local supermarkets, for example.

The second stream appears to be linked to differentiated quality models (cf. Chapter 3), representing a rejection of the agro-industrial model, and a quest for something different in terms of ecology, ethics, locality and community. To some extent it connects with wealthier consumers who attach a value to traceability, locality, being organic, and having a 'story'. Geographically, those pursuing this kind of lifestyle are pushed to the fringes, to the less fertile land between mountains and away from cities – land unattractive for commercial purposes (Rosenberger, 2017).

But there is also a middle ground, which looks surprisingly like the past. The silkworms have gone, and so have the factories, but tourists are now finding their way to rural Japan in increasing numbers, and there are hopes that they can provide new sources of supplementary income that are interesting enough to attract young people back to rural villages. And the two new streams may not deviate far from this middle ground. The progeny of the rebirth of Japanese agriculture may not be so alien after all.

Notes

1 This research has received funding from the EU's Horizon 2020 research and innovation programme under the Marie Skłodowska-Curie grant agreement No 645,763. It also draws on the following works: Whittaker and Scollay (2017); Whittaker et al. (2013), commissioned by Fonterra (dairy co-operative Group), New Zealand.

2 Roughly 700 cooperatives are the constituent units of prefectural and national peak organisations that have been the organisers and defenders of rural interests. JA's activities span agricultural inputs, advice, retail, banking, insurance, and so on. With roughly 10 million members, less than half of whom are full members directly engaged in agriculture, they exert considerable political muscle.

3 Saito (2015: 409) estimates the relative household income ratio of samurai, merchant and farmer incomes in the 1840s as 1.2:1.1:1.0, as compared with 7.3:2.4:0.8 for the gentry, middle and lower class in pre-industrial (1688) England.

4 'While the shogunate regime strictly protected the property right of individual farmers, farmers were strictly regulated and protected against the commodity market risk, the financial market risk, and the natural weather risk. This resulted in an egalitarian society' (Nakabayashi, 2017: 4).

5 Despite some parallels, Dore and Ouchi (1971) caution against superimposing the rise of German fascism on Japan's case.

6 Davis was comparing the postwar growth of South Korea and Taiwan – former Japanese colonies– with Mexico and Argentina.

7 MAF Minister Akagi declared in 1964: 'The time has come to reform the farmland system to enable farmers to expand their operations. We should enable a public corporation to buy, sell, and lease farmland for farmers who want to expand their farm size' (cited in Yamashita, 2008a). A bill was drawn up and submitted twice, but scrapped both times.

8 In 2012, roughly 40%, or 1.1 million hectares of paddy land were subject to diversion schemes. A reversal of considerable symbolic significance happened in 2011, when household spending on bread overtook that of rice, reflecting changing diets and lifestyles: *Asahi shinbun*, 16 August 2012.

9 In 1980 MAFF's budget accounted for 8.4% of the General Account, of which 65% went to subsidies, the bulk of it for rice (Yoshimatsu, 1998: 340).

10 Those votes counted more than urban votes because electoral boundaries had failed to keep up with urbanisation, and JA was able to mobilise farmers and rural non-farmers alike.

11 Feed accounts for some 30% of production costs in Hokkaido; it is higher elsewhere in Japan.

12 Cf. Nōrin suisan chōki kin'yu kyōkai, 2005.

13 Under the scheme, farmers received a fixed payment of ¥15,000 per 0.1 hectares of planted rice, designed to cover the gap between the assessed standard production cost of ¥13,700 per 60 kg and a standard selling price of ¥12,000 per 60 kg. An additional price-contingent payment would be triggered if the average producer price in any current year fell below the standard selling price.

14 As its name suggests, this was a vehicle for trans-Pacific integration involving the United States and its NAFTA partners Canada and Mexico, as well as with the resource-rich economies of Australasia, Australia and New Zealand, with Chile and Peru in South America, and with Singapore, Malaysia, Viet Nam and Brunei Darussalam in Southeast Asia. Formal TPP negotiations were launched in March 2010.

15 This rationale was expressed even more strongly by Japanese food processing company executives we interviewed; they might benefit from Japan joining TPP, but it would mean fundamentally changing their business models, premised on 'solidarity of producers, processors and retailers' (sei-sho-han ittai). They were still 'part of the village'. Although large in terms of sales, most of those sales are generated domestically.

16 The first two arrows comprised, respectively, monetary expansion and fiscal stimulus. The third arrow, given the shorthand title of 'structural reform' consisted of a 'growth package' aimed at stimulating private investment by deregulation, internationalisation and other initiatives (Aso, 2013), which critics say has been partial at best (Posen, 2014).

17 In fact, agriculture is no longer the main business of many of the roughly 700 individual cooperatives. A 2014 survey of 482 cooperatives found that two-thirds directed less than 10% of their loans to agriculture. Profits from the sale of agricultural goods exceeded those from other activities in only nine cooperatives (*Shūkan daiyamondo*, 29 November 2014, pp. 68–75).

18 *Asahi shinbun*, 30 October 2017; *Jiji.com* 28 November 2017.

19 The number of 'corporation management entities' more than doubled from 8,700 to 18,857, and their share of total agricultural sales increased from 15% to 27%, between 2005 and 2015 (Ministry of Agriculture, Forests and Fisheries (MAFF), 2017: 6), although these figures include conversions from non-corporate entities as well as new entrants.

20 *Nikkei shimbun*, 11 August 2013.

21 *Shūkan daiyamondo*, 29 November 2014. This is one of the attractions of the possible corporatization of JA Zennō.

22 In part, this has been prompted by interest in agriculture shown by Google, Microsoft and Amazon, as well as Phillips.

23 *Nikkei shimbun*, 5 April 2014; 25 January 2015.

24 *Nikkei shimbun*, 5 March 2015.

25 *Nikkei Weekly*, 4 June 2012.

26 Ministry of Agriculture, Forests and Fisheries (MAFF) (2017: 7). This trend is partly from concentration, and partly from the formation of community-based farm cooperatives. The number of commercial farm households (outside Hokkaido) cultivating paddy land of 10 hectares or more almost doubled over the same period (ibid.).

References

Allaire G., Boyer R., 1995. Régulation et conventions dans l'agriculture et les ISS. In Allaire G. and Boyer R. (Eds.), *La grande transformation de l'agriculture*. Paris, INRA_Economica.

Allaire G., Wolf S., 2004. Cognitive Representations and Institutioal Hybridity in Agrofood Innovation. *Science, Technology and Human Values*, 29(4), 431–458.

Aso T., 2013. *What is Abenomics: Current and Future Steps of Japanese Economic Revival.* Mimeo.

Davis D., 2004. *Discipline and Development: Middle Classes and Prosperity in East Asia and Latin America.* Cambridge, Cambridge University Press.

Dore R., 1965. Land Reform and Japan's Economic Development: A Reactionary Thesis. *Developing Economies*, 3(4), 487–496.

Dore R., Ouchi T., 1971. Rural Origins of Japanese Fascism. In Moreley J. (Ed.), *Dilemmas of Growth in Prewar Japan*. Princeton, Princeton University Press.

Francks P., 2016. *Japan and the Great Divergence: A Short Guide*. London, Palgrave.

Fukutake T., 1989. *The Japanese Social Structure: Its Evolution in the Modern Century*. 2nd edition, Tokyo, University of Tokyo Press.

George Mulgan A., 2006. *Japan's Agricultural Policy Regime*. London, Routledge.

George Mulgan A., 2017. Loosening the Ties that Bind: Japan's Agricultural Policy Triangle and Reform of Cooperatives. *The Journal of Japanese Studies*, 42(2), 221–246.

Hall D. (2015), 'The Role of Japan's General Trading Companies (Sōgō Shōsha) in the Global Land Grab', paper presented to Land-grabbing, conflict and agrarian-environmental transformations: Perspectives from East and Southeast Asia' conference, Chiang Mai University, 5–6 June.

Harada Y. (2011), 'TPP to Nihon nōgyō' (TPP and Japanese Agriculture) for the Tokyo Foundation, accessed at www.tkfd.or.jp/research/project/news.php?id=865 on 22 August 2012.

Honma M., 2012. 'Nihon nōgyō 2020nen ni muketa seidō kaikaku no hōkō' (The Direction of Reform Towards Japanese Agriculture in 2020). In The 21st Century Public Policy Institute (Ed.), *Nōgyō saisei no gurando dezain (Master Plan for Agricultural Reform)*. Tokyo, The 21st Century Public Policy Institute pp. 106–123.

Jentsch H., 2017. Tracing the Local Origins of Farmland Policies in Japan: Local-National Policy Transfers and Endogenous Institutional Change. *Social Science Japan Journal*, 20(2), 243–260.

Kawashima H. (2011), 'TPP o ki ni "sentaku" to "shuchu" Nihon wa Orandagata nogyo yushutsukoku ni nareru' (With TPP as an Impetus 'Selection and Focus' Japan can Become a Dutch-style Exporting Country). *The Economist Report*, March, 97–99.

Kobari M. (2012), 'Nōgyōhōjin ni okeru jinzai ikusei no torikumi' (Measures to Nurture Human Resources in Agriculture Businesses), Nōrin kin'yū, July.

Ministry of Agriculture, Forests and Fisheries (MAFF), 2010. *Hōkatsuteki keizai renkei ni kansuru shiryō (Data on Trans-Pacific Strategic Economic Partnership Agreement)*. Tokyo, MAFF.

Ministry of Agriculture, Forests and Fisheries (MAFF), 2011a. *FY2010 Report on Food, Agriculture and Rural Areas in Japan – Summary*. Tokyo, MAFF.

Ministry of Agriculture, Forests and Fisheries (MAFF), 2013. *Policy Deployment of "Active Agriculture, Forestry, and Fisheries"*. Tokyo, MAFF, February 2013.

Ministry of Agriculture, Forests and Fisheries (MAFF), 2017. *FY2016: Summary of the Annual Report on Food, Agriculture and Rural Areas in Japan*. Tokyo, MAFF.

Nakabayashi M. (2017), 'Dual Structure of Emerging Japan: Revisit to Economics of Peasantry,' ISS Discussion Paper Series F-144, Institute of Social Sciences, University of Tokyo.

OECD 2006. *OECD Economic Surveys: Japan*. Paris, OECD.

Pomeranz K., 2000. *The Great Divergence: China, Europe and the Making of the World Economy*. Princeton, Princeton University Press.

Posen A., 2014. What Can We Learn from Abenomics. *East Asia Forum Quarterly*, 6(3), 3–6.

Rosenberger N.N., 2017. Young Organic Farmers in Japan: Betting on Lifestyle, Locality, and Livelihood. *Contemporary Japan*, 29(1), 14–30.

Rothacher A., 1989. *Japan's Agro-Food Sector: The Politics and Economics of Excess Production*. London, Macmillan.

Ryan P., 2017. *The Sōgo Shōsha: An Insider's Perspective*. Tokyo, Marubeni Research Institute.

Saito O., 2015. Growth and Inequality in the Great and Little Divergence Debate: A Japanese Perspective. *The Economic History Review*, 68(2), 399–419.

Sakuyama (2014), 'The Grand Design for Japan's Agriculture, Forestry and Fisheries Policies,' http://ap.fftc.agnet.org/ap_db.php?id+295&print=1, accessed 6 January 2015.

Shogenji S., 2010. *Nihon no nōsei kaikaku: Genba shiten no nōsei tenkai (Reform of Japanese Agriculture Policy: Changes in Agriculture Policy from the Viewpoint of the Field)*. Tokyo, Tokyo zaidan kenkyūsho.

Smith T., 1959. *The Agrarian Origins of Modern Japan*. Stanford, Stanford University Press.

Waswo A., 1977. *Japanese Landlords: The Decline of a Rural Elite*. Berkeley, University of California Press.

Whittaker D. H., R. Scollay (2017), La renaissance de l'agriculture japonaise? in G. Allaire and B. Daviron (eds), *Transformations et transitions dans l'agriculture et l'agro-alimentaire*. Versailless, QUAE.

Whittaker D.H., Scollay R., Gilbert J., (2013). TPP and the Future of Food Policy in Japan, NZAI Occasional Paper 13-01, in association with the Auckland, APEC Studies Centre, University of Auckland.

Yamashita K., 2008a. 'The Issues in the Farmland System' at www.tokyofoundation.org/en/articles/2008/the-issues-in-the-farmland-system accessed 8 December 2012.

Yamashita K., 2008b. 'The Perilous Decline of Japanese Agriculture' at www.tokyofoundation.org/en/articles/2008/the-perilous-decline-of-japanese-agriculture-1 accessed 8 December 2012.

Yamashita K., 2015. *Evaluating the Trans-Pacific Partnership Pact*. Tokyo, Research Institute of Economy, Trade and Industry (RIETI).

Yasuba Y., 1996. Did Japan Ever Suffer from a Shortage of Natural Resources before World War II? *Journal of Economic History*, 56(3), 543–560.

Yoshimatsu H., 1998. *'Japan's Keidanren and Political Influence on Market Liberalization'* in *Asian Survey*, 38(3), 328–345.

Yoshino M., Lifson T., 1986. *The Invisible Link: Japan's Sogo Shosha and the Organization of Trade*. Cambridge MA, MIT Press.

12 Transforming the dairy sector in post-communist economies

Actors and strategies

Pascal Grouiez and Petia Koleva

Introduction

Since the beginning of the 1990s, economies of the former Soviet Union and Eastern Europe have experienced deep structural changes referred to as the 'transition from plan to market'. The legacy of the socialist era left the agricultural sectors of Russia and Bulgaria highly concentrated around collective farms (*kolkhozes*) and State-owned farms (*sovkhozes*). Private plots, averaging 0.5 hectares (ha) and totalling less than 1.5% of arable land in each country, were the only form of private farming authorised before 1990. As with most agricultural segments, the dairy industry was managed by the agri-food industry and organised around highly integrated complexes on a regional basis: collective farms, State-owned farms and combinats. The comprehensive agricultural sector reforms (e.g. land restitution, dismantling of existing organisational structures, opening up to competition, etc.) challenged the legacy compromises between actors and opened up a field for strategic action, leading to new compromises about production, distribution and consumption of agricultural goods. This chapter endeavours to analyse the integration of the Russian and Bulgarian agricultural sectors into the globalised agri-food industry by focusing on the dairy industry. Which actors are involved in defining these new compromises? To what extent have they mobilised local institutional legacies, and how have they enabled these economies to adapt to the conditions for redeployment of the agricultural sector?

To answer these questions, we adopt an evolutionary and institutionalist perspective based on the dialectic between legacy and innovation. This path-dependency/path-shaping approach differs from both the neoclassical mainstream economics insights and some socio-economic analyses in terms of path-dependency. It implies that actors can intervene in the historical process and reshape the historical path of the sector. We build on the concept of 'transaction' (as developed by John R. Commons) to identify the strategies of the main dairy sector actors and their impact on manufacturing and marketing practices. From the manufacturing point of view, we underline that two risks are managed by the processing industries: supply disruption risk on the one hand, and the risk of not meeting the required quality standards on the other. The first risk management ensures the continuation of the process of production, while the second allows mass-market suppliers to reduce the capability of weaker suppliers (milk-producers) to develop a 'nested market' for milk products (Van der Ploeg et al., 2012). From a marketing point of view, some processing industries can use their brands as a quality label tool that meets the expectations of a specific category of local consumers. In parallel with these industries, dairies succeed in being niche players through the development of specific labels. Consequently, domestic dairy industries both in Russia and Bulgaria have restructured in two relatively autonomous markets for milk products: one with global and the other with local quality standards.

Path-dependent path-shaping post-communist change

Since 1990, two theoretical conceptions have dominated the post-communist economic agenda. The most influential, formulated by the so-called Washington Consensus, stressed the need for privatisation, liberalisation and stabilisation policy in order to break with the communist economic past. It was criticised by many economists because of its teleological and deterministic notion of change oriented to a final equilibrium state (i.e. the 'market economy') and based on the same reforms for all countries notwithstanding their institutional and economic heritage (see Chavance, 2011, for a summary of the debates). The alternative perspective underscores the dynamic properties of economic systems, which are made up of heterogeneous agents, tend to evolve in a non-linear fashion, and are characterised by multiple trajectories with path-dependence. However, some of these alternative approaches – e.g. McDaniel (1996) and Stark (1995) – tend to overemphasise the role of the path effect (negative or positive) and could be also seen as deterministic.

Since the mid-1990s, a few heterodox analyses of the post-socialist transformation have attempted to provide a more balanced combination of legacy factors with those related to the deliberate introduction of innovations. The ambition is to analyse both the recent past and the near future by adopting an approach in terms of path-dependency/path-shaping. According to Nielsen et al. (1995: 7) this approach 'implies that within specific, historically given, and potentially malleable limits, social forces can redesign the "board" on which they are moving and reformulate the rules of the game'. While the uncertainty resulting from the reshaping of an economic system may encourage individuals and organisations to reproduce past behaviour, this uncertainty can also trigger – for 'instituted' agents more or less tightly grouped together – deliberate actions driven by anticipations of the emerging order. The fatalism associated with path-dependency is thus rejected because social forces can interact with current circumstances and make new trajectories possible (Federowicz, 2000). In other words, within their path-dependent development, dairy sector actors make strategic choices and choose their path. This possibility enables a 'path enforcing' evolutionary development of the sector based on previous choices (e.g. productive specialisation, privatisation policy etc.). Inversely, it leaves also room for 'path-breaking' development (e.g. the emergence of nested markets).

Hence the path-dependency/path-shaping perspective could be seen as a modern reinterpretation of the dialectic between routine and strategic transactions, as highlighted in the 1930s by John R. Commons (1934), one of the founders of institutional economics. According to him, in resolving ordinary, familiar problems, individuals adopt habitual behaviour in routine transactions, governed by existing institutions. However, whenever a new problem arises, the behaviour patterns are no longer given and anticipations are disrupted: individuals are then involved in strategic transactions. These transactions are 'strategic' insofar as they develop new behaviours that can lead to the building of new routines. In order for the new practices discovered on an individual level to become institutions governing organisations and society, these practices must be selected. In capitalism, this selection process mainly takes place through collective bargaining (at the society level) and the choices of decision-makers (at the organisation level). When a compromise is found, strategic transactions become routine ones. The new institutions are the result of a synthesis between the new situations with new behaviours, and the old institutions.

In this chapter, we argue that individual actions are the expression of economic institutions, and individuals are continually placed in hierarchical relationships with the rest of society (or a community) as a group of active individuals. Thus, the institutions provide the rules of conduct, but these rules are constantly evolving due to individual and collective actions (which Commons calls 'working rules'). Moreover, Commons' conception of a socialised individual enables him to consider collective action as more than simply coercive. Indeed, collective action restricts, liberates or expands individual action. Commons considers three different types of transactions that specify the modes of negotiating working rules. A *bargaining* transaction is an interaction involving the exchange of produced wealth. It ensures recognition of the principle of property. The resulting transfer of property occurs between actors with comparable powers. A *managerial* transaction involves the organisation and control of production, it ensures the recognition of the principle of productive efficiency. A *rationing* transaction involves the distribution of wealth created (principle of sharing), within a specific institutional context. The three categories of transactions are differentiated according to their implicit power relations: equal parties (bargaining), unequal with the existence of a higher authority (managerial), or collective action (community or society) restricting individual action (rationing). Moreover, the ethical dimension of decisions is guided by the community's working rules, while helping these rules to evolve. The purpose of collective action is to ensure that the outcomes of transactions are not socially undesirable or unreasonable – two notions that are dependent on the institutional and historical context. The actors take part in the evolutionary dynamic of rules and are constantly acting by combining the economic, social and political dimensions of their decisions (Barthélemy and Nieddu, 2007). The social and political dimensions (i.e., the distributive dimension) of an instituted economy (Gislain, 2010) can highlight the territorial aspects of any economic activity.

The transposition of this grid to the analysis of a sector such as dairy offers the opportunity to integrate both its productive (sector-specific) and distributive (territorial) dimensions. Our objective is to cast light on the new setup of the dairy sector by going beyond a purely economic explanation for its evolution and change.

We will study changes in the Russian and Bulgarian dairy sectors. In each case, we will describe the institutional legacy, in order to better grasp the context in which strategic transactions (bargaining, managerial and rationing) occur. We will show how the transitional history forces private-sector operators to adapt to the new institutional context in order to successfully carry out the transformations needed to re-industrialise the dairy sector.

Towards a new model of cooperation in the Russian dairy industry: the path-shaping role of international investors

Gorbachev's 1987 and 1989 reforms on individual labour activity paved the way for the creation of the first 'peasant farms', a new form of individual, market-oriented farming operation. The 'Law on peasant farms' of February 1990 confirmed the right for a physical person to possess (not 'own')[1] a plot of land for life, and to transmit it as an inheritance to heirs. It also confirmed that the local soviets were the land lessors. This enabled the individual entrepreneur to gain access to land without going through the collective farm's management. Finally, it gave kolkhoz and sovkhoz members the possibility to leave the collective farm and be allocated a plot of land. The creation of peasant farms thus gave rise to a political conflict regarding two possible trajectories for the transformation of the agricultural sector. The first trajectory, at the national level (liberal party), promoted 'individual farms'. The second trajectory, defended by some regional actors (conservative party), aimed to adapt the collective farms to the new economic and political context. Indeed, conservative politicians were assisted in their task to support former collective farms by the shock therapy itself.[2] The latter had accelerated the scissor effect crisis (i.e. the difference between the price of agricultural commodities and the price of industrial goods), which caused decapitalisation of the agricultural sector (in particular, contraction in the size of cattle herds[3]), whereas the collective farms were the only entities to finance social infrastructures in rural areas (e.g. schools, hospitals and roads). As a response, certain conservative regions such as Oryol Oblast were aspiring to greater autonomy and began to develop a new agricultural model. Thus, while certain regions turned their agricultural sectors over to market forces, through wide-scale dismantling of collective farms and their transformation into a large number of individual farms (known as the 'Nizhny Novgorod' model), others such as Oryol Oblast imagined a new organisational structure for the agricultural sector: the public agro-holding (the 'Oryol' model). These public agro-holdings were based on the principle of grouping together kolkhozes and sovkhozes, with the objective of securing the food supply and ensuring continued agricultural production across the entire region.

After the devaluation of the rouble in 1998, the agro-holding model developed in Oryol Oblast was adopted in several regions of the Federation, but on the initiative of private-sector actors. The emergence of these private-sector verti-cally integrated agro-holdings merely reflected the existence of a new economic and social compromise between local institutions and private investors. This compromise, which is still relevant today, has enabled the recapitalisation of large-scale farming operations, their integration within the processing industry, and the preservation of certain social services for rural populations. However, the agribusiness investors' expectation of a rapid return on investment led them to ignore the dairy cattle farms and the dairy sector. Indeed, the latter requires considerable investments in equipment while the expected returns are measured in years.

The land property was gradually transferred from the State to the farm workers after 1991. Each employee of a kolkhoz became the owner of a plot of land, held in shared ownership with other employees that form a collective of landowners (on average, five to ten families share ownership of a plot of 100–200 hectares). The *payshik* (shareholder) has a coupon indicating the surface area of farmland that he or she owns in the collectively owned plot (generally a section measuring 6–10 hectares), but does not know its location. Consequently, the communities of collective landowners were important actors involved in maintaining the balance between investors and rural populations or more generally in influencing the future path of former collective farms. Paradoxically, these groups of collective landowners were able to influence the negotiations between regional institutions and private investors in order to ensure the recognition of the need for private agro-holdings to finance social services for rural populations (including hospitals, schools, roads, maintaining milk cattle in multi-activity farms, etc.; see Grouiez, 2013).[4]

The legacies of the Russian transitional processes include differences that are attributable to the heterogeneous actors that took part in the conflicts and determined the compromises. The conflicts mainly involved the distribution of land and which characteristics should be favoured. These conflicts did not lead to a selection of operations based solely on economic performance criteria.

In the case of the dairy segment, because Russian investors paid no attention to this sector, international actors appeared in the market during the transitional period These include international investors such as Danone, which became the number one actor in the Russian dairy industry by acquiring a 57% share in Russian dairy industrial group Unimilk.

To adapt the dairy industry in the face of international competition, Danone sought to protect dairy production from two risks: the first related to supply breakdowns, and the second to the quality of the milk. However, the institutional environment (i.e. the institutional legacy) influenced the behaviour of this operator. The need to re-industrialise the milk supply chain became a major issue for Danone. The objective was to bring dairy products to the Russian market that complied with international norms and could meet the increasingly demanding expectations of Russian urban westernised consumers. Danone, whose presence in Russia goes back to 1992, was not originally a producer of agricultural products and for many years it partnered with Russian producers (mainly large agricultural operators). However, these large operators called on smaller operators (individual farms and plots of land) in order to meet Danone's demand and this production chain was not compatible with the traceability criteria required for dairy production. Danone took advantage of its Unimilk acquisition to conduct strategic transactions with heterogeneous actors (producers, local politicians, international and national partners, etc.).

Beginning in 2008, notably during the national Russian agricultural exhibition 'Golden Autumn', Ministry officials stated that cooperation between small and large agricultural operators was not compatible with the modernisation of dairy production. The Ministry of Agriculture imposed a new norm that required labelling for the milk tanks used in the collection process. This labelling requirement made it impossible, for practical reasons, to collect milk from individual farmers. Meanwhile, Danone developed a training programme to adapt to the Russian institutional environment, by targeting large-scale farming operations and individual farmers, and by seeking to fill in the gaps in dairy production in terms of the intrinsic quality of the milk and the way it is collected. In 2011, Danone worked with one of its long-term partners, the 'Verbilovskoye' corporate farm, to offer a series of training courses for its animal scientists. In 2012, Danone signed an agreement with the Ministry of Agriculture and dairy industry experts to create the 'Milk Business Academy' (MBA), a school to train both the livestock farmers under contract with Danone and other farms that looked forward to new partnerships. Danone dedicated a budget of €1.3 million to this project in 2012.[5] Nearly 150 livestock farmers took classes the first year, and Danone trained approximately 986 livestock farmers in the first three years.[6] According to 'investinrussia' medium an additional €0.51 million (RUB36 million) was allocated for expanding and developing the business academy's instruction programmes in 2015. Danone's stated objectives were:

- To increase Russian farmers' long-term productivity.
- To ensure a high-quality milk supply and increase production by 7–10% in the operations involved, by developing sustainable operating practices.
- To secure the dairy industry's downstream segment (i.e. dairy plants).
- To provide access to high-quality expertise for communities of mid-sized livestock farmers.

To achieve these goals, an educational centre operates a 100-cow demo-farm and provides in situ farm training. The creation of a training centre is aimed at diffusing rules for livestock farming to a population of farmers with varying levels of knowledge, through:

- Sustainable practices that are promoted in areas such as animal welfare, feed systems, breeding and good milking practices, based on a team of trainers and university professors mainly from France.
- Increased awareness for decision-makers, managers and owners of large-scale and mid-sized livestock farms.
- An intensive one-week course for animal scientists.[7]

What makes this training centre unique is that it adapts to the Russian institutional environment by taking into account the two main ways in which production is organised in Russia (large-scale operations and independent farms, part of the institutional legacy described above) and by adapting its teaching programme to the specific features of each of these operating models. However, this policy implies a transformation in firms' strategic orientations, especially for the large-scale operations. The latter have been preserved for political reasons and to secure the food supply, but must now rethink their position in a market that is heavily influenced by the demands of Russian consumers.

Indeed, two strategic transactions enable Danone to operate as a leader of the milk-product market in Russia. An initial *bargaining* transaction took place with several corporate farms, e.g. 'Verbilovskoye', and individual farms based on a price/quality trade off. Danone conducted, simultaneously, a *rationing* transaction (principle of sharing) with the Ministry of Agriculture to ensure the sustainability of the *bargaining* transaction. The *rationing* transaction allowed the creation of a training centre in order to generate knowledge transfer within the milk industry. Meanwhile, Danone was ensuring a high level of skill for its own business partners. These combined transactions ensure the growth of dairy farms' productivity without using any *managerial* transaction that would consist in vertical integration of corporate farms as did the agro-holding investors for other food products.

This choice is not systematic, and currently, both large-scale operations and individual farms are inventing some avoidance strategies to bypass the production norms spearheaded by Danone. For example, one operator we interviewed in Vologda Oblast in 2012 had decided to terminate its contract with Danone, which it regarded as too restrictive on sanitary standards, and to sign a strategic partnership – a *bargaining* transaction – with a regional dairy plant that offered a lower purchase price for milk production (RUB11 per litre vs RUB15 per litre for Danone) but did not push for any additional investment. Furthermore, the trade journal *Agroinvest* noted that after the milk price dropped to RUB7–8 per litre in 2009, the following year some independent farmers invested in transformation facilities for their own milk in order to boost their margins. For example, in the Moscow region, the 'APK-Nepecino' dairy farm, which has pasteurised a portion of its milk production since it was created in 1990, established its own transformation facilities in 2008 in order to process milk into fresh cheese and kefir. These independent farms sell their products on the regional market, mainly to retailers located near their production sites, at prices lower than the dairy plants while still generating margins of 20–30%.

Both dairy plants and some independent corporate farmers as 'APK-Nepecino' used the local milk production argument to earn the confidence of the rural consumers. Thus, a *nested market* (Van der Ploeg et al., 2012) emerges in which suppliers sell their products to non-westernised consumers. These production rationales are in frontal opposition with the aim of *mass-market* actors such as Danone to secure their supply from Russian producers. In the dairy industry, we are currently witnessing a process similar to the one identified during the early phases of the transition from communism. Heterogeneous actors initiate competing strategies for organising production, and these different strategies lead to selecting and/or maintaining various kinds of operations (and eliminating others). A similar process is observed in Bulgaria, as we shall now demonstrate.

The trajectory of the Bulgarian milk industry: reactivating past standards in the context of European integration

The changes in the Bulgarian dairy industry demonstrate the disastrous effects of the reforms of the early transitional period on dairy production, which had long been associated with Bulgaria's brand image abroad. Beginning in 1992, the 53 highly concentrated dairy processing combinats, inherited from the planned economy, were privatised as a part of the shock therapy economic reform, along with the launch of the land restitution policy. As in Russia, this policy was marked by a political conflict between the partisans of collective structures and the defenders of individual farms. The conflict was exacerbated by the strong political instability at the beginning of the transition (five different governments in six years). The strategy of left-wing governments was to protect the legacy of farming organisations by authorising the restitution of land and capital only to collectives of individuals belonging to a cooperative. Conversely, centre-right governments, motivated by the desire for a definitive break with the communist past, opted for land restitution to individuals, in hopes that the free-market forces would then allow for consolidation of land plots and the emergence of large-scale, modern, individual farms (Koleva, 2004).

The lack of consensus delayed the formation of a land market that would allow for the transfer of property rights (*bargaining* transactions) and consolidation of land plots. At the same time, liberalisation in the dairy business opened up the sector to many Bulgarian private-sector operators, drawn to the large dairy live-stock population (not only dairy cattle, but also sheep, goats and buffalo), low commodity prices, and hopes for turning a quick profit. In 1995, there were already 826 dairy plants in Bulgaria.

From that period onwards, the sector disorganisation resulting from the lack of a clear path-shaping restitution strategy began to impact the supply chain for dairy plants. Indeed, given a severe shortage of capital, the private cooperatives (the legacy of the old collective farms) began to halt the least-profitable and most labour-intensive activities, such as milk production. Meanwhile, the individual farmers who had received small plots of land focused on subsistence livestock breeding. As a result of this process, one-third of the total dairy cattle herd disappeared over a few years, before stabilising in the mid-1990s.[8] The path-dependency effect of early reforms is still visible today: at present, farms have an average of four cows, and only 8% of all farms have more than 100.

Through the various instruments of the common agricultural policy (CAP), as well as quality standards and classification of producers, the EU has been a major actor in shaping the path of the Bulgarian dairy industry over the past few years. Pressure is more focused on industrial groups, which are directly subject to procedures for obtaining a CAP export licence so they can export products to the EU market. Note that just before Bulgaria's EU accession, only 9% of Bulgarian dairy producers held export licences. The process of bringing producers up to standard came at a hefty price, with bankruptcies and dramatic closures for the structures that were unable to modernise. Although EU funding was available for building new facilities and increasing herd sizes, only the major industrial groups were able to benefit due to the co-funding requirements. In the end, the number of operators was divided by four in the span of 15 years, and the 227 firms in the sector in 2009 all held EU export licences.

Not all industry actors could face the stiff competition for access to a now-scarce local resource, while the market for the finished products was internationalised – especially after Bulgaria joined the EU in 2007 – and increasingly demanding (in terms of quality, product differentiation, delivery lead times, packaging, etc.). Therefore, various strategies for cooperation or avoidance were implemented.

As in Russia, international industrial leader Danone, which opened an office in Sofia in 1993, engaged in *bargaining* transactions with large farms spread across the country based on a price/quality trade off. Some *rationing* transactions such as covering transport costs and paying farmers every two weeks show Danone's willingness to retain farmers and establish a long-lasting relationship in order to widen its lead in the *mass-market* segment. However, progressively the focus on profit seeking in a competitive European context led mass-market investors to favour *bargaining* transactions in order to optimise their portfolio at the transnational level. Thus, in 2014 Danone transferred the ownership over its dairy manufacturing capacities in Bulgaria, the Czech Republic and Portugal to Schreiber Foods, an American company specialised in the production of private labels for various retail chains. This strategy differs from the motivation to enter a new market that led Danone to invest in post-socialist countries in 1990s. Consequently, Danone's brands are still present on the Bulgarian market through the milk products produced by Schreiber Foods, which appears a less costly strategy after the post-subprime crisis in the EU.

Some Bulgarian dairy plants have also been able to combine *bargaining* and *rationing* transactions, with a greater emphasis on territorial anchoring; unlike Danone, these dairy plants work with several hundred small farmers from their own regions. For example, Dimitar Madjarov Ltd, a company located in Plovdiv with more than 300 employees, differentiates itself by:

- investing in the purchase of collective equipment (milking machines used collectively by several farms that are located far from any market outlet);
- making advance payments to farmers to offset the high seasonal effect of milk production;
- organising twice-monthly discussion groups and training sessions on new EU regulations;
- assisting farmers in preparing applications for farm subsidies.

In parallel, Dimitar Madjarov Ltd has signed long-term contracts with the major food retailers in Bulgaria; these retail chains purchase 60% of its production (including cheeses, butter and traditional dairy specialities), with the rest being sold on the wholesale market (10%) or abroad (30%).

Nonetheless, this example is far from typical because the vast majority of Bulgarian milk processors are small. On the whole, the sector shock created by the requirements of EU membership was made worse by the fact that the industrial groups were unable, on their own, to assist farmers in meeting the two-fold demands of quality and supply security. In the face of multinationals relying on the westernisation of consumption and trying to promote global standards of dairy products, different avoidance strategies regarding the mass-market were observed, some with speculative purposes and others oriented to the creation of *nested* markets. One of these speculative avoidance strategies consisted of purchasing imported raw materials (powdered milk, vegetable fats) and manufacturing 'dairy' products without using animal milk, for sale on the domestic market. This approach merely worsened the situation of small farmers. Unable to sell to dairy plants or forced to sell their milk production for very low prices, some dairy farmers turned to local food networks. Others attempted to grow their herd size by leasing land for forage, but then they faced the obstacle of incoherent national legislation. According to the law on land concessions, tenants (i.e. farmers) must invest in maintaining land at their disposal for a period of five years, but if the land is located in mountainous areas (which are mostly suitable for livestock), then the municipality can change tenants every year! Thus, micro-farms can only expand with support from town councils.

The avoidance speculative strategy chosen by certain local industrial groups, by reducing the quality of dairy products, ultimately created dissatisfaction among Bulgarian consumers. In Bulgaria, yogurt and cheese are staples of the daily diet.[9] Under pressure from consumers, the Bulgarian Institute for Standardisation (BDS, 2006) successfully issued a national seal of quality in 2012, initially for dairy products, then for other food products, by using norms that had been defined in 1980s. Admittedly, these norms were associated with the production of the large combinats of the socialist era. However, consumers perceived them as being more in line with their expectations for authenticity, quality and taste. By sanctioning the excesses of some speculative avoidance strategies, Bulgarian consumers unwittingly encouraged individual farmers to adopt territorialised, nested solutions that boosted the value of local products. It is noteworthy that some domestic players such as LB Bulgaricum and Dimitar Madjarov Ltd are currently trying to use European norms of food packaging as a support for their local products' expansion strategy. They point to the need to modify the reactivated BDS regulation on yogurt with respect to packaging, as the only three materials authorised in the 1980s for contact with this product (glass, ceramics and polypropylene) are considered to be inadequate with the technological, processing and logistic challenges in 2017. In fact, behind this argument we can discover the issue of the greater profitability that some other, cheaper materials (such as the polystyrene) could help to achieve.

Conclusion

The analysis of the Russian and Bulgarian trajectory of the dairy sector during the last 25 years offered original insights on the effects of path-dependence and path-shaping factors that transformed the agricultural landscape in Eastern Europe. Starting from the common Soviet heritage of large agro-industrial structures, Russia and Bulgaria experienced radical economic and institutional change, including similarities (shock therapy at the beginning of the 1990s), but also specificities (land reform, and integration into the global or regional economies). These path-shaping reforms at the national level produced divergent evolutions in term of domestic farm structure: large-farm operators (agro-holding) in Russia and small individual farms in Bulgaria.

In this context, some local operators seized the opportunity to implement their own path-shaping economic strategies. Russian agro-holdings motivated by fast return in investment abandoned the dairy sector, highly intensive in qualified labour and capital. In Bulgaria, the lack of a national strategy for the future of the sector combined with the EU integration reforms created strong uncertainty that gave priority to survival strategies. This situation left room for the deliberate action of international actors (including Danone).

The use of a Commonsian perspective helped us to highlight the nature of the strategic transactions in which they engaged. They entered the domestic markets with different size and future opportunities while having to deal with constraints in term of financial viability. Given the specific configuration of large-farm operators in Russia and disinterest in milk products, Danone re-industrialised the dairy sector by introducing global norms supposed to meet the expectations of westernised consumers representing a huge market (17 million in Moscow and its region). By doing so, Danone led strategic transactions with several actors in order to introduce new quality standards in the dairy industry (creation of the Milk Business Academy supported by the Ministry of Agriculture and quality trade-offs with different corporate farms). In Bulgaria, this strategy did not pay because of the limited national market (population of 7 million) and the chaotic restructuring of farms that did not offer the opportunity for bargaining transactions with them. Consequently, Danone preferred using its label as a quality norm considering it as the only strategy compatible with its financial constraints.

In contrast with the above strategies that contributed to shape the mass markets for milk products, local operators in both countries engaged in strategic transactions leading to the creation of a variety of niche markets. In Russia, the strong market power of Danone left little room for local operators. However, the latter were able to implement two different niche strategies: the first was based on the investment of corporate farms in milk manufacturing facilities, while the second consisted in commercial contracts between commercial farms and dairy plants. In both cases, the local origin of the products is put to the fore, thus targeting non-westernised consumers. In Bulgaria, domestic actors had to deal with several norms imposed by the EU CAP. While these norms were considered by some actors as favouring standardisation of milk products to the benefit of the mass-market, other actors used different EU norms (e.g. on yogurt packaging) to defend their strategy based on the development of a nested market.

The Commonsian grid of analysis facilitates the understanding of processes occurring in post-communist Europe. We have shown that even if the planned economy's heritage matters, we cannot decipher the current structure of the dairy industry without taking into consideration the way the operators having complex and sometimes contradictory strategies interpret it and shape future paths.

Notes

1 The first law was a reform of land ownership. This law gave farmers the right to 'possess' (not 'own') land. The second law created a legal status for individual farms.
2 Shock therapy refers to a specific package of policy reforms callously adopted (contrasting with gradualism): 'immediate deregulation of prices, liberalisation of external trade; quick macroeconomic stabilisation through fiscal and monetary restraint and drastic cuts in subsidies; rapid privatisation and other structural and institutional reforms' (Popov, 2007).
3 In 1999, the OECD estimated the total cattle herd population, which in Russia mainly includes dairy cattle, was just 40% of the 1990 level. Likewise, many dairy plants with direct ties to dairy cattle farms went bankrupt. According to data from Ikar (a firm specialised in agri-food market expertise), in 2007 only 1.5–2% of livestock operations still had their own milk pasteurisation and transformation facilities. Most dairy plants that survived were supplied with milk from abroad.
4 The local press often acts as a spokesperson for complaints of communities of landowners in cases where the new investors in private agro-holdings fail to respect their commitments. Such complaints sometimes lead to legal conflicts that are decided at the federal level. This was the case in the conflict between the director of the agro-holding Nobel-Ojl and the landowners in Oryol Oblast; the community of landowners won their lawsuit and Nobel-Ojl had to resume the financing of social services (see reports in the newspaper *Orlovskaâ Pravda*, 2 June 2009).
5 The same year, Danone and Cargill developed an aid programme for livestock farming, targeted to large-scale operations under contract with Danone. This programme aims to improve the quality of cattle feed. Danone supplies the livestock farmers with cattle feed in order to increase the yield and quality of milk. Lastly, it takes part in programmes to help operations finance the purchase of modern production equipment.
6 See: www.milkacademy.ru/results_iba.html.
7 The animal scientists, trained at agricultural universities, are highly specialised but do not have good knowledge about raising dairy breeds that have been genetically selected for their level of productivity. Conversely, the owners of individual farmers have much more general knowledge than the animal scientists working on large operations, but sometimes have no knowledge of the international health standards for collecting and preserving milk.
8 The decline in sheep herds was even more dramatic, with the population stabilising at one-quarter of its size before the transition.
9 According to official statistics, the average Bulgarian consumer buys 27 kg of yogurt per year (vs 21 kg in France), along with 23 kg of milk and 14 kg of cheese. These figures do not include the substantial quantity of self-produced dairy products. Yellow cheese and yogurt are traditionally eaten at breakfast; ayran (a yogurt beverage) and brined white cheese (similar to feta) are eaten at every meal.

References

Barthélemy D., Nieddu M., 2007. Non-Trade Concerns in Agricultural and Environmental Economics: How J. R. Commons and Karl Polanyi Can Help Us. *Journal of Economic Issues*, 41(2), 519–527.

Bulgarian Institute of Animal Science, 2006. Milk and Meat in Bulgaria: Scattered Farms, Scattered Segments, Special Issue. *Economics of Animal Science*, 362, November.

Chavance B., 2011. The Postsocialist Experience and the Resistible Learning Process of Economic Science. *International Journal of Management Concepts and Philosophy*, 5, 2.

Commons J.R., 1934. *Institutional Economics*. New Brunswick, Transaction, 1990.

Federowicz M., 2000. Anticipated Institutions: The Power of Path-Finding Expectations. In Dobry M. (Ed.), *Democratic and Capitalist Transitions in Eastern Europe*. Alphen-sur-le-Rhin, Kluwer, 91–106.

Gislain J.J., 2010. Pourquoi l'économie est-elle nécessairement instituée? Une réponse commonsienne à partir du concept de futurité. *Revue Interventions Economiques [Online]*, 42.

Grouiez P., 2013. Understanding the Puzzling Resilience of the Land Share Ownership in Russia: The Input of Ostrom's Approach. *Revue de la régulation [Online]*, 14, Autumn.

Koleva P., 2004. *Système productif et système financier en Bulgarie (1990–2003)*. L'Harmattan, Collection "Pays de l'Est".

McDaniel T., 1996. *The Agony of the Russian Idea*. Princeton, Princeton University Press.

Nielsen K., Jessop B., Hausner J., 1995. Institutional Change in Post-Socialism. In Hausner J., Jessop B. and Nielsen K. (Eds.), *Strategic Choice and Path Dependency in Post-Socialism*. Cheltenham Glos, Edward Elgar, 3–34.

Popov V., 2007. Shock Therapy versus Gradualism Reconsidered: Lessons from Transition Economies after 15 Years of Reforms. *Comparative Economic Studies*, 49(1), 1–31.

Stark D., 1995. Not by Design: The Myth of Designer Capitalism in Eastern Europe. In Hausner J., Jessop B. and Nielsen K. (Eds.), *Strategic Choice and Path Dependency in Post-Socialism: Institutional Dynamics in the Transformation Process*. Aldershot, Edward Elgar, pp. 67–83.

Van der Ploeg J.D., Jingzhong Y., Schneider S., 2012. Rural Development through the Construction of New, Nested, Markets: Comparative Perspectives from China, Brazil and the European Union. *The Journal of Peasant Studies*, 39(1), 133–173.

13 Large-scale land investments and financialisation of agriculture

An analysis based on agro-financial *filières*

Antoine Ducastel and Ward Anseeuw

Introduction

The '*Global AgInvesting Europe* 2014' conference held on 1–3 December 2014, brought together diverse organisations including the Teachers Insurance and Annuity Association of America—College Retirement Equities Fund (TIAA-CREF),[1] the Bill and Melinda Gates Foundation, Rabobank[2] and the International Finance Corporation[3] to discuss investment prospects in agriculture. For a minimum entrance fee of $2,895, participants could attend around 30 panels on topics such as 'Precision agriculture and the development of Big Data', 'Agricultural opportunities in Australia and New Zealand' and 'What is responsible agricultural investment?'.[4] Various agents organise such conferences several times a year[5] and across all continents.

Agriculture has recently been described as 'green gold'; it had been abandoned by investors and external financing for many years (Lipton, 1977; Bezemer and Headey, 2006) but has made a major comeback[6]. On a macro-economic level, these investments are based on a Malthusian perspective (Ducastel and Anseeuw, 2011) according to which agricultural production cannot keep pace with the continuously increasing demand (due to the growth of the world population, the emergence of a middle-class in emerging countries, or the development of the biofuel sector). But in addition to these macro-economic factors, purely financial motives are also invoked to justify this turnaround in fortune. Hence, following the 2008 financial crisis, the finance industry began to seek 'alternative assets' unrelated to stock markets that would constitute an inflation hedge (HighQuest Partners, 2010). Several publications highlighted the excellent historical financial performance and the contra cyclic nature of agricultural assets, which justified their inclusion in a financial portfolio, along with shares and bonds (Chen et al., 2013).

These agricultural investments cover a diverse series of financial products. Golberg et al. (2012) attempted to clarify the matter by proposing a typology of these agriculture-related assets and establish a distinction between:

- derivatives and index funds on the futures market in agricultural products;
- stocks and shares of agricultural holdings and agri-food companies;
- farmland investment funds.

All these financial products have developed rapidly over the last decade. First, the futures markets experienced a boom characterised by an influx of speculators, in the wake of financial innovations such as funds indexed to the price of specific products (Kerckhoffs et al., 2010; Clapp, 2014).[7] Second, the purchase of company securities, either through stock exchange transactions or investment capital transactions for non-listed companies (Burch and Lawrence, 2013), have profoundly modified the corporate capital structure of agri-food companies, and particularly of multinationals, to the advantage of powerful professional share-holder conglomerates.[8] Third, land investment, either directly or more often via a Specialised Investment Fund (SIF), is part of a broader dynamic of *land grabbing* whereby public or private, national or foreign investors are acquiring agricultural land.[9]

Over the last few years, their financial 'land grabbing' propensities have been the subject of many scientific works that have attempted to characterise this social phenomenon (Cotula and Blackmore, 2014); however, their efforts have often been hindered by a lack of empirical data. Based on two case studies related to land investment funds in South Africa, this article presents a study of the 'bottom-up' financial *land grabbing* phenomenon (Bayart et al., 1992) based on an analysis of agro-financial filières. In the first section of this study, we present a critical review of the literature on this subject, and then outline the relevance of an approach based on the agro-financial filières. In the second section, we describe an empirical study following the direction of the capital flow, from the investors to the agricultural holdings. Finally, we examine the changes wrought on agriculture by the development of these filières.

The financialisation of agriculture and farmland

The role played by financial actors in the *land grabbing* process has attracted much attention. A series of publications first focused on identifying investment funds and companies engaged in this land acquisition process (Grain, 2010; Daniel, 2012; Buxton et al., 2012); then some further literature proposed a more structural analysis by describing land acquisition as part of the general financialisation of agriculture and agricultural land.

Since the year 2000, the notion of financialisation has been used to describe a variety of social phenomena (expansion and proliferation of financial assets, changes in the shareholder structure of companies, household indebtedness), which has resulted in a lack of conceptual clarity (Fine, 2012). In order to characterise these different studies, French et al. (2011) differentiate between three 'financialisation schools'. The first is in keeping with the regulation school tradition and considers financialisation as a specific accumulation regime on a national scale (Boyer, 2000). The second is a 'cultural political economy' approach that analyses the growing influence of financial markets on companies – via the shareholder value – and households (Froud et al., 2002). The third is a socio-cultural approach focused on the dissemination of financial mechanisms and rationale in daily life (Langley, 2008; James, 2014).

The studies that specifically concern the financialisation of agriculture are often aligned with a political economy tradition condensed in a description by Epstein (2005):

> Financialisation means the growing role played by financial motives, financial markets, financial actors and financial institutions in the functioning of domestic and international economies.
>
> (Epstein 2005: 3)

In other words, financialisation should be understood as the growing influence of 'financial capital' over various aspects of the economy.

This approach has a number of advantages. First, it establishes a link between a plurality of spatially dispersed financial practices by connecting them to the global changes in contemporary capitalism. From this perspective, the proliferation of index funds on the futures market in Chicago and the acquisition of over 3,000 hectares of arable land for a British investment fund in Zambia (Chu, 2013) are both part of the same global agricultural financialisation process. Secondly, this approach opens up new avenues of reflection on the upheavals generated by the 'penetration' of finance into agriculture, and specifically on the relations and power play between 'financial capital' on the one hand, and the actors, social groups and interest coalitions that participate in the sectoral political economy on the other, be it organised by large-scale distribution (Burch and Lawrence, 2013) or labourers and farmers (Isakson, 2014). Third, the current development of investment funds and companies dedicated to agricultural land is an opportunity for us to examine the temporality of financialisation and the dynamics of contemporary capitalism – why does financial capital invest in agricultural land today? Fairbairn (2014) considers that the financialisation of agricultural land illustrates the recent changes in financial capital, marked by 'a desire to return to the real economy'. In this regard, the financialisation of agriculture is the last frontier, or 'spatial fix' (Harvey, 2006) of a global financial capital expansion process into new activities and sectors with a view to the continual renewal of the conditions of its accumulation. This process plays out in agriculture according to modalities and mechanisms comparable with previous sectoral financialisation operations. In this context, the financialisation of agriculture is associated with the emergence of a new 'food regime' (Burch and Lawrence, 2009), which in turn leads to the renewal of capital accumulation forms within the sector as the financial income of non-financial actors increases.

Nonetheless, this perspective has significant limitations and should be supplemented by other studies on the process of financialisation (Ouma, 2014). Firstly, this approach views financialisation as a homogeneous process that does not capture the multiple local and national variations, the forms of resistance it meets, or its failures. It tends to ignore the diversity of the actors engaged in the process and uses vague, ill-defined, homogenising categories, starting with the 'financial capital' category, which ignores the power relations and conflicts that structure and organise financial space (Williams, 2014). Generally speaking, this approach masks the entire chain of intermediaries – investors, asset managers, consultants, national and local governments, farmers – who interact within the framework of land financialisation implementation. Lastly, finance on the one hand and agriculture on the other are objectified as closed spaces, confined to distinct spatial scales (global and local), which are mutually exclusive. This opposition between the global and local dimensions has been criticised by Bayart (2004) in his analysis of bottom-up globalisation which includes re-appropriation, subjectification and extraversion mechanisms. Similarly, instead of being a top-down process, financialisation is mainly a negotiated process on all levels as has been demonstrated in a number of publications, whether in regard to companies, through management accountants (Pezet and Morales, 2010), or in the German automobile industry (Kadtler and Sperling, 2002).

A number of publications have sought to go beyond such criticism, firstly, by undertaking more detailed empirical studies in order to characterise the different aspects of financialisation, and examining it in the context of historical developments and national or local long-term land dynamics. These works reveal the existence of an original map of the phenomenon by focusing on the 'emerging' countries of Egypt (Dixon, 2013) and Russia (Visser et al., 2012), and more recently, on the United States (Gunnoe, 2014). By drawing on other schools of thought and analysis frameworks relating to the financialisation process (economic sociology, social studies of finance, or Actor-Network-Theory), a new perspective has developed in regard to the interaction between actors and the mechanisms (technical and cognitive) that enable and embody the financialisation of farmlands. Thus, a number of authors have examined the conditions that led to the conversion of farmlands into financial assets via innovating securitisation methods (Fairbairn, 2014) or the introduction of new assessment methods (Williams, 2014).

Other publications are more interested in investigating the daily work that drives the financialisation of farmlands. All these studies provide insight into practices that underpin the capital funding allocation for agriculture (Ortiz, 2014), such as intermediation (Ducastel and Anseeuw, 2015) or the cognitive and discursive resources that make such investments possible and legitimate (Chu, 2013), always insisting on the trial-and-error approach, and experiments that accompany the entire process (Ouma, 2014). A final series of publications focuses more on the power play and alliances between actors, which underpin the financialisation of land, by examining the process in the context of national elite reproduction (Dixon, 2013), for instance. The aim of these diverse publications is to 'open the black box of finance' (Ouma, 2014) by studying the interfaces between finance and agriculture, which are gradually transforming farmland into a financial asset. The 'agro-financial filière approach' that we propose in this chapter has a similar purpose.

As pointed out by Raikes et al. (2000), the concept of a filière first appeared in the field of French agronomic research in order to characterise the production-trade chains of the major agricultural products of colonial farming (coffee and cocoa). The original concept of a filière puts special emphasis on the interdependence of actors in production-trade chains, and traces the continuous flow between the metropolis and the colonies. Subsequently, the regulation school became interested in this concept as a regulatory space. Allaire (1995) defines it as 'a technological space and area of coordination between actors' the development of which is part of a process of desectoralisation and the proliferation of local rules and agreements. Filières therefore, allow for the emergence and institutionalisation of transnational and 'cross-cutting' regulatory frameworks (Gras and Hernández, 2014), for a specific product.

The analysis of filières in terms of socio-technical networks (Callon et al., 2000), can be used to complement this regulatory approach in the context of an economy of qualities. In this regard, the attachment between agents is forged during the successive qualification and requalification stages in the construction of the product. Hence the filière and the product are constructed simultaneously. This approach emphasises the role of objects, or instruments (Lascoumes and Le Galès, 2004), in the establishment of relationships and coordination between the actors.

The success or failure and the stabilisation of a given filière are based on the mobilisation of agents who are individually engaged in social relations and power structures in their respective domains (Bourdieu, 2000). The causes and forms of enrolment of the actors in the filières are largely influenced by their social positions within their respective domains and, hence, to the structure of their capital; conversely, the participation of the actors in the filières tends to modify the structure of their original domain.

The concept of an agro-financial filière is put forward in this article in order to characterise the entrenchment of stabilised exchanges and interdependencies between multiple agents from diverse social backgrounds (institutional investors, professional asset managers, farmers, etc.) for the construction of a specific financial product. Rather than opposing finance and agriculture and analysing how the former tends to colonise the latter, an approach based on the agro-financial filière allows for the progressive autonomisation of a specific social space that 'straddles' both these sectors. Thus, the notion of the agro-financial filière is akin to that of the 'AG space' proposed by Williams (2014, 423), namely, a hybrid space at the intersection of agriculture and finance.[10] However, our approach has a different dimension, as it only concerns a single financial product.

It is not a matter of objectifying these filières, which are not directly perceived and organised as such by the various actors who are stakeholders in them. Rather, we will must use this concept to account for the complexity of creating an original financial product, by integrating the non-financial dimension of this productive work and gaining a better understanding of how the enrolment of actors from the agricultural sector (South African in this case), starting with the farmers, affects the organisation of farm work. It now remains for us to illustrate this approach empirically.

South African agricultural holdings as a basis for creating financial instruments

In the remainder of this chapter, we will attempt to analyse two specific agro-financial filières dedicated to producing financial assets from South African agricultural holdings. The operators in these filières are engaged in constructing innovating financial products, such as the Alpha and Beta investment funds.[11] The differences and similarities of these two products enable us to analyse the stabilisation of mechanisms specific to an emerging 'agro-financial industry', and the differentiation and experimentation strategies implemented in the various filières.

Institutional investors and professional managers are the two groups that play a decisive role in the creation of both the financial products. The coordination of these actors is based on a 'special purpose vehicle'[12] – a temporary financial mechanism that makes it possible to convert economic goods or activities into financial instruments[13] and to formally delegate the management of the investors' capital to the manager. Once the investment fund has been formalised, managers engage in the business of acquiring agricultural holdings in South Africa on behalf of the fund, then set up management and valorisation strategies for these holdings by mobilising a whole range of players and mechanisms. When the fund expires, the manager resells the farm to remunerate the investors who are unitholders in the fund.

Aglietta and Rigot (2009) define institutional investors as 'specialised financial institutions that collectively manage contractual savings schemes on behalf of third parties with specific objectives in regard to risks, yield and maturity'.[14] Their investment policy is largely determined by the structure of their balance sheets:[15] they invest in diversified equity-based or debt-based financial instrument 'portfolios' but also potentially in agriculture-related financial assets.

Professional asset managers manage a portfolio of securities on behalf of a third party.[16] In the case of agriculture-related asset management, these management companies act as financial entrepreneurs by engaging in an 'uncertain translation process' (Bessy and Chauvin, 2013) between investor demand for financial products, and the actors and institutions in the agricultural sector. These intermediaries are not performing a 'literal translation' but they play an active part in matching the supply to the demand; all the more so because they connect physically and culturally distant objects and actors.

The first investment fund, known as the Alpha fund, was set up in 2008, and is exclusively financed by a large American university endowment[17] fund.[18] Towards the end of the 1990s and as part of its 'natural resource' investments, this endowment fund invested massively in forest plantations throughout the world. More recently it has turned its attention to agricultural holdings, first in the United States and then in other countries like South Africa. The management of its capital was transferred to a South African company initially dedicated to coverage and speculation activities on the agricultural futures market, known as the South African Futures Exchanges. Gradually, the company diversified its activities, with a focus on managing agricultural (agricultural holdings) and agro-industrial (crushing plants) physical assets on behalf of third parties.

The second fund, known as the Beta fund, was set up in 2010. Over 85% of the fund units[19] are held by a South African Public Service Pensions Fund and by one of the oldest insurance companies in the country.[20] Both of these funds made their first investments in South African agricultural land as part of a drive to promote an 'impact investment' policy. This framework, developed by a network of investors and financial institutions, establishes a correlation between the capital invested and the environmental and social 'production' of the company, which then have to be measured using a standard methodology. Tthe allocation of capital is therefore based on the financial and 'social profitability' of the company (Chiapello, 2015). Lastly, an Open-ended investment company (Sicav), founded in Luxembourg and which comprises several institutional investors and European individuals, holds the remaining Beta fund units. The management of this fund has been transferred to a joint venture involving the insurance company and entrepreneurs specialised in the export of South African fruit. Specially created for the occasion, this management company has since then set up a second fund for the acquisition of agricultural holdings in South Africa, and a similar fund in Swaziland.

This brief overview highlights several characteristic traits of agro-financial filières. First of all, investors are all world leaders in their respective areas (endowment fund, pension funds and insurance companies). Hence, financial innovation, or rather financialisation as an 'extension of the area of finance' (Deffontaines, 2013), is mediated by a few global actors who concentrate financial, as well as technical and human capacity in order to produce a new class of assets. These long-term investors (Aglietta and Rigot, 2009) are involved in agriculture with the aim of diversifying their portfolio in order to preserve the long-term market value of their capital. In this regard, it is interesting to note that the financial products analysed in this chapter were launched at the height of the financial crisis. Hence, the interest in agricultural holdings must be understood in its interdependence with other asset classes – in other words, these agriculture-related assets are only of value to institutional investors in relation to the shares and bonds in their portfolio. But engagement in agriculture is part of a rhetoric and general discourse concerning the social responsibility of finance, particularly among South African investors, the latest expression of which is their adherence to the global network of 'impact investment'.

Both the management companies are implanted in South Africa, even if one of them employs mostly foreigners. These management companies consider themselves intermediaries between investors on a global level and the national agricultural land market. This local anchorage illustrates the substantial influence of local networks and careers in ensuring the transit from one level to another and constructing a financial product. Furthermore, these companies illustrate the permeability between the various agro-financial filières because they pass from the management of one financial product to another (from futures contracts to agricultural holdings) and from one country to another (from South Africa to Swaziland). These management companies employ small teams of around five to ten people, which can be typified in general terms as a unit responsible for the interaction with investors and other actors in the agricultural sector; a group of financial analysts (responsible for the identification, assessment and negotiating of financial transactions); a back-office that supervises the accounts and management of agricultural holdings; and technical consultants make up the rest of their number. The accounting and financial functions are by far the most important; financial assessments figure prominently both as a primary basis for decision-making and as a communication vector between managers and investors.

The special purpose vehicle constitutes the third link in these agro-financial filières. This specific financial mechanism supports a commissioning relationship and a securitisation process. The meeting between investors and managers results in the creation of a legal entity that exists independently of the first two entities, but is also formatted to suit this particular relationship. Hence, it is an ad hoc structure that only exists to facilitate investments by the investor–manager partnership and disappears as soon as this collaboration ceases. In our two case studies, these entities have a limited life span, ranging from 10 to 12 years,[21] during which period the investors' capital is frozen. The shares of these vehicles are distributed between the investors in proportion to their capital investment; the management of this capital is entrusted to the manager. The status of this type of structure may vary from one jurisdiction to another; however, in this case, they are all legally classified as partnerships. This means that the manager has an obligation of means towards the investors and not an obligation of result (absolute obligation), which Montagne (2006) has termed a 'procedural accountability'.

Thus, the operation and governance of these funds, such as capital disbursement modalities, are strictly regulated by a series of contractual procedures with which managers have to comply.[22] These legal documents include an 'investment thesis' which defines potential target companies in terms of minimum and/or maximum investments, locality, etc. Lastly, an investment committee, comprising independent members and representatives of the investors and the manager, constitutes the final decision-making authority that validates the investments. In the present instance, these two financial vehicles have been registered in Mauritius and not in South Africa. In the year 2000, Mauritius developed a very advantageous fiscal policy to attract foreign companies, particularly financial companies, by exempting them from paying dividend tax and capital gains tax. At the same time, Mauritius included international anti-money laundering and financial regulation mechanisms in its national jurisdiction[23] which put it in an ideal position to capture financial flows towards Africa. A new geography of agro-financial filières is emerging between the offices of the big institutional investors concentrated in global financial hubs of the United States, Europe, South Africa, Mauritius, which is accumulating an increasing number of investment vehicles for the African continent; and the South African agricultural holdings.

The structuring of the agro-financial filières must therefore be understood in the light of a given investor–manager relationship, but also in relation to the mobilisation of a set of stabilised legal and financial mechanisms: the trust, the Mauritian jurisdiction, the investment thesis, etc. Agro-financial filières are in fact largely modelled on the institutional architecture of other financial filières, particularly private equity and real estate assets, whose 'underlying' filières (urban and peri-urban companies and real estate) are considered to be similar.

These institutional arrangements enable actors to move in spaces with which they are familiar, or that they recognise. Although they invest in a specific economic activity, namely agricultural production, its promoters mobilise a series of bench-marks (Karpik, 2007) in order to make it into an innovative product and a legitimate asset on the financial product markets. For instance, these fund promoters set financial return targets that make them comparable with equity returns or real estate investments of an identical duration; the Beta fund, for instance, has set a target yield return of 10% plus inflation.

Once the investment funds had been formalised, their managers made every effort to put their respective investment theses into practice, namely to acquire agricultural holdings in South Africa in order to develop them. Fund investment policies have proved to be markedly different as regards the selection of agricultural holdings and their valorisation strategies.

In 2013, the Alpha fund held five agricultural holdings, which were all the product of farm consolidation, namely 22,000 hectares located in four different provinces.[24] These farms used to produce mainly cash crops, i.e. maize, corn and soya crops on irrigated and non-irrigated plots. In 2014, the Beta fund held four consolidated farms, comprising 5,500 hectares located in the same four provinces. These were all fruit-growing farms using irrigated plots, where the main varieties of fruit were lemons and table grapes, as well as pears, apples and peaches.

The investment practices of these two funds have some common traits, however. First, they only invest in agricultural holdings situated in South Africa. Their decision to concentrate exclusively on the South African land market is based on the managers' attachment to their local roots, and on the comparative advantages of the country on the global market[25] of farm-related assets. In this regard, the existing financial infrastructures in South Africa, such as the agricultural commodity futures market in Johannesburg or the strict legal regulations surrounding private land ownership, provide them with a competitive advantage over other countries in the region. Furthermore, their holdings are located in several different provinces that are sometimes thousands of kilometres apart. This provincial diversification strategy is an attempt to mitigate the natural risks inherent to farming activities, particularly by acquiring holdings in zones that have different ecological characteristics.[26] The agricultural holdings possessed by the funds are in both cases the result of a farm consolidation process that merged several family holdings and reflect their commitment to generate economies of scale. Moreover, in most cases, these holdings were in financial trouble before being taken over.[27]

Lastly, both entities rely on their ability to increase the value of the agricultural holdings, that is to create value from the invested capital, which in turn, will remunerate the investors and the manager. These value creation strategies can take advantage of two levers: the merchant value of the farm and the productive income flows. As we shall see, these strategies rely heavily on an advanced financial rationalisation, the implementation of economies of scale, the integration of upstream or downstream activities, and the increase in the productive capacity of the farms.[28] Nonetheless, farmland development techniques are far from being standardised and fund managers are engaged in an experimentation process as they seek an optimal development 'formula'. The differences between these two funds, as regards the farm management methods, for instance, are part of this quest for value creation.

Agricultural production on the holding is directly managed by the Alpha fund manager. He is responsible for hiring labourers and sub-contractors, supervising the operations, purchasing inputs, 'hedging' his production on the futures market, and then selling his production. Consequently, his income is based both on operational income and on the appreciation in the price of the farm during the holding period. The Beta fund, on the other hand, immediately leased its farms to operators for ten-year periods, for an annual rent equal to 8% of the farm value, thus separating the ownership of the land and the production activities. This second fund is structured on the real estate investment trust model or equivalent trusts in the forestry sector (Gunnoe, 2014). It earns income from the rent received and the appreciation of the farm price on the final resale date.

The construction of both the investment funds that we have studied requires the mobilisation of actors and mechanisms from multiple social spaces. It is this multi-positioning that leads to the success or failure of the agro-financial filières. Hence it is interesting to examine the forms of implantation within a specific social space, the South African agricultural sector. Our aim is to gain insight into the social origins of this expansion in the area of finance on one hand, and the disruptive effect on the economic and social structures on the other.

Towards a transformation of South African agrarian structures due to financialisation?

In acquiring South African farms and mobilising the staff and range of skills and know-how in the local agricultural sector, the funds are confronted with the institutions and actors of the South African agricultural sector; they also contribute to the introduction and dissemination of new ways of doing and thinking about agriculture and farmland in South Africa.

Both funds employ dedicated staff to carry out their development strategies on the South African farms they have acquired. Their capacity to recruit and retain this staff depends, as we shall see, on the current dynamics within a changing sector. The gradual integration of these farms into agro-financial filières goes hand-in-hand with an original type of organisation as regards production, and the productive and non-productive work carried out on and around the farms.

Both the funds employ on-site managers to run the farms, who are either hired directly by the fund manager or by the operator. These managers are white Afrikaans farmers, who are often former owners, or the sons of family farm owners. Hence, they have often gone from being independent farmers to being salaried managers, sometimes on the same agricultural holding. In pursuing an on-site manager career, they take advantage of the opportunity to continue working on an agricultural holding, but also sometimes to break away from family supervision. It is interesting to note that this choice does not only involve the man himself, but also his household, as management companies often hire the farmer-manager's wife to carry out the administrative tasks. Thus, the creation of financial products is based on the mobilisation of a specific social and 'ethnic' group, in this case the rural Afrikaans households who are in danger of losing their social status, by giving them the possibility of maintaining their social position.

On-site managers have to strike a balance between the economic and social reality of the farm and the financial reality of the manager. Indeed, fund managers centralise the running of the development and/or management activities of their geographically dispersed holdings. This type of 'remote' control requires a series of specific techniques and skills. Consequently, the funds have developed advanced monitoring and evaluation procedures that on-site managers must adhere to: daily reports on the activities, validation by the head office of the operational and investment costs, monitoring of agricultural input stocks. Thus, a reporting chain is developed along the agro-financial filière, from investors to on-site managers, which fosters remote control and coordination between physically distant actors. Furthermore, the fund managers use the latest technology, such as Google Earth Pro, to monitor specific farms, plot by plot, and record their results. For the management of farm labourers, the funds rely increasingly on sub-contractors and/or introduce profit-sharing management methods.

The control procedures that on-site managers have to follow regulate and restrict their potential activities and transform their duties by imposing on them an ever-increasing load of administrative tasks (Hibou, 2012). Farm managers are therefore not only subject to a change in status but also a change in career, which requires different procedures and working patterns (to those practised in the commercial agriculture model). Hence, the paternalistic relationship, a legacy of apartheid, whereby labourers lived with their families on the farm, is gradually being replaced by a market relationship via the recourse to sub-contracting. From a business perspective, the responsibility of farmer-managers is limited to the achievement of performance goals within the budget determined by head office.

However, the retraining of farmer-managers to become employees has not been achieved without difficulty, as the strict procedural control over agricultural practices is sometimes felt to be a waste of time and detrimental to truly productive activities. Recurrent conflicts arise between head office and the on-site managers in regard to the proper use of capital and the prioritisation of tasks on the farm. In both cases we studied, the high turnover rates point to the difficulty of importing a managerial rationale into a sector that was historically structured around independent farms.

Agricultural engineers also have a distinct role to play in managing agro-financial filières and developing agricultural holdings. The Alpha and Beta funds both employ an agricultural engineer either as a consultant or directly within the organisation, whose role is to assess the ecological, soil and climate parameters before the fund acquires the agricultural holding. The engineer actively contributes to defining an agronomically motivated development plan for the farms (soil improvement plans, selection of the best varieties, introduction of no-till crops, etc.), and they conduct physical monitoring by regularly visiting the sites on behalf of head office. Thus, the training and experience of the agronomist enables them to move with ease between the reference frame and standards specific to the agricultural sector on the one hand, and the business environment on the other. They are, therefore, a major actor in the shift towards agro-financial filières.

The developers of agro-financial filières disseminate specific practices and organisational forms in order to reach their financial profitability goals. They are 'organisational weapons' (Foureault, 2014) that transform the forms of productive organisation for the benefit of institutional investors.

When funds acquire agricultural holdings, they also import their own rationale, which differs substantially from the mind-set that shaped the South African agricultural sector structured around the independent commercial farmer. Firstly, they only commit to specified periods of time (10–12 years), and so have a specific time frame that differs greatly from the 'life-long' commitment prevalent among the farmers.

Secondly, the funds differ from commercial farmers, even in their ultimate goals. Indeed, whereas farmers view their agricultural holdings from an economic and cultural perspective, as a production tool and a place to live, or from a financial perspective, as a source of savings, investment funds only take into account the financial rationale of farm development. The goal of the latter is to increase the value of the property by using a number of levers, of which the income generated by farming activities is only one component. As agricultural holdings are viewed from a financial perspective in which development-related activities are uppermost, the new forms of organisation differ from the historical form of the agricultural holding; this is what is meant by an 'agro-financial' holding with its own set of job qualifications.

We have already mentioned some of the levers specific to the development strategies implemented by the funds, and in particular the conversion of family farms to an entrepreneurial mind-set, including the creation of economies of scale. However, these strategies also have an impact on the choice of crops and their varieties. Each agricultural product is in fact associated with a more or less stable and significant risk/return variable. Thus, maize and wheat are the subject of futures contracts on the Johannesburg market, which guarantee a regular income for Alpha; conversely, the margins generated by lemon production are greater, but there is no sophisticated risk management tool, which may explain why the Beta fund outsourced the production to a third party. Consequently, the choice of crop is crucial in order to maximise the value of the agricultural income and of the farm itself, and this choice is dependent on a specific investment model. Both the managers also developed plans to change the crops or to introduce more productive varieties.

The funds also rely extensively on the services of a panel of consultants in order to create value on the farms. Firstly, in the legal arena, they draw on corporate and fiscal legal expertise in order to optimise the structures of their holdings and subsidiaries.[29] Furthermore, they increasingly turn to the experts for social and environmental assessments, and for impact investments. This type of investment, requires a metrology system, embodied in a series of specific assessment practices, the object of which is to elucidate and quantify their positive repercussions (Chiapello, 2015). Thus, the Beta fund sends investors a specific yearly report, which measures the fund's performance against the social and environmental indicators developed by the International Finance Corporation (IFC).[30] A specialised consultant is hired on a part-time basis to prepare this report, but also to formalise a set of social and environmental standards and procedures that the funds must comply with on their agricultural holdings. This reference framework is 'put together' for each fund based on a composite set of good governance codes (IFC indicators, 'Guidance for Responsible Investment in Farmland') and external certification (GlobalGap). Hence, the environmental and social dimension is increasingly acknowledged as a separate lever for development.

The assessment method used by these funds constitutes the last dimension in the farmland development process. Assessment and development are, in fact, two sides of the same coin (Vatin, 2013). Both funds developed assessment matrices largely based on the discounted cash flow method (DCF). These matrices calculate a future capital return rate – the internal rate of return – based on a series of projections and expectations. Although this method is widely used in the finance sector (Middelberg, 2014), the South African farmland market used the 'comparable sales' method in the past. The latter focuses on the evolution of land transactions, hence on market dynamics, whereas the DCF method focuses on the productive value of the farm. By using this method, the funds can evaluate the capacity of the managers to enhance the value of the property over time; and by assessing the available cash flow they can also measure the level of return on investment provided to investors (Redon, 2014). The change in the assessment method is not just a shift in perspective, but has repercussions on the work on and around the agricultural holdings, which is now geared towards the maximisation of the future value of the holding, and also influences the forms taken by the agricultural holdings, as instruments for the valorisation of capital.

The financial product, the asset, is thus neither the farmland, nor the production extracted from agricultural holdings, but the value of the shares of the investment funds holding the land. The value of these shares and their return rates are linked, but not limited to the value of the farmland and of the agricultural products. Numerous other factors must be taken into consideration (entrepreneurial and fiscal architecture, environmental and social impacts, and assessment practices) in order to understand the underlying coding specific to this financial product. Agrofinancial filières do not, therefore, contribute to the commodification of farmland, but rather, engage in a financial 'metamorphosis' of the said farmland.

The possession of farmland and the organisation of production implemented by investment funds, although a relatively marginal phenomenon in South Africa, nevertheless has an impact on the sectoral architecture of agriculture. Indeed, it may be assumed that the production of these financial assets goes hand-in-hand with a 'desectoralisation' (Muller, 2010) of South African agriculture shown by the disappearance, or at least the hybridisation, of sectoral institutions, and results in a new social configuration and new power relations.

Firstly, thanks to their substantial capital injections,[31] investment funds select and legitimise certain actors and certain agricultural practices to the detriment of others (Ortiz, 2008). The Beta fund manager, for instance, only hires out its farms to large fruit exporting and marketing companies seeking to secure their supply. The fund seeks to protect itself from the non-payment of rent by the tenant, so prefers to use multinational, highly integrated companies. Hence, in this particular situation, the development of agro-financial filières strengthens the position of the dominant agro-industrial actors, and more specifically, the entrenchment of 'agribusiness' in South Africa (Hervieu and Purseigle, 2009).

Moreover, these transformations favour certain social and racial groups more than others. Thus, white Afrikaans rural households are affected on two levels: first, many farmers sell their farms to funds, which results in career changes or a move to the city or abroad; second, only a minority is able to engage in a manager-farmer career. Consequently, the emergence of these agro-financial filières contributes to the demise of the independent commercial farmer model, which is a cornerstone of the sectoral architecture and thus plays a part in weakening the position of this sectoral elite. Moreover, they provide career change opportunities to a section of this elite, whereas blacks and 'coloureds', the rural populations that were discriminated against under the apartheid government, are still confined to their roles as farm labourers. The development of agro-financial filières seems therefore to be accelerating the transformation of the sectoral structures inherited from apartheid, but not always in areas where changes were expected (Anseeuw, 2013).

Lastly, funds and their managers take no part in sectoral institutions, such as AgriSA, the South African union of commercial farmers. In other words, they do not engage with the political, economic and social issues of the agricultural profession. However, they do engage in collective initiatives on a global level, international forums such as 'Global AgInvesting' or the Global Impact Investing Network, where investors, fund managers, farmers and regulators congregate. These spaces contribute to the dissemination and legitimisation of practices and cognitive frameworks[32] and are thus akin to emerging regulatory forums. In this regard, agricultural holdings are territorial enclaves that appear to be less involved in the South African agricultural sector and increasingly part of these global agro-financial spaces.

However, several elements tend to relativise this 'desectoralisation', starting with the resistance from and adaptive capacities of the former sectoral elites. In this regard, the example of former producer cooperatives is of some interest (Ducastel and Anseeuw, 2014). During apartheid, these cooperatives were in the hands of the farmers and played a major role in the single marketing channel policy. In spite of the conversion of these cooperatives to private sector companies, sector deregulation policies in the 1990s[33] and, more recently, the offensive launched on their capital via investment funds, they are still controlled by the more affluent farmers. In fact, over the last few years these former cooperatives have adopted financial valorisation strategies for their assets[34] and have achieved a 'reversal of financialisation' (Burch and Lawrence, 2009) led by the sectoral agricultural elites.

It is, therefore, relevant to ponder on the ability of these sectoral elites to re-appropriate financialisation mechanisms in order to maintain their position between the global and the sectoral dimensions. This question is all the more legitimate when we consider that the financial vehicles have a limited life span of 10–12 years, at the end of which the fund managers have to resell the agricultural holdings. Who, then, will the next purchasers of these farms be?

Conclusion

We have attempted to analyse the social dynamics that lead numerous financial and non-financial actors to engage in the production of a financial product, which in this case is an investment fund dedicated to South African farmlands. The production process requires specific cognitive and technical mechanisms, both to coordinate investors, managers, consultants and sub-contractors and to increase the value of the underlying asset. The linkage of actors and the progressive empowerment of these filières are gradually transforming the actors' forms of organisation as well as their territorial and sectoral anchorage.

Through this 'agro-financial filière approach', tangible capital circulation modalities are emerging. The circulation of capital is based on the construction of a financial product, i.e. the transformation of an economic activity, agricultural production, and/or of a good, the farmland, into a financial security. Here we see another aspect of the financialisation of agriculture that goes beyond the opposition between agriculture and finance and highlights the ramifications that are unfolding. Moreover, financialisation is not the all-inclusive and homogenising process that is sometimes described, but rather the product of spatially and temporally situated social configurations. For instance, the ability of fund managers to hire and retain local farmers is a major factor in determining the successful outcome of their farmland development operations.

The South African agricultural sector case is of particular interest in that it expands the scope of financialisation studies. The social and political foundations of the sector were consolidated and stabilised during the apartheid era. The development of agro-financial filières is based on the ability of managers to circumvent or take advantage of these sectoral structures. At present, South African agricultural elites alternately challenge or cooperate with the investment funds and their subsidiaries. The fractured nature of the agricultural sector will determine to a large extent the restructuring of agriculture both as a social space and as an economic activity.

Notes

1 TIAA-CREF (Teachers Insurance and Annuity Association—College Retirement Equities Fund) insures and manages pension schemes for over 3 million people and 15,000 establishments in the fields of education, research and health.
2 A Dutch bank formed from the merger of cooperative bank networks.
3 A subsidiary of the World Bank Group that supports the private sector.

4 See the Global AgInvesting website, www.globalaginvesting.com/Conferences/
   Agenda?eventId=25 (consulted on October 10, 2016).
5 In addition to Global AgInvesting, other instances include the Ag Innovation Show-
   case and the Terrapinn Agriculture Investment Summit.
6 CCFD, 2006. 'Soya against life', http://ccfd-terresolidaire.org/infos/souverainete/agri
   culture/campagne_613 (consulted on 10 October 2016).
7 The increased participation of speculative funds played a major role in the 2007–
   08 food crisis (Clapp, 2014) as the increase and acceleration of transactions on
   these markets have a tendency to increase the price volatility of agricultural
   products.
8 Thus, Palpacuer et al. (2006) in the case of European multinational agribusiness
   companies, and Baud and Durand (2011) in the case of large-scale distributors,
   highlight the growing importance of these shareholders and their predominant influence in the corporate governance of these companies.
9 See Cotula et al. (2009), World Bank (2010), ILC (TBP).
10 'A distinct social and institutional space, a particular field of engagement, visualiza-
   tion and intervention that exists somewhere between finance and farming, sharing
   features of both while being irreducible to either one' (Williams, 2014).
11 This case study draws extensively on participant observations conducted by one of the
   authors in two of the South African asset management companies. One of the
   conditions that made it possible to carry out this field work was the anonymity
   granted to all the companies and the investors involved.
12 Also known in other jurisdictions as a joint securitisation fund or a '*special purpose
   vehicle*'.
13 See Jorion (2007) on American real estate securitisation, or Deffontaines (2013) on
   the securitisation of public procurement financing.
14 Such as pension funds, insurance companies, endowment funds, etc.
15 Namely, the features of their deposits (duration of the commitment, financial compensation, etc.).
16 Unlike institutional investors, they often specialise in a single asset category.
17 The endowment fund is an investment fund held by a single organisation, often a non-
   profit organisation, that invests donations in the organisation and whose profits are
   intended to finance its own activities.
18 In 2013, this endowment fund had a portfolio of over US$30 billion allocated to several
   asset classes: shares (49%), *hedge funds* (15%), bonds and fixed income securities
   (11%), and tangible assets (25%). This last category includes real estate investments,
   commodity futures contracts and lastly, a fraction called 'natural resource'.
19 In 2013, this pension fund held over US$100 billion in management assets in South
   Africa and on the rest of the continent; 5% of this capital is allocated to 'investments
   for development', such as the Beta Fund.
20 In 2013, this company held the equivalent of US$385 billion in assets.
21 Which is the standard for investment funds, and particularly private sector equity
   funds.
22 In compliance with this legal documentation, the Beta fund may not invest less than
   50 million Rand (or €3.8 million) per agricultural holding.
23 Thus, Mauritius is one of the only African countries to be a member of the Global
   Initiative for Fiscal Transparency.
24 Limpopo, KwaZulu-Natal, Western Cape and Northern Cape.
25 'Land in South America is seven times more expensive than in South Africa and is
   utilized for the same underlying agricultural activity with the same soils, climates and
   yields. This provides the ideal opportunity to leverage this arbitrage opportunity and
   simultaneously generate above average cash flow return.' (Alpha Fund, presentation
   brochure)

26   The diversification of assets and the use of probabilistic models for capital allocation are based on the 'modern portfolio theory' (Markowitz, 1952), which contends that the asset's risk should not be assessed by itself but by how it contributes to a diversified and complementary portfolio's overall risk; it is based on a mean-variance analysis.
27   One of the properties was purchased at an auction after the winding-up order had been issued by the court.
28   The Beta fund directly funded the development of new plots, the introduction of more productive citrus fruit varieties and the installation of protection nets, particularly anti-hail nets.
29   With the Alpha fund, each group holding is incorporated as a subsidiary of the parent company based in Mauritius.
30   There are eight 'environmental sustainability performance standards' that are considered authoritative in this field.
31   The Beta fund is worth 462 million Rand, which is around €35 million.
32   For instance, when they discuss the question 'what is a responsible agricultural investment?', as stated in the introduction.
33   Marketing of Agricultural Products Act (1996).
34   Through the creation of specialised investment funds (Ducastel and Anseeuw, 2014), for instance.

References

Aglietta M., Rigot S., 2009. *Crise et rénovation de la finance*. Paris, Odile Jacob.

Allaire G., 1995. De la productivité à la qualité, transformation des conventions et régulation dans l'agriculture et l'agro-alimentaire. In Boyer R., Allaire G. (dir.), *La grande transformation de l'agriculture: lectures conventionnalistes et régulationnistes*. Versailles, Inra éditions/Economica, pp. 381–410

Anseeuw W., 2013. Between Subsistence and Corporate Agriculture. The Reinforcement of South Africa's Agricultural and Territorial Dualisms. In *Conference Land Divided: Land and South African Society in 2013, in Comparative Perspective*, 24–27 mars, Le Cap, p. 12.

Bayart J.-F., 2004. *Le gouvernement du monde: une critique politique de la globalisation*. Paris, Fayard.

Bayart J.-F., Mbembe A., Toulabor C., 1992. *Le politique par le bas en Afrique noire: contribution à une problématique de la démocratie*. Paris, Karthala, Collection Les Afriques.

Baud C., Durand C., 2011. Financialization, Globalization and the Making of Profits by Leading Retailers. *Socio-Economic Review*, 10(2), 241–266.

Bessy C., Chauvin P.-M., 2013. The Power of Market Intermediaries: From Information to Valuation Processes. *Valuation Studies*, 1(1), 83–117.

Bezemer D., Headey D., 2006. Something of a Paradox: The Neglect of Agriculture in Economic Development. Gold Coast, Australia. In *International Association of Agricultural Economists Conference*, 12–18 août, Queensland (Australia).

Bourdieu P., 2000. *Les structures sociales de l'économie*. Paris, Le Seuil.

Boyer R., 2000. Is a Finance-Led Growth Regime a Viable Alternative to Fordism? A Preliminary Analysis. *Economy and Society*, 29(1), 111–145.

Burch D., Lawrence G., 2009. Towards a Third Food Regime: Behind the Transformation. *Agriculture and Human Values*, 26(4), 26–79.

Burch D., Lawrence G., 2013. Financialization in Agri-Food Supply Chains: Private Equity and the Transformation of the Retail Sector. *Agriculture and Human Values*, 30(2), 247–258.

Buxton A., Campanale M., Cotula L., 2012. Farms and Funds: Investment Funds in the Global Land Rush. *IIED Briefings Papers*, en ligne http://pubs.iied.org/17121IIED/ (consulté le 10 octobre 2016).

Callon M., Meadel C., Rabeharisoa V., 2000. L'économie des qualités. *Politix*, 13(52), 211–239.

Chen S., Wilson W.W., Larsen R., Dahl B., 2013. *Investing in Agriculture as an Asset Class.* Agribusiness and Applied Economics Report 711, Department of Agribusiness & Applied Economics, Fargo, North Dakota State University.

Chiapello E., 2015. Financialisation of Valuation. *Human Studies*, 38(1), 13–35.

Chu J.M., 2013. Creating a Zambian Breadbasket: "Land Grabs" and Foreign Investments in Agriculture in Mkushi District. *LDPI Working Paper*, 33, 27 p.

Clapp J., 2014. Financialization, Distance and Global Food Politics. *The Journal of Peasant Studies*, 41(5), 1–18.

Cotula L., Blackmore E., 2014. *Understanding Agricultural Investment Chains: Lessons to Improve Governance.* Rome, FAO/IIED.

Cotula L., Vermeulen S., Leonard R., Keeley J., 2009. *Land Grab or Development Opportunity? Agricultural Investment and International Land Deals in Africa. rapport de recherche.* Londres, IIED/FAO/IFAD.

Daniel S., 2012. Situating Private Equity Capital in the Land Grab Debate. *The Journal of Peasant Studies*, 39(3–4), 703–729.

Deffontaines G., 2013. *Extension du domaine de la finance? Partenariats public-privé (PPP) et 'financiarisation' de la commande publique.* Thèse de doctorat en sociologie, Paris, Université Paris-Est.

Dixon M., 2013. The Land Grab, Finance Capital, and Food Regime Restructuring: The Case of Egypt. *Review of African Political Economy*, 41(140), 232–248.

Ducastel A., Anseeuw W., 2011. La libéralisation agricole post-apartheid en Afrique du Sud: nouveaux modèles de production et d'investissement. *Afrique contemporaine*, 237, 57–70.

Ducastel A., Anseeuw W., 2014. The Financialisation of South African Grain Cooperatives. What Room for the Agrarian Capital? In *5th Annual Conference of the International Initiative for the Promotion of Political Economy "The Crisis: Scholarship, Policies, Conflicts and Alternatives"*, 16–18 septembre, Naples, IIPPE.

Ducastel A., Anseeuw W., 2015. Agriculture as an Asset Class: Reshaping the South African Farming Sector. *Agriculture and Human Values*, à paraître, 1–11, http://dx.doi.org/10.1007/s10460-016-9683-6 (consulté le 10 octobre 2016).

Epstein G.A., 2005. *Financialization and the World Economy.* Northampton (Mass.), Edward Elgar Publishing.

Fairbairn M., 2014. "Like Gold with Yield": Evolving Intersections between Farmland and Finance. *The Journal of Peasant Studies*, 41(5), 777–795.

Fine B., 2012. La financiarisation en perspective. *Actuel Marx*, 51, 73–85.

Foureault F., 2014. *Remodeler le capitalisme. Le jeu profond du Leverage Buy-Out en France, 2001–2009.* thèse de doctorat en sociologie, Paris, Institut d'études politiques de Paris.

French S., Leyshon A., Wainwright T., 2011. Financializing Space, Spacing Financialization. *Progress in Human Geography*, 35(6), 798–819.

Froud J., Sukhdev J., Williams K., 2002. Financialisation and the Coupon Pool. *Capital & Class*, 26(3), 119–151.

Goldberg R., Segel A.I., Herrero G., Terris A., 2012. Farmland Investing: A Technical Note. *Harvard Business School Background Note*, pp. 211–222.

Grain, 2010. *Les nouveaux propriétaires fonciers: les sociétés d'investissement en tête de la course aux terres agricoles à l'étranger. rapport.* Barcelone, Grain, p. 13

Gras C., Hernández V., 2014. Agribusiness and Large-Scale Farming: Capitalist Globalisation in Argentine Agriculture. *Canadian Journal of Development Studies*, 35(3), 339–357.

Gunnoe A., 2014. The Political Economy of Institutional Landownership: Neorentier Society and the Financialization of Land. *Rural Sociology*, 9(4), 478–504.

Harvey D., 2006. *The Limits to Capital*. Londres, Verso.

Hervieu B., Purseigle F., 2009. Pour une sociologie des mondes agricoles dans la globalisation. *Études rurales*, 183, 177–200.

Hibou B., 2012. *La bureaucratisation du monde à l'ère néolibérale*. Paris, La Découverte.

HighQuest Partners, 2010. Private Financial Sector Investment in Farmland and Agricultural Infrastructure. *OECD Food, Agriculture and Fisheries Papers*, 33, Paris, OECD Publishing.

Isakson S.R., 2014. Food and Finance: The Financial Transformation of Agro-Food Supply Chains. *The Journal of Peasant Studies*, 41(5), 749–775.

James D., 2014. *Money from Nothing: Indebteness and Aspiration in South Africa*. Stanford (Californie). Stanford University Press.

Jorion P., 2007. *La crise: des subprimes au séisme financier planétaire*. Paris, Fayard.

Kadtler J., Sperling H.J., 2002. The Power of Financial Markets and the Resilience of Operations: Argument and Evidence from the German Car Industry. *Competition & Change*, 6(1), 81–94.

Karpik L., 2007. *L'économie des singularités*. Paris, Gallimard.

Kerckhoffs T., van Os R., Vander Stichele M., 2010. *Financing Food: Financialisation and Financial Actors in Agriculture Commodity Market. rapport*. Amsterdam, Somo.

Langley P., 2008. *The Everyday Life of Global Finance: Saving and Borrowing in America*. Oxford, Oxford University Press.

Lascoumes P., Le Galès P. (dir.), 2004. *Gouverner par les instruments*. Paris, Les Presses de Sciences Po.

Lipton M., 1977. *Why Poor People Stay Poor: Urban Bias in World Development*. Cambridge (Mass.), Harvard University Press.

Markowitz H., 1952. Portfolio Selection. *Journal of Finance*, 7(1), 77–91.

Middelberg S., 2014. Agricultural Land Valuation Methods Used by Financiers: The Case of South Africa. *Agrekon*, 53(3), 101–115.

Montagne S., 2006. *Les fonds de pension entre protection sociale et spéculation financière*. Paris, Odile Jacob.

Muller P., 2010. Secteur. In Boussaguet L., Jacquot S., Ravinet P. (dir.), *Dictionnaire des politiques publiques*. Paris, Les Presses de Sciences Po, pp. 591–599

Ortiz H., 2008. *Anthropologie politique de la finance contemporaine : évaluer, investir, innover*. Thèse de doctorat en anthropologie, Paris, EHESS.

Ortiz H., 2014. Financial Valuation and Investment: Techniques, Organizations, Politics. In *Finance, Food and Farming Workshop*. 25 janvier, La Hague, International Institute of Social Studies (ISS).

Ouma S., 2014. Situating Global Finance in the Land Rush Debate: A Critical Review. *Geoforum*, 57, 162–166.

Palpacuer F., Pérez R., Tozanl S., Brabet J., 2006. Financiarisation et globalisation des stratégies d'entreprise: le cas des multinationales agroalimentaires en Europe. *Finance, contrôle, stratégie*, 9(3), 165–189.

Pezet A., Morales J., 2010. Les contrôleurs de gestion, 'médiateurs' de la financiarisation. *Comptabilité contrôle audit*, 16(1), 101–132.

Raikes P., Jensen M.F., Ponte S., 2000. Global Commodity Chain Analysis and the French Filière Approach: Comparison and Critique. *Economy and Society*, 29(3), 390–417.

Redon M., 2014. *L'influence des Directeurs administratifs et financiers dans le processus de financiarisation.* mémoire de master en gestion, Paris, Université Paris-Dauphine.

Vatin F., 2013. Valuation as Evaluating and Valorizing. *Valuation Studies*, 1(1), 31–50.

Visser O., Mamonova N., Spoor M., 2012. Oligarchs, Megafarms and Land Reserves: Understanding Land Grabbing in Russia. *The Journal of Peasant Studies*, 39(3–4), 899–931.

Williams J.W., 2014. Feeding Finance: A Critical Account of the Shifting Relationships between Finance, Food and Farming. *Economy and Society*, 43(3), 401–431.

World Bank, 2010. *Rising Global Interest in Farmland. Can It Yield Sustainable and Equitable Benefits?* Rapport de recherche, Washington DC, World Bank.

14 Conclusion

Alternative sketches of a second Great Transformation

Benoit Daviron and Gilles Allaire

Through different and complementary approaches, the chapters gathered in this volume have developed novel contributions in three domains: (i) new perspectives on the global history of agri-food ecology in relation to social structures; (ii) an institutionalist approach to agri-food ecology and economy; and (iii) an overview of the shifting organisation of agri-food capitalism and the diversity of agriculture across world structures and through specific studies of various issues addressed in various national contexts.

The issues we want to discuss in this concluding chapter relate to a second Great Transformation announced by various authors as initiated in the 1990s. All the chapters describe and analyse what look like very ambivalent historical processes, in the sense developed in Chapter 1. Various studies referring to Regulation Theory (RT) (Chapter 3) or Convention Theory (CT) (Chapter 4) have stressed the differentiation within quality standards and conventions and the diversity of production models, which can be related to the two alternative cognitive paradigms of decomposability and identity (Allaire and Wolf, 2004; see Chapter 9). Chapter 5 sees a threat of autocracy in the new regime of standardisation related to world mass markets. While acknowledging changes, Chapter 6 stresses that no effective transition from the global industrial regime (i.e. the use of fossil energy) is appearing. There is today a media-focused discussion around the importance of climate change. However, the engagements announced by the governments to reduce CO_2 tend more towards discourse than accountability. Chapter 7 shows the multi-polarity of the global food system, which was revealed in truly transparent terms by the so-called 'global food crisis' of 2006–08 and its aftermath. The technological convergence of the food, feed and other land-based sectors as general sources of biomass prompts debate around the analytical relevance of a distinct political economy of food. Chapter 8, related to the emblematic device of the dairy industry modernisation worldwide – the Holstein cow – shows an increasing commoditisation process of genetic resources, while these processes maintain the status of public and common resources. Chapter 9 analysed bio-economy projects with two opposing paradigms: the dominant one relating to the life sciences and the other to agro-ecology. Chapter 10 explores alternative paradigms in the development of organic farming and food. Chapter 11 shows how new participants are now entering Japanese agriculture, with diverse agendas, and the non-stabilisation of Japanese agricultural policy. Chapter 12 shows how enterprises in the Russian and Bulgarian dairy industry develop avoidance strategies from the technological norm and quality standards prevailing in mass markets governed by multinational firms; while some small companies take advantage of local sources of raw material and proximity with consumers, others take advantage of the international market's speculative opportunities to reduce production costs. Finally, Chapter 13 gives a critical appraisal of financialisation of land production in South Africa by showing how advanced agronomic and financial technologies, as well as systematic monitoring and management and production certification, contribute to guaranteeing the valorisation of financial investments. Considering all these contradictory and diverging orientations in economic and political practices and representations, it appears that recent decades did not give birth to coherent modes of regulation stabilised in relation to a relatively steady world order.

If the era of the First Great Transformation has ended, followed by a new period of liberalism, a second Great Transformation cannot yet be fully or even coherently characterised with respect to the agri-food economy and global ecology. Globalisation, as it was called in the 1990s, generated important expectations in terms of economic growth and the extension of welfare, but – as we now know – what actually happened was the reverse effect, at least in OECD countries. The liberalisation of finance caused a large transformation in commodity production and distribution, including agri-food, but the series of changes identified in the chapters of this volume do not lead to a unique model without resistances or failures.

Money and finance play a globally systemic role and threaten the world with financial crises that have social and economic impacts, as shown by several episodes in the 1990s. The world 'dominated' by finance is in fact very heterogeneous in economic, social and power terms. Finance accelerates the circulation of credit and thus contributes to an acceleration in all domains of economic life: flexibility of industrial chains and obsolescence of industrial products, and increasing scales of sourcing for processing industries (including food), come with trade liberalisation and shifting behaviour of consumers. We include in that mosaic the increasing production of waste and its impact on climate. While long-term relations remain stable within food chains and in consumers' choices, change seems to now be a 'normal' behaviour, going with the diffusion of competition logics in all social relations, or in other words, the triumph of neo-liberal ideologies. As Bonanno and Wolf (2017: 285–286) argue, the power of this ideology 'derives from its openness and flexibility' and it is possible to describe elements of both crisis and resilience of neoliberalism, to analyse both the forms of its domination and the 'resistance that it engenders'; thus the authors claim that 'we may find ourselves in a world in which an organised regime and a dominant discourse may not necessarily emerge'. According to Raymond Aron (1969), this renews the dialectic of socialisation characterising modern society.[1] Socialisation is not just the building of so-called 'social capital', but the mediation between the individual and social structures that feed social cohesion and renew the nature of inequalities. In recent decades, social rights have created increasing inequalities. The development of the welfare state, which in the past created social cohesion, became more an engine producing bitterness and exclusion. While Aron wrote his book when the crisis of the nation/state, as the main socialisation framework, was at its onset, today the dialectic of socialisation (autonomy vs interdependence) takes on new features related to the governance of common goods, the sovereignty of people's lives and social routes, in a context where various and diverging ways are offered to building capabilities

Several visions of the future are competing, and if much information on the state of the world leads to pessimism, *le pire n'est pas sûr*. We do not share the vision of a global and final crisis of capitalism revealed by the ongoing political and social disorder and even chaos in many countries. We prefer to identify the contradictions and alternative directions, in distinguishing, as we did in the introduction, on the one hand the topics of industrialisation and socialisation historical process and on the other hand liberal and transformation periods; for more on this, see Table 14.1.

So far, globalisation has resulted in a multi-polar world

The world agriculture that came out of the two world wars, the depression of the 1930s, and of the independence of former colonies, was organised on a national basis (possibly extended in the case of the EEC) through the intervention of the US. From the end of the 19th century global markets were split into isolated national markets between which residual trade occurred with US food aid constituting the archetype. After a runaway[2] phase of this model, the collapse of international prices on agricultural markets triggered by the 1982 debt crisis soon led to growing pressure for a major overhaul of agricultural policies. The report on world development published by the World Bank in 1986 can be seen as the moment where that is clearly explicit: too many subsidies in 'developed countries', too much taxation in the so-called 'developing' countries, such was the summarily formulated diagnosis. Therefore, agricultural markets should be liberalised and competitiveness become the first goal of public intervention.

However, liberalisation was conducted according to procedures and a pace clearly differentiated depending on the country. In indebted 'developing' states, it was implemented quickly within the context of the structural adjustment programmes aiming to restore macroeconomic balances. For the OECD countries, whose objectives were both to reduce agricultural spending and meet the pressures of self-proclaimed 'fair trade' states, liberalism was organised more gradually, as an accompaniment and subsequently to negotiations conducted under the General Agreement on Tariffs and Trade (GATT) and the World Trade Organization (WTO). In the end, in both cases, domestic prices were reconnected to international prices – public market interventions having been eliminated. However, in Europe or the United States, direct aid programmes have had the same role of supporting and stabilising agricultural revenues.[3]

Table 14.1 Industrialisation and socialisation of agriculture since the 19th century.

	Industrialisation	Socialisation
19th-century liberalism	Frontiers (pioneer fronts) Mechanisation	Colonialism International division of labour Futures market Gold standard Accumulation oriented family farm
First Great Transformation	Injection of massive amount of fossil energy/accumulation of physical capital in agriculture (and the economy in general) • Specialisation on food • Motorisation/Mechanisation • Spread of basic agronomic knowledge • Chemical fertiliser • Pesticide • Disintegration of agriculture and livestock	Extension of wage society and the social state Invention of national agricultures: • Building of agricultural policies: stabilisation of prices and revenues, producers' rights and status, etc. • Nationalisation of commodity markets Reduction of social (at country level) and political (at international level) inequalities
Neoliberalism	Renew of frontiers, land grabbing GMOs and colonisation of life entities New information communication technologies	Neoliberalism/globalisation/ financialisation Quality turn, Transnational normalisation Liberalisation Food security standards Concentration in industries and trade Market media regime Privatisation of common-pool resources Renew of inequalities
Second Great Transformation	*Bio-economy versus agro-ecology* Integrative biology and bio-informatics	*Dematerialisation versus Alternative food* Cities' food policies and new rural policies New drivers of socialisation (as social networks) Mass market avoidance strategies

	Bio-economy	Agro-ecology	Dematerialisation of market devices	Alternative food networks
	• Healthy food • Green chemistry	• Healthy food • Energy saving and auto production	Trans humanism Gated communities	Transnational social movement Ambivalence of the commons

A third food regime?

Following Philip McMichael (2005), many authors (Pritchard, 1998; Pechlaner and Otero, 2008; Holt Gimenez and Shattuck, 2011) announced the emergence of a *third food regime* qualified often as the 'corporate' or 'neo-liberal' food regime (Burch and Lawrence, 2009). This idea was part of a renewal of the analyses carried out in terms of French Regulationism theorising, its revival signalled by the publication of several special issues of journals such as the *Journal of Agrarian Change*, the *Journal of Peasant Studies* and *Agriculture and Human Values*.

Markets and competition are seen as the organising principles of this food regime, in contrast to the previous two. Its end point would be the creation of a truly global agriculture, seen as a transnational space integrated by commodities markets:

> In contradistinction to previous food regimes constructed by hegemonic British and US states, the food regime under neoliberalism institutionalizes a hegemonic relation whereby states serve capital. This, to me, is the distinctive organizing principle by which corporate rights have been elevated over the sovereign rights of states and their citizens—the World Trade Organization (WTO) rules (among other, ongoing trade agreements) made this clear.
>
> (McMichael, 2016: 649)

Corporations therefore substitute for States in the management of agriculture. This move is particularly illustrated by agricultural research showing how agrochemical and seed companies now control a large part of the activities even in public institutions (Howard, 2016).

The 2007/08 international food price spike has been debated at length by the proponents of the notion of food regimes. Described as a global food crisis, this episode has often been presented as a harbinger of the crisis of the third food regime and its deleterious effects: destruction of ecosystems, destruction of the smallholder farmers, outrageous market speculation. It has also often been analysed in connection with 'land grabbing', which was itself presented as the result of policies promoting agrofuels.

However, 10 years later, the analyses of the causes and the consequences of the 2007/08 price spike evoked questions surrounding many of the published arguments, leading to a reconsideration of the notion of food crisis itself.[4] The FAO, which first announced a sharp rise in the numbers of malnourished and hungry, published new data showing that the downward trend had continued. The increase in international prices has been transmitted to domestic prices in very different ways depending on the country and at times has not been transmitted at all. This transmission has been much stronger in France than in Niger, China or India, for example. Nevertheless, in countries or regions where it takes place, as in the European Union and Latin America, the influence of the instability of international commodities markets on domestic markets has two consequences: on the one hand, agricultural producers and food and distribution firms seek coverage of financial risk, thus reinforcing the financialisation of the agri-food economy; on the other hand, they seek to escape competition, developing private standards and contractual arrangements. This is one of the reasons for the current development of organic farming and short chains in the countries of the North, and fair trade labels in the South. The magnitude of land grabbing has also been seriously reconsidered. The NGOs *Grain*, which published a report published in 2008 that largely triggered the debate and the wave of work on the question, has shown more recently an important number of failures of acquisition projects, including some of the most ambitious (Grain, 2016). Yet, the link of land grabbing with biofuels has not been confirmed. Biofuel production, whose growth slowed strongly, still remains concentrated in Brazil, the United States and Europe.

Tipping over in Asia

The global agri-food landscape is being upset by the shift of the world economy towards Asia. It has become increasingly difficult to ignore the reconfigurations caused by the accelerated growth of China. Becoming (or rather once again) the manufacturer of the world, this country generates a prodigious primary commodity demand. Over the past two decades this demand has been mainly oriented towards energy and minerals. In 2014, China's share of global consumption was 69% for iron ore, 51% for nickel, 45% for lead, 47% for copper (Chaumet and Pouch, 2017: 107). In many ways, China is walking in the footsteps of Japan and has taken on a number of its model elements, including a highly internationalised economy based on the massive mobilisation of distant mineral resources (Bunker and Ciccantell, 2007). Japan also imports large quantities of biomass. At first this was mostly non-food agricultural products and particularly fibre, but – while maintaining the protection of rice, a food that is culturally important but is declining in the diet – Japan has been importing corn and soy intended for animal feed since the 1960s. From the 1990s, it became an importer of animal products as well.

From 2000, Japan entered into crisis at the same time as Chinese growth accelerated. Certainly, Japan then met a number of domestic and specific problems (ageing, over-indebtedness, insularity), but the two phenomena are closely linked. In 2000, China adopted the 'Go Out strategy' (*zou chuqu*), with its accession to the WTO. A number of flagship industries in Japan then moved to China: steel, shipbuilding and electronics, as predicted by the 'flying geese theory' (Akamatsu, 1962). In the international markets for biomass, the displacement of Japan by China is spectacular. Measured as a percentage of world imports, the large decline of the Japanese position, whose share fall from 18% to 6% between 2000 and 2015, is in line with the Chinese rise, whose share leapt from 2% to 14%. In a few years China has become the number one importer of soybeans (two-thirds of world imports), cotton, rubber and wood.

For food, self-sufficiency remains the Chinese official line. The objective of 95% self-sufficiency in grains (cereals and soya), established in 1996, has been reaffirmed in 2008 in the framework of the medium-term food security strategy (2008–20). For some observers China's food security strategy is now called a 'dual' strategy, founded on national production, complemented by 'moderate imports' (Zhang and Cheng, 2017). The debate on the future of this policy seems open. Some advisers of the Government (see Cheng and Zhang, 2014; Zhang and Cheng, 2017) argue publicly for an opening to imports and for the adoption by China of a 'global agricultural policy'.

Will China re-enact in the coming year, the United Kingdom strategy from the 19th century (i.e. workshop of the world, imports of primary commodity and free exchange), or will it follow the path of Japan with a complete end to its ongoing growth? The future of the landscape of the global food industry will depend a great deal on the answer to this question. As stated in the conclusion to Chapter 7 (Wilkinson and Goodman): The 'perspective of resource access and security appears to offer a fruitful point of entry for the analysis of the political economic restructuring of contemporary agriculture and food systems'.

Changes in quality regime: from 'industrial' regime to media regime

By the end of 1980s, it was argued that modern capitalism was becoming increasingly disorganised (Offe, 1985; Lash and Urry, 1987) based on five arguments: a decline in the corporatist regulation of sectoral national markets; a restriction of the power of nation states to control transnational economic forces; the expansion of a service economy and that of intangible assets (fictitious capital); the appearance of new social movements (differing from worker and industrial unions); and cultural fragmentation. The same arguments can be found in the current descriptions of the neo-liberal era. Appadurai (1996) adds that 'the complexity of the current global economy has to do with certain fundamental disjunctures between economy, culture, and politics' (1996: 33). Recognising a 'new media order' he was aware to not confuse 'some ineffable Mc-Donaldization of the world and the much subtler play of indigenous trajectories of desire and fear with global flows of people and things', thus 'each time we are tempted to speak of the global village, we must be reminded that media create communities with "no sense of place" (Appadurai, 1996: 29).

The new agri-food economy is related to changes in the set of institutions that make markets exist, in particular the working rules regarding food quality construction on both the production and consumption sides. This is shown by the success of notions such as the 'quality turn', '*économie de la qualité*', 'alternative food networks', etc. Institutional approaches have introduced the concept of 'quality regime' to embrace all the social interactions relating to qualities ordering, including actors of the domestic, industrial, administrative, science and media worlds. The quality regime in general refers to the relations within various spheres where inquiries and representations of quality ordering of methods, people, products and other social realities develop across domestic, professional, legal, politic, media and social networks. Depending on historical and geosocial contexts, one can distinguish a variety of quality regimes, according to the type of social actors responsible for the formations of quality.

In Fordist agri-economies the quality of products sold by large retail networks was determined by the constraints of agriculture industrialisation and food processing. But given the social modes of eating, in this period the quality values were still in relation with the domestic and community family style of life and the responsibility in the domain of food was at the level of the consumer, i.e. the family structure and within it, women.

From the 1990s, changes intervene in all the listed spheres that progressively overwhelm the quality regime characterising Fordist agri-food economies. We have already mentioned individualisation and related changes in the formation of demand with the development of eating out and food services, globalisation of a 'tri-standards regime' structuring markets (see Chapter 5), and so-called 'market-oriented' public policies. We could add, primarily, changes in the legal and public sphere related to intellectual property rights (cf. Chapter 1), financial devices (cf. Chapter 13), and responsibility regimes, characterised by the ascendance of liability for the industry (sanitary norms and the 'sanitary crisis') and retail; and, secondly the accelerating development of the media and the scale of the influence of written and numeric media (i.e. Big Data) and of 'social networks' defining new generations.

Because of the predominance of supermarkets, and because of the role of the media in the circulation of quality representations, we proposed to call this new configuration the *media regime* (Allaire and Daviron, 2006). This new situation of food markets developing since the 1990s corresponds to Busch's concept of tripartite standard regime (TSR) (Chapter 5). At the same time, overt individualisation is a deep change in lifestyles. It corresponds to transformations in the whole food complex through convenience food, 'foodism', nutrition services and food clubs. So-called 'alternative food networks' inscribe themselves in the global change of the quality regime, integrating a bundle of qualities, from organic food to fortified food.

Analysing financial markets, Orléan (1999) argues (following Keynes) that financial evaluation is not the outcome of a rational (impossible) calculation of the future, but of a 'logic of opinion' (mimetism). The evaluation of qualities in the media regime is likewise a logic of opinion (Allaire, 2002). The media regime brings to bear a whole technology of market observation and configuration, with transaction traceability on one side and on the other the globalisation of public issues that goes together with the proliferation and opening-up of quality forums. The TSR model adapts to different domains of standard setting organisations. The TSR introduces in the post-Fordist mass market, non-trade values as organic, fair trade, geographical indications,[5] local and ethnic food, etc. This is not simply the result of a technical and procedural standardisation, but of a process of 'second coding' that translates these values in monetary hierarchies of values (Jessop, 1990). Knowledge in the form of symbolic and political claims crosses and moves into these domains of 'transcendental resources', as described by Allaire and Wolf (2004). For example, Fouilleux and Loconto (2017) show relations and moves within the organic TSR and sustainability TSR. Chapter 13 shows how important the TSR is for the financial profitability of land investments.

Market actors adapted by developing the use of supply contracts specifying quality characteristics. That provided an argument for the development of so-called 'market-oriented' agricultural policies directed more by the market and quality standard setting then became a crucial issue. On the demand side, crises of trust arose with the 'food scares' of the 1990s. These led to various initiatives from various origins, whether professional bodies, distributors, nation states or international forums, where both the main multinational firms are involved and non-governmental organisations are represented. The development of the TSR rests on the development of a new cultural environment, both in the sense of complementary changes in quality conceptions (i.e. intellectual or cultural frame-works) and in the sense of the development of new institutional arrangements, in terms of cultural commons, property rights and governance instruments.

The media regime is characterised by a new form of crisis: crises of opinion or crises of quality that spark professional crises, whereas the industrial regime was characterised by crises of under- or over-production. Under the new regime, the media space, which is no longer industry-fragmented, is heterogeneous; it is materialised in the form of recourse to varied expert recommendations and in the form of pluri-cultural social spaces. This new form of sectoral delimitation emerged corresponding to identity quality standard markets. This includes food safety standards promoted by big international retail consortia and alternative food labelling.

Third party certification (TPC) is not only a change in market governance structures, but a change in regulatory regimes at the level of global value chains. Hatanaka, Bain and Busch (2004) argued that TPC 'reflects a broader shift from public to private governance' and the growing power of supermarkets to regulate the global agro-food system. These authors notice that TPC 'also offers opportunities to create alternative practices'; as a matter of fact 'alternatives' tend to integrate the TSR. In this governance shift, 'social movements' have to be taken into account, especially if by movements we mean the social dynamics of the spreading of knowledge related to quality judgements and the social critique. AFNs, according to the situation, can be presented as a springboard for the construction of more equitable and sustainable food systems or as diversification of markets, duality in quality regimes and social re-stratification of consumption (Verhaegen, 2012).

Competition includes avoidance strategies. They are behaviours to protect the actors from competition on a mass market governed by TSR as claimed above. According to the cases analysed in Chapter 12, there are two avoidance strategies with different time horizons: (i) taking opportunist short-term advantage on a speculative market[6]; and (ii) targeting protected market segments ('nested markets'), alongside the general mass market. These characteristics can be generalised. For example, Chapter 10 shows two types of avoidance strategies from the conventionalisation of the organic standard regime, by multiplying private labels and by radical change of the guarantee system (e.g. Participative Guarantee Systems).

While TSR in its various domains, including sustainability, can be extended to conform to commodities in US and European markets, China is one of the few countries to have developed specific environmental and social guidelines relating to its outbound investment and finance. While alternative food networks are mainly a western social invention, Appadurai (2000) sees a 'series of social forms has emerged to contest [and] interrogate' the current workings of capital on a global basis; they 'rely on strategies, visions, and horizons for globalisation on behalf of the poor that can be characterised as "grassroots globalisation"', with one example as the development of Participative Guarantee Systems' (Appadurai, 2000).

Forces for change

Where are the capacities to resist social and ecological destruction that define neo-liberal forces? To resist and reverse booming intra-national inequalities? To make existing and realist orientations for 'the good life' in sharing resources? To help to address these political issues, we conclude by proposing the opening of debate in critical social sciences along three points we see as important to address the future of food and landscape.

Towards a general economy approach or socio-ecology

In the 1980s and 1990s, Nature was not present as a real actor in the developing social sciences regarding agri-food economies. In an article published 20 years ago, Goodman (1999) analyses the consequences of this erasure of nature in agro-food studies and considered the several theoretical perspectives developed in the 1990s, such as post-structuralism and actor-oriented rural sociology, to which we can add the RT and CT research collected in Allaire and Boyer (1995). He argues that these theoretical approaches are

> restricted by their unexamined methodological foundations in modernist ontology. The nature-society dualism at the core of this ontology places agro-food studies, and their "parent" disciplines in the orthodox social sciences, outside the broad intellectual project that is advancing the greening of social theory and militates against effective engagement with the bio-politics of environmental organizations and Green movements.
>
> (Goodman, 1999: abstract)

To escape the impasse of ontological polarities, Goodman refers to Latour's conceptualisation 'in terms of heterogeneous collective associations of elements of Nature and elements of the social world' (Latour, 1993: 1007) (Goodman, 1999: 26). With this view, economic objects and the goal of economics have to be completely reconsidered.

Several methods can be considered to enlarge economic thinking and to reintroduce Nature as an actor in the social world. In the 1970s, when ecology and political economy rarely overlapped, Georgescu-Roegen (1971) introduced the notion of the 'bioeconomic', and Passet (1979) proposed a conceptualisation of three spheres whereby the economic sphere lies in a sub-system of the social sphere, itself a sub-system of the biosphere (see Chapter 8). More recently, researchers at the Institute of Social Ecology in Vienna have developed a conceptual approach of the notion of social metabolism, initially formulated by Karl Marx, to characterise historic metabolic regimes according to different uses of energy and matters and different productions of waste. However, very few authors, if any, provided a radically wider perspective by placing life and human societies within the dynamic of the cosmos, as did the philosopher Bataille in a 1932 article ('The Notion of Expenditure'), considering that living matter (over the course of ages) and human societies (over the course of centuries) appropriate an increasing flow of energy (see Chapter 6). Bataille's general economy does not refer to 'needs' or utilitarism but to the 'consummation' or burning of energy, which is the *raison d'être* of Nature and thus of human societies, while needs are social formations (see Chapter 2). These perspectives raise a tricky issue: Is it possible to escape ongoing economic downturns without increasing the productivity of capital and the ecological surpluses?

Beyond the modernist ontology: the commons

Behind 'alternative food networks' and 'nested markets' are the concepts of 'the commons' – in the sense of Ostrom's common-pool resources – and of common goods. The concept of the commons has in recent years attracted agri-food scholars, from the issue of biological resources and agro-ecology to food itself, considering notably in the perspective of the right to food.

The commons, politically, is characterised by 'two opposite attractors' (Latour, 2017): the local, exposed to the risk of reactionary identification or excluding communitarianism, and the global, conveying the one hand a logic of enlargement of areas of competition and on the other hand a discourse about 'global public goods'. The global public good perspective emerged in the 1990s with the recognition of ecological upheavals, the Rio Conference (1992) and its sequels (Thomas, 2005) and the Convention of the United Nations on climate change, bringing together the countries' participants once a year since 1995. Global public goods are 'ideal goods' (Allaire, 2013) with ambiguous and conflicting operational definitions. However, as Latour (2017) argues, the opposition of global vs local commons remains locked in the 'arrow of time of the modern' – from the 'reactionary' local to the 'progressive' global – which we must escape to design a future. This leads to defining a new focus outside of the conception of the modern global perspective linking social progress and economic growth. Thus, this re-definition might very well lead to the deployment of the notion of *the commons* if it seen as a set of resources that we must take care of in a multitude of domains left to explore and connect.

As argued by Muller (2010), since the beginning of the 21st century, sustainable development has entered into the discourse of public policy. While questioning the impact of economic activities on Nature, this concept does not offer a representation of what would be a durable global response to environmental crises. At the same time, in Europe and the USA, the politics of organic agriculture settled on a political consensus surrounding a coexistence of technical models. This thesis, however, allows for a 'green' intensive agriculture – including organic farming practised on an industrial scale – and multiple variants of agricultural models claiming ecological credentials. It allowed recognition of a gradient of alternatives to the intensive industrial farming model known for its negative impact, and the reform of agricultural policies by targeting the 'production of public goods', limited to a list of so-called environmental goods, and the public funding for the development of a variety of the alternatives according to the principle 'public money for public goods'. But it also weakens the more critical politics of alternatives such as organic agriculture, which is thus recognised as one contributor among others, including adaptations of dominant systems for the production of environmental public goods.

There is a sort of cognitive dissonance between the presentation and valorisation of a correct lifestyle in the media: organic farming, solidarity and short food circuits and 'local' or 'natural' products, and the agricultural policy issues discussed in intergovernmental bodies. While the efficiency and directions of public policy at national and international levels are contested, local governments, cities and regions, not only in Europe, invest in agriculture of proximity, thus representing urban consumers. But proximity does not necessarily refer to family farming or small-scale producers and new national and regional agricultural policies include production systems that can rely on competing social values. A similar situation can be described in the field of energy, with 'transition' policies based on an energy mix that results in some noticeable global changes (Cahen-Fourot and Durand, 2016). Today, in the middle of 2018, nuclear energy has been de-emphasised (after Fukuyama) but probably not definitively defeated. In addition, there is strong pressure to reduce the use of fossil fuel energy in land transport, which notably allows the revival of competition in industrial sectors, including in China.

Growth or sufficiency?

The current intellectual consensus to denounce the damages of productivism in the agri-food sector does not extend to the condemnation of growth or the promotion of degrowth. It is impaired by a call to realism citing the growing food needs of world population and other uses of biomass. Nevertheless, the debate on the growth imperative has shifted the orientation of agronomic research. Thus, for example, the European Commission's Standing Committee on Agricultural Research (SCAR) highlights a new intellectual and political situation in which the productivist paradigm opposes a paradigm of 'sufficiency'. It introduced the idea that productivity itself is perhaps not a reasonable goal. This report (Freibauer et al., 2011) distinguishes two major narratives: one centred on productivity, with massive investments in R&D and the removal of 'barriers to markets', i.e. more competition and intellectual property rights, and the other on 'sufficiency'. What is significant is the recognition by the SCAR of possible scenarios which consider that the challenge for the future is not increased intensification but changes in institutional setting and more social responsibility for scientific research.

At the global level at this contemporary moment, it seems as if the crises of the industrialisation of agriculture are much less of a problem than they should be. Indeed, is the crisis in this current model itself or is this a crisis of the effects of the dissemination of this model? Humanity seems to have overcome the energy constraints embedded in agriculture and food production and consumption through industrialisation, but we might have also postponed these crises by passing them on to future generations. Yet, the alarm bells ringing about ecological crises in food and that of climate change across the globe might very well silence the voices of social movements and novel social practices in current and future political processes.

Notes

1 Raymond Aron's book, *Les désillusions du progrès. Essai sur la dialectique de la modernité*, published in 1969, after its publication in English in 1965 (*Encyclopaedia Britannica*), analyses modern society through three dialectics (contradictions): dialectic of equality (distinguishing social and political equality), which engenders frustrations; the dialectic of socialisation of which the engine is the tension between the inevitable rise of an ideal of individual autonomy and the continuity of a dynamic of socialisation; and the dialectic of universality. Means of mass socialisation develop at the same time where the autonomy of the individual is asserting itself as an aspiration. Everything happened as if 'the disillusionment of progress', created by the dialectics of modern society, and, as such, inevitable, had been proven by the young generation of the 1960s with such intensity that the rampant dissatisfaction expressed in revolt.
2 This 'runaway' has taken various forms depending on the country. In Europe and the United States it resulted in the formation of growing surpluses, making exports increasingly necessary. For some countries said 'developing', as the petroleum-producing countries or African countries, this rather led to increasing taxation of the agricultural sector and the appearance of a large food deficit.
3 For the European case, see Chapter 1.
4 The construction of the 'food crisis', in which numerous international organisations, Government and NGOs participated, deserves a special analysis (see Bricas N., Daviron B., 2008. De la hausse des prix au retour du 'productionnisme' agricole: les enjeux du sommet sur la sécurité alimentaire de juin 2008 à Rome. *Hérodote*, (4), 31–39).
5 According to new European agricultural regulation, consortia managing GIs have the competency to control supplied quantity.
6 In the example Bulgarian local dairy plant turning to the international market of milk powder, when the prices are low; another convincing example can be found in the international expansion of the Chinese industry of tomato paste, see Malet (2017).

References

Akamatsu K., 1962. A Historical Pattern of Economic Growth in Developing Countries. *The Developing Economies*, 1(s1), 3–25.

Allaire G., 2002. L'économie de la qualité, en ses secteurs, ses territoires et ses mythes. *Géographie économie société*, 4(2), 155–180.

Allaire G., 2013. Les communs comme infrastructure institutionnelle de l'économie march-ande. *Revue de la régulation. Capitalisme, institutions, pouvoirs*, (14).

Allaire G., Boyer R., 1995. Régulation et conventions dans l'agriculture et les ISS. In Allaire G. and Boyer R. (Eds.), *La grande transformation de l'agriculture*. Paris, INRA_Economica.

Allaire G., Wolf S., 2004. Cognitive Representations and Institutional Hybridity in Agrofood Innovation. *Science, Technology & Human Values*, 29(4), 431–458.

Allaire G., Daviron B., 2006. Régimes d'institutionnalisation et d'intégration des marchés: le cas des produits agricoles et alimentaires. *Les nouvelles figures des marchés agro-alimentaires: apports croisés de l'économie, de la sociologie et de la gestion. Journées d'études du GDR-CNRS, Montpellier, France*, 23, 113–125.

Appadurai A., 1996. *Modernity Al Large: Cultural Dimensions of Globalization*. Mine-apolis,University of Minnesota Press.

Appadurai A., 2000. Grassroots Globalization and the Research Imagination. *Public Culture*, 12(1). 1–19.

Aron R., 1969. Les désillusions du progrès. Paris, Gallimard.

Bonanno A., Wolf S.A., 2017. *Resistance to the Neoliberal Agri-Food Regime: A Critical Analysis*. Abingdon-on-Thames, Routledge.

Bunker S.G., Ciccantell P.S., 2007. *East Asia and the Global Economy: Japan's Ascent, with Implications for China's Future*. Baltimore, JHU Press.

Burch D., Lawrence G., 2009. Towards a Third Food Regime: Behind the Transformation. *Agriculture and Human Values*, 26(4), 267.

Cahen-Fourot L., Durand C., 2016. La transformation de la relation sociale à l'énergie du fordisme au capitalisme néolibéral. Une exploration empirique et macro-économique comparée dans les pays riches (1950–2010). *Revue de la régulation. Capitalisme, institutions, pouvoirs*, (20).

Chaumet J.M., Pouch T., 2017. *La Chine au risque de la dépendance alimentaire*. Rennes, Presses universitaires de Rennes.

Cheng G., Zhang H., 2014. *Chinal's Global Agricultural Strategy: An Open System to Safeguard the Country's Food Security*. Singapore, The S. Rajaratnam School of Interna-tional Studies (RSIS), RSIS Working Paper 19.

Fouilleux E., Loconto A., 2017. Voluntary Standards, Certification and Accreditation in the Global Organic Agriculture Field. A Tripartite Model of Techno-Politics. *Agriculture and Human Values*, 34(1), 1–14.

Freibauer A., Mathijs E., Brunori G., Damianova Z., Faroult E., Gomis J.G., O'brien L., Treyer S., 2011. *The 3rd SCAR (European Commission–Standing Committee on Agricul-tural Research) Foresight Exercise*. Brussels, European Commission

Georgescu-Roegen N., 1971. *The Entropy Law and the Economic Process*. Cambridge, (Mass.), Harvard University Press.

Goodman D., 1999. Agro Food Studies in the 'Age of Ecology': Nature, Corporeality, Bio Politics. *Sociologia Ruralis*, 39(1), 17–38.

Grain, 2016. *The Global Farmland Grab in 2016: How Big? How Far?* Barcelona, GRAIN, p. 11.

Hatanaka M., Busch L., 2008. Third-Party Certification in the Global Agrifood System: An Objective or Socially Mediated Governance Mechanism? *Sociologia ruralis*, 48(1), 73–91.

Holt Gimenez E., Shattuck A., 2011. Food Crises, Food Regimes and Food Movements: Rumbling of Reform or Tides of Transformation. *Journal of Peasant Studies*, 38(1), 109–144.

Howard P., 2016. *Concentration and Power in the Food System: Who Controls What We Eat?* London, Bloomsbury.

Jessop B., 1990. *State Theory: Putting the Capitalist State in Its Place.* University Park, Pennsylvania, Penn State Press.

Lash S., Urry J., 1987. *The End of Organized Capitalism.* Madison, University of Wisconsin Press.

Latour B., 1993. *We Have Never Been Modern.* Trans. C. Porter. Cambridge, MA, Harvard University Press.

Latour B., 2017. *Où atterrir? comment s' orienter en politique.* Paris, La Découverte.

Malet J.B., 2017. *L'Empire de l'or rouge: Enquête mondiale sur la tomate d'industrie.* Paris, Fayard, p. 288

McMichael P., 2005. Global Development and the Corporate Food Regime. In Buttel F.H. and McMichael P. (Eds.), *New Directions in the Sociology of Global Delelopment – Research in Rural Sociology and Development Volume 11.* Amsterdam, Elsevier, pp. 265–299.

McMichael P., 2016. Food Regime for Thought. The Journal of Peasant Studies 43(3), 648-670.

Muller P., 2010. Les changements d'échelles des politiques agricoles. Introduction. In Hervieu B., Mayer N., Muller P., Purseigle F. and Remy J. (Eds.), *Les mondes agricoles en politique. De la fin des paysans au retour de la question agricole.* Paris, Les Presses de Sciences Po.

Offe C., 1985. *Disorganized Capitalism: Contemporary Transformations of Work and Politics.* Cambridge : Mass., The MIT Press.

Orléan A., 1999. *Le pouvoir de la finance.* Paris, Odile Jacob.

Passet R., 1979. *L'économique et le vivant.* Paris, Economica.

Pechlaner G., Otero G., 2008. The Third Food Regime: Neoliberal Globalism and Agricultural Biotechnology in North America. *Sociologia Ruralis*, 48(4), 351–371.

Pritchard W.N., 1998. The Emerging Contours of the Third Food Regime: Evidence from Australian Dairy and Wheat Sectors. *Economic Geography*, 74(1), 64–74.

Thomas F., 2005. Droits de propriété industrielle et 'communs' agricoles. Comment repenser l'articulation entre domaine public, biens collectifs et biens privés? In Vanuxem S. and Guibet-Lafaye C. (Eds.), *Repenser la propriété, essai de politique écologique.* Marseille, Presses universitaires d'Aix-Marseille, pp. 171–190.

Verhaegen E., 2012. Les réseaux agroalimentaires alternatifs: Transformations globales ou nouvelle segmentation du marché. In *Agroécologie entre pratiques et sciences sociales.* Dijon, Educagri editions, pp. 265–279.

Zhang H., Cheng G., 2017. China's Food Security Strategy Reform: An Emerging Global Agricultural Policy. In Wu F. and Zhang H. (Eds.), *China's Global Quest for Resources: Energy, Food and Water.* London, Routledge, pp. 23–40.

Index